Second Generation General System Theory: Perspectives in Philosophy and Approaches in Complex Systems

Special Issue Editors

Gianfranco Minati
Eliano Pessa
Ignazio Licata

MDPI • Basel • Beijing • Wuhan • Barcelona • Belgrade

MDPI

Special Issue Editors

Gianfranco Minati
Italian Systems Society, Milan
Italy

Eliano Pessa
University of Pavia
Italy

Ignazio Licata
ISEM Institute for Scientific Methodology
Italy

Editorial Office
MDPI AG
St. Alban-Anlage 66
Basel, Switzerland

This edition is a reprint of the Special Issue published online in the open access journal *Systems* (ISSN 2079-8954) from 2016–2017 (available at: http://www.mdpi.com/journal/systems/special_issues/sggst).

For citation purposes, cite each article independently as indicated on the article page online and as indicated below:

Author 1; Author 2. Article title. *Journal Name* **Year**, *Article number*, page range.

First Edition 2017

ISBN 978-3-03842-440-6 (Pbk)
ISBN 978-3-03842-441-3 (PDF)

Table of Contents

About the Special Issue Editors

Gianfranco Minati graduated in Mathematics from the University of Milano. He has switched from an executive position in a large industrial–financial Italian group (1979–1984) to Research. He is the founder (1996) and president of the Italian Systems Society http://www.airs.it; president of the European Union for Systemics http://www.ues-eus.eu/; doctoral lecturer at the Polytechnic of Milan http://www.abc.polimi.it/fileadmin/abc/images/news/PHD_event/phdcourse0170209.pdf. In the US, he was a member of the Consulting Faculty at the Saybrook University, San Francisco, and adjunct faculty at the OHIO State University (OSU). His research activities deal with collective behaviour; dynamic usage of models; emergence; ergodicity; logical openness; mesoscopic coherence; meta-structures; multiple-systems; quasi-systems; systems science; theoretical incompleteness. Author of 40 articles, 15 books, 32 chapters in edited books, and editor of seven books. For more information, see http://www.gianfrancominati.net/.

Eliano Pessa holds both a graduate and a postgraduate degree in Physics. After working at the University of Rome, La Sapienza, as an associate professor of Mathematics and Artificial Intelligence, currently he is a full professor of General Psychology and Cognitive Modelling at the University of Pavia (see http://psicologia.unipv.it/didattica/docenti.html?chronoform=dettagliodoc&rec=208). He is also a member of the managing committee of the Italian Association for Systems Research. His research activities deal with Computational Neuroscience; Theoretical Physics; Complex Systems; Cognitive Science; Artificial Intelligence; Philosophy of Science. In particular, he contributed to the theory of phase transitions and emergence processes as well as to a number of foundational problems of Quantum Field Theory and to the building of a theory of consciousness and of a model of depressive disorder. He is the author or co-author of 10 books, co-editor of six volumes and author of more than 100 papers in international and national journals.

Ignazio Licata theoretical Physicist, is Scientific Director of Inst. for Sci. Methodology (ISEM), Palermo. He is professor at the School of Advanced International Studies on Theoretical and non-Linear Methodologies of Physics, Bari, Italy, and at the International Institute for Applicable Mathematics & Information Sciences (IIAMIS), B.M. Birla Science Centre, Adarsh Nagar, Hyderabad 500, India. He is Editor of the Electronic Journal of Theoretical Physics (EJTP), member of the editorial board of Quantum Biosystems and Il Nucleare and of the association Computability in Europe (CiE). His research areas include Foundation of Quantum Mechanics; Quantum Cosmology; Dissipative Quantum Field Theories; Quantum Information; and Physics of Emergence and Organization. His recent books are entitled Quantum Potential. Physics, Geometry, Algebra (with D. Fiscaletti), Springer 2013, and Beyond Peaceful Coexistence. The Emergence of Space, Time and Quantum (editor), ICP 2015. His recent work is on sub and super Turing Systems, with particular emphasis on Quantum Pattern Recognition.

Preface to "Second Generation General System Theory: Perspectives in Philosophy and Approaches in Complex Systems"

Following the classical work of Norbert Wiener, Ross Ashby, Ludwig von Bertalanffy and many others, the concept of System has been elaborated in different disciplinary fields, allowing interdisciplinary approaches in areas such as Physics, Biology, Chemistry, Cognitive Science, Economics, Education, Engineering, Social Sciences, Mathematics, Medicine, Artificial Intelligence, Philosophy, and Simulation Science.

The new challenge of Complexity and Emergence has made the concept of System even more relevant to the study of systemic aspects with high contextuality.

This conceptual shift in System concepts runs through the entire area of natural philosophy and epistemology, and leads to the questioning of old and new science words in a new conceptual archipelago.

In his essay entitled American Lessons (1985), the Italian writer Italo Calvino proposed six key words for the new millennium:

- Lightness
- Quickness
- Exactitude
- Visibility
- Multiplicity
- Consistency

We think that these words, with others emerging from the contributions in this book, can conceptually represent significant, fresh approaches to systems research dealing with complexity.

This Special Issue focuses on the nature of new problems arising from the study and modelling of complexity, their eventual common aspects, properties and approaches—already partially considered by different disciplines—as well as focusing on new, possibly unitary, theoretical frameworks. In particular, this issue is also devoted to the philosophical and structural aspects of the complexity and emergence theories.

Research is based on two crucial requirements: the introduction of new approaches and results, in addition to problems and proposals that are non-linearly deducible from previously available knowledge. The robustness of approaches and results makes them pillars of science. Problems and proposals should be innovative rather than simply correct. Too often, non-questionability relates to obviousness and guarantees the publication of very well documented articles. This is the case for Systems science too. This Special Issue, as well a previous one entitled Towards a Second Generation General System Theory (Editors: Gianfranco Minati, Eliano Pessa, 2014) http://www.mdpi.com/journal/systems/special_issues/second-generation-general-system-theory, and other publications aim to introduce fresh impetus into systems research when the possible detection and correction of mistakes require the development of new knowledge—not just the detection of inconsistencies with previously available knowledge. This book contains contributions presenting new approaches and results, problems and proposals. The context is an interdisciplinary framework dealing, in order, with electronic engineering problems; the problem of the observer; transdisciplinarity; problems of organised complexity; theoretical incompleteness; design of digital systems in a user-centred way; reaction networks as a framework for systems modelling; emergence of a stable system in reaction networks; emergence at the fundamental systems level; behavioural realization of memoryless functions.

Gianfranco Minati, Eliano Pessa and Ignazio Licata
Special Issue Editors

systems

MDPI

Article

Formal Proof of the Dependable Bypassing Routing Algorithm Suitable for Adaptive Networks on Chip QnoC Architecture

Hayat Daoud [1,2], Camel Tanougast [2,*], Mostefa Belarbi [1], Mikael Heil [2] and Camille Diou [2]

[1] Laboratory of Informatics and Mathematics (LIM), University of Tiaret, Tiaret 14000, Algeria;
     hayat.daoud@hotmail.fr (H.D.); mbelarbi@univ-tiaret.dz (M.B.)
[2] Laboratoire de Conception, Optimisation et Modélisation des Systèmes (LCOMS),
     ASEC Team, Université de Lorraine, Metz 57070, France;
     mikael.heil@univ-lorraine.fr (M.H.); camille.diou@univ-lorraine.fr (C.D.)
*   Correspondence: camel.tanougast@univ-lorraine.fr

Academic Editors: Gianfranco Minati, Eliano Pessa and Ignazio Licata
Received: 31 October 2016; Accepted: 20 January 2017; Published: 22 February 2017

Abstract: Approaches for the design of fault tolerant Network-on-Chip (NoC) for use in System-on-Chip (SoC) reconfigurable technology using Field-Programmable Gate Array (FPGA) technology are challenging, especially in Multiprocessor System-on-Chip (MPSoC) design. To achieve this, the use of rigorous formal approaches, based on incremental design and proof theory, has become an essential step in the validation process. The Event-B method is a promising formal approach that can be used to develop, model and prove accurately SoC and MPSoC architectures. This paper proposes a formal verification approach for NoC architecture including the dependability constraints relating to the choice of the path routing of data packets and the strategy imposed for diversion when faulty routers are detected. The formalization process is incremental and validated by correct-by-construction development of the NoC architecture. Using the concepts of graph colouring and B-event formalism, the results obtained have demonstrated its efficiency for determining the bugs, and a solution to ensure a fast and reliable operation of the network when compared to existing similar methods.

Keywords: network on chip; switch and adaptive-routing; event-B formalism; formal proof; correct-by-construction

1. Introduction and Related Work

The growing chip complexity and the need to integrate more and more components on the chip (e.g., processors, DSP cores or memories) imposes the trend for embedded systems moving towards Multiprocessor System-on-Chip (MPSoC). Consequently, in the new SoC paradigm, the network centric approaches, called Networks-on-chip (NoCs), are progressively becoming the main on-chip communication mediums and are the major issue in MPSoC. Indeed, integrating a NoC in the SoC provides an effective way to interconnect several Processor Elements (PEs) or Intellectual Properties (IPs) (processors, memory controllers, etc.) [1], with high levels of modularity, flexibility, and throughput. Generally, a NoC consists of routers and interconnections allowing communications between PEs and/or IPs. Communication on NoC relies on data packet exchanges. The paths used by the data packets between a source and a destination through the routers are defined by the routing algorithm. Hence, on-chip communications must have a high degree of availability; that is, a high probability of correct and timely provision of requested services. To achieve this, correctness of such communicating structure should be ensured. Formal methods can be used to verify complex MPSoC in order to ensure that these systems satisfy their functional and communication requirements [2].

Indeed, the application of formal methods helps increase confidence to building correct hardware systems. Therefore, formal methods are of essential importance to the development of such novel and complex platforms. Consequently, several previous studies propose new system formalisms based on generic and hierarchical connectors for handling the complexity of on-chip communications and data flows [3–9]. However, few previous studies have focused on fault tolerant communicating systems, in particular for proof validation of the on-chip communication reliabilities [10–14]. With an increasing complexity and reliability evolution of MPSoC, where NoCs are becoming more sensitive to phenomena generating permanent, transient, or intermittent faults, several solutions have been proposed and formalized [15,16]. The aim is to define mechanisms allowing the bypasses of the faulty nodes or regions to achieve reliability of NoC performing on-chip communications of the designed SoC. Generally, such routing schemes, before being designed, are behaviourally verified by simulation with created stimuli. This allows quick detection of the coarse errors. However, simulations cannot find all possible errors to ensure reliable functionalities before their design, and are often considered insufficient for improving the dependability of NoCs [1]. Indeed, unlike formal verification, simulations arouse a computer model by input stimuli, which is, unfortunately, not exhaustive because no extensive testing can ensure a large error coverage. Furthermore, experience has shown that if half of the errors come from carelessness during the design, the other half are from the level of the specifications.

Formal methods have the ability to produce critical systems for large industrial projects through the generation of original mathematical models that can be formally refined at various levels until the final refinement containing accurate implementation and verification details. Typically, verification by simulation does not allow the detection of all possible design errors [17]. In this paper, we propose to use the Event-B formal method, especially the correct-by-construction paradigm [18], to specify hardware systems. This paradigm offers an alternative approach to prove and define correct systems and architectures, for the reconstruction of a target system using progressive refinement through validated methodological techniques [19]. Our goal is to complete the simulation time in the design flow with a formal proof method. The preconditions for the formal development of microelectronic architectures lead to the description and/or the design of the architecture.

A large body of works has focused on the use of formal methods to verify communication systems and protocols, and model-checking methods or its composition with proving theorems are the most widely used. The work of Clarke et al. described in [20] proposed checking the temporal properties of parameterized ring networks and binary trees. A first step consists of using a free-context grammar to model the network communication systems while temporal properties are verified using a model checker. In [21], Amjad used a model checker implemented in HOL to verify AMBAAHB protocols and PDB. Bharadwaj et al. [22] proposed a broadcast protocol in a binary tree network using the SPIN model checker demonstrator. In [23], Curzon developed a structural model for Fairisle ATM switch and compared its behavioural specification using HOL. The free deadlock in the network was verified by Gebre Michael et al. using the PVS tool [24]. Some works, which are based on semi-formal methods, were also proposed to detect and debug failures. For example, Chenard et al. proposed assertion listeners PSL [25,26] which were synthesized using NGC tool [27] in a NoC. Analytical approaches do not cover the dynamic behaviour and performance of a system, but rather analyse it statically. Model checking, which has been adopted in this paper, is an automated technique to verify each model of a system satisfies its specifications [28]. The model is described as a state machine and the specification is described in a temporal logic. A model control algorithm uses the transition function associated with the state machine to explore the state space and define the states that do not meet the specifications. If a state is found, its traces leading to this state are reported. If such a state is not found, the system is proved correct. Model checking is widely adopted by academic and industry communities, mainly because it is a fully automatic model. The major problem is its combinatorial blow-up of the number of states that have to be explored, usually called state space explosion. This severely limits the scalability of model checking. Theorem proving is a technique where the evidence of a mathematical theorem is formalized so that a computer program can guarantee its accuracy. The main advantage of the theorem is the ability to deal with the parametric systems.

The verification of SoC communications [29] describes the main challenges in the design of NoCs [1] and discusses some aspects of audit networks where formal methods are useful. Dynamic reconfigurable NoCs are adequate for FPGA-based systems, where the main problem arises when IP components must be defined dynamically at run time. Given the rapidly changing and highly complex MPSoCs, the constraints related to the complexity and the increasing number of interconnected modules or IP such as the cost and performance must be resolved. Current communications of NoCs implement the data transmission between the interconnected nodes. Sometimes the communication of this kind of networks is difficult or even impossible. This is the main reason why XY fault-tolerant routing algorithms have been developed [30]. These algorithms allow bypassing faulty or unavailable regions through the error detectors introduced inside the network in order to ensure that the data packets are not loss. Routers can control the miss-routing of previous detectors (e.g., packet on the path XY, etc.). In addition, new techniques and adaptive fault-tolerance routing with error detection and path routing based on the well-known XY turn model have been introduced. In the case of these algorithms, zones corresponding to already detected faulty nodes or unavailable regions in the NoC are defined. As routers can control if previous switches have performed routing errors (e.g., packet out of the XY path, etc.), the neighbour routers of these zones must not send data packets toward these faulty routers or unavailable regions. To achieve this, chains or rings around the adjacent faulty nodes or regions are formed in order to delimit rectangular parts in the NoC covering all the faulty nodes or unavailable regions. In these chains or rings of switches, the routing tables are modified and differ from the standard tables related to the XY routing algorithm. These specific switches integrate in their tables additional routing rules allowing to bypass the faulty zones or regions dedicated to dynamic IP/PE instantiations, while avoiding starvation, deadlock, and livelock situations [19]. On the other hand, formal studies have only focused on NoC performance [31,32], latency [33], bandwidth [34], estimation of consumption [34], error detection and correction [35,36], and required and used logic area. Others propose methods of free-deadlock routing [37,38] to characterize the traffic.

This paper aims to investigate event-B concept and its application to specify and verify formally the correctness of the SoC communications. The contributions are twofold: to demonstrate the NoC behaviour; and to prove the dependability of a proposed fault tolerant routing algorithm used in a specific NoC. More precisely, in our formal approach, by considering the architecture of routers and the defined communication rules of the specific routing algorithm under normal and fault conditions, we ensure a correct transmission of data in the network and a reliable functionality. This is done from the checking criteria being invariants of behavioural of logic and communication blocks constituting the routers, and designed for bypassing faulty or unavailable regions in the NoC. We also aim to demonstrate the step-wise construction of the adaptive NoC, which can be useful for a pre-silicon verification by testing the behaviour of the NoC architecture in a virtual environment with a formal verification tool.

The remainder of this paper is organized as follows. Section 2 presents an overview of the event-B method showing the NoC architecture studied with the audit results of the formal verification. In this section, the behavioural modelling of the NoC with its basic element switches are detailed by describing the architecture of the fault tolerance and the model description. This section also details the proofs of the proposed fault tolerant modified XY routing algorithm associated to the modelled NoC to overcome some faulty events encountered in the network in order to maintain the dependability during the communication operations. Section 3 discusses the results obtained from the various levels of formal modelling detailed in Section 2. Finally, Section 4 concludes this paper along with the future work.

2. Stepwise Specification of the System Using Event-B

2.1. Definition of Event-Band and Proof Obligations

The main reason for selecting Event-B as the modelling language is its the refinement feature which allows a progressive development of models. Moreover, Event B is also supported by a complete RODIN toolset [38] providing features including refinement, proof obligations generation, proof assistants and model-checking facilities. The Event B modelling language can express theorems or safety properties, which are invariants, in a machine corresponding to the system [39]. Event B allows a progressive development of models through refinements. The two main structures available in Event-B are:

-   Contexts, which express the static information about the model.
-   Machines, which express dynamic information about the model, invariants, *safety properties*, and events.

An Event B model is defined either as a context or as a machine. A machine organizes events (or actions), which modify the state variables and uses static information defined in a context. The refinement of models provides a mechanism for relating an abstract model and a concrete model by adding new events or variables. This feature allows to develop gradually Event-B models and to validate each decision step using the proof tool. The refinement relationship should be expressed as follows: a model M is refined by a model P, when P simulates M. Thus, from a given model M, a new model P can be built and asserted to be a refinement of M describing the architecture. Model M is an abstraction of P, and model P is a refinement (concrete version) of M. Likewise, context C, seen by a model M, can be refined to a context D, which may be seen by P (see Figure 1). The final concrete model is close to the behaviour of the real system that executes events using real source code. The refinement of a formal model allows us to enrich the model via a step-by-step approach and is the foundation of our proposed correct-by-construction approach. Refinement provides a way to strengthen invariants and to add details to a model. It is also used to transform an abstract model to a more concrete (real) version by modifying the state description. This is done by extending the list of state variables (possibly suppressing some of them) and by refining each abstract event to a set of possible concrete versions, and by adding new events.

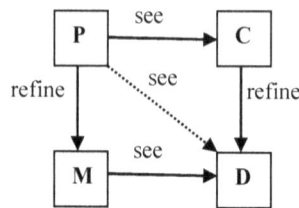

Figure 1. Illustration of machine and context relationships.

The proofs obligation defines what is to be proved for an Event-B model. These proofs are automatically generated and concern *Invariant, Preservation, Feasibility, Fusion*, and so on. The RODIN Platform tool [40], called Proof Obligation Generator, decides what is to be proved in order to ensure the correctness of the model. Therefore, just to check contexts and machine texts and to decide what is to prove in these texts, there are eleven rules for the proof obligation all defined and labelled within the Robin platform. Actually, the refinement-based development of Event B requires a very careful derivation process, integrating possible tough interactive proofs for discharging generated proof obligations, at each step of development. The automatic prover is designed in order to drive the heart's mode in which the evidence Prove (pr) command is tried on each proof obligation. An automatic mode simply stores the maximum tempted strength for each obligation, which is the level of automatic proof obligation. On the other hand, the interactive demonstration allows the operator to decide him/her self what proof commands are applied. The sequence of commands that a user has chosen to demonstrate an obligation is stored along with the state of proof.

2.2. NoC Architecture Description and Modeling

The network transmission is realized through routers constituting the network, and by using switching techniques of data packets constituted of messages and routing rules [1]. Usually, the network has a grid-like form and is built with on-chip routers characterized by a 2D mesh topology, Ack/Nack flow control, and store-and-forward buffering strategy, which avoid deadlock situations. Figure 2 illustrates the mesh-network (4 × 4 mesh topology) where peripheral switches are associated to a *PE*. Therefore, boundary switches are connected to one PE and with two or three neighbours, whereas other nodes are connected to four neighbours. Each NoC element (*PE* or *router*) possesses a specific address but only *PEs* act as emitter and receiver of messages through the network. Thus, in a $n \times n$ 2D mesh, a node or switch K is identified by a two element vector (k_x and k_y), $1 \leq k_x, k_y \leq n$, where k_x and k_y are the coordinates of dimension x and y, respectively. The routers communicate with other neighbours in the four possible directions through a fixed number of packets. For this purpose, each switch has four incoming and outgoing ports from which it can receive and send the packets (see Figure 1). Four pairs of ports towards other adjacent nodes are named, respectively, *West*, *North*, *East* and *South*. In our case study, we consider a Quality-of-Service NoC (QNoC) switch (see Figure 2b) [1] which consists of routing logic and control logic with inputs/outputs for each direction. This microelectronic architecture communicates with four neighbouring elements.

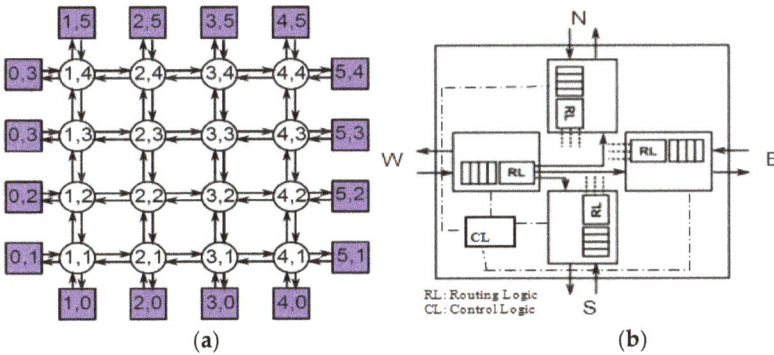

Figure 2. (**a**) Mesh topology; and (**b**) general structure of the QNoC switch architecture.

The computing elements associated with the NoC network communicate through messages. A message consists of a fixed number of packets and is based on a wormhole flow control. The number of packets Np in a message is not fixed by the network (Np $\geq$ 1). The format of a packet is shown in Figure 3. The first field denotes the destination address whose size depends on the total number of units constituting the NoC communication network. The second field contains the information about the size of the packet, while the third field contains the (user-ID) identification number of the packet. We consider the size of the ID field to be less than or equal to the size of the message, and it is never zero. The last field contains the data to be transmitted. Packets can flow in different directions depending on the status of the router (free busy or failed) but the communication units in a single packet (conflicts) must follow the same path.

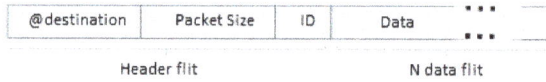

@destination	Packet Size	ID	Data	•••

Header flit   N data flit

Figure 3. Format of a packet used in NoC.

In the next subsections, we detail the formal specification that has been validated from the RODIN Event-B software. The refinements of the formal proof demonstrating the dependability of the proposed fault tolerant routing algorithm suitable for adaptive NoC is also given. Therefore, we give the axioms and invariant express results obtained under RODIN toolset environment proving the efficiency of the proposed fault-tolerant adaptive routing algorithm based on the XY and turn model routing schemes.

2.3. NoC Formal Development and Discharge Obligations

An incremental development of a NoC architecture using the Event-B formalism [41] and the formalization of the architecture is presented from an abstract level down to a more concrete level in a hierarchical way (see Figure 4).

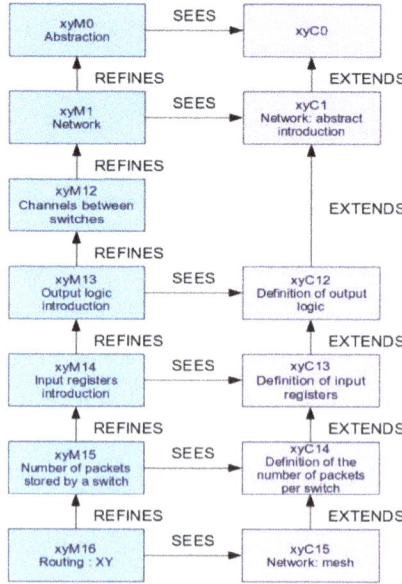

Figure 4. Step-by-step Modelling of NoC Architecture.

- The first model xyM0 is an abstract description of the service offered by the NoC architecture (see Figure 5): the sending of a packet (p) by a source switch and the receiving of (p) by a destination switch.

Figure 5. Abstraction Level.

A set of switches (NODES); a set of packets (MSG); a function src, associating packets and their sources; and a function dst, coupling packets and their destinations, are defined in the context xyC0. The machine xyM0 uses (*sees*) the contents of context xyC0, and with these, describes an abstract view of the service provided by the NoC architecture:

- An event SEND presents the sending of a packet (m), by its source (s), to a switch destination (d).
- An event RECEIVE depicts the receiving of a sent packet (m) by its destination (d).

Moreover, the model xyM0 allows us to express some properties and invariants:

| ran(received) ⊆ran(sent) |

This invariant ensures that each packet received by a destination switch has been sent by a source switch.

- The machine xyM1 refines xyM0 and introduces a network (*agraph*) between the sources and destinations of packets (see Figure 6). Some properties on the graph are defined in context xyC1: graph is non-empty, non-transitive and is symmetrical.

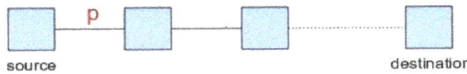

Figure 6. Adding Network.

The events in xyM0 are refined:

- Event SEND: When a source sends a packet, it is put in the network.
- Event RECEIVE: A packet is received by its destination, if it has reached the destination.

New events are also introduced by xyM0:
Event FORWARD (see Figure 7): in the network, a packet (p) transits from a node (x) to another node (y), until the destination (d) of packet (p) is reached.

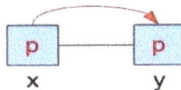

Figure 7. Transfer of a Packet (p) between Switches.

- Event DISABLE: A node is *disabled*. The node is not allowed to communicate with its neighbours (*failure*, etc.). During the *disabling* of some nodes, we ensure that the packets transiting in the network will eventually reach their destinations (either after a reconfiguration of the network or by always keeping a path to destinations available).
- Event RELINK: This event models the reconfiguration of the network. *Disabled* nodes are re-*enabled*: the links between them and their neighbours are restored, therefore allowing communications and packets transfers. The reconfiguration of the network helps in demonstrating the safety of data transmission between a source switch and a destination switch.

The machine xyM1 also presents some properties of the system:

| ran(received) ∩ ran(store) = ∅ |

This invariant demonstrates that a packet (p) sent by a source is either traveling in the network (store) or is received by a destination.

- The second refinement decomposes the event FORWARD of xyM1 into two events:

- A refinement of the event FORWARD depicts the passing of a packet (p) from a switch (x) to a channel (ch), leading to a neighbour (y) (see Figure 8).
- An event FROM_CHANNEL_TO_NODE models the transfer of a packet (p) from a channel (ch) to a connected switch (n) (see Figure 8). The machine xyM12 also defines some properties:

$$\text{ran(c)} \cap \text{ran(switch)} = \varnothing$$

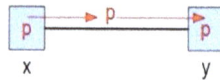

Figure 8. Channel Introduction.

The invariant expresses that each transmitted packet is either in a channel or in a switch. A sent packet cannot be in a channel and in a switch at the same time.

- The third refinement allows us to introduce the structure of a switch gradually. We express, in xyM13, that the switches possess output ports (see Figure 9). The abstract event FORWARD is further decomposed:

- The refinement of event FORWARD adds the fact that a packet (p), which is leaving a switch (x) and heading for a neighbour (y), first enters the output logic (op) of the switch (x) leading to (y).
- A new event OUTPUT_BUFFER_TO_CHANNEL models the transition of a packet (p) from an output port (op) to a channel (ch) leading to a target switch (n).

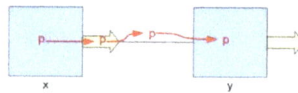

Figure 9. Adding Output Ports.

Moreover, new properties and invariants are defined in xyM13:

$$\begin{array}{l} \text{inv1 : ran(chan)} \subseteq \text{ran(sent)} \\ \text{inv2 : ran(outputbuffer)} \subseteq \text{ran(sent)} \\ \text{inv3 : ran(outputbuffer)} \cap \text{ran(chan)} = \varnothing \end{array}$$

The invariant inv1 expresses that each packet transiting in a channel (ch) has been sent by a source (s); inv2 demonstrates that each packet transiting in an output port (ch) has been sent by a source (s); and inv3 presents the fact that a packet is either in an output port or in a channel, the packet cannot be in an output port and a channel between two switches at the same time.

- The fourth refinement (xyM14) adds input ports to the structure of a switch (see Figure 10).

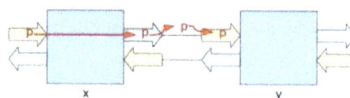

Figure 10. Adding Input Ports.

The event SEND is refined: when a switch source (s) sends a packet (p), the packet (p) is put in an input port (ip) of the switch (s). The actions described by the abstract event FORWARD are decomposed:

- The event SWITCH_CONTROL, a refinement of FORWARD, models the passing of a packet (p), from an input port (ip) of a switch (x), to an output port (op) leading to a switch (y).
- The event OUTPUT_BUFFER_TO_CHANNEL presents the transition of a packet (p), from an output port (op), to a channel (ch) leading to a target switch (n).
- The event FROM_CHANNEL_TO_INPUT_BUFFER demonstrates the transition of a packet (p) from a channel (ch) to an input port (ip) of a target switch (n).

The machine xyM14 also presents properties and invariants:

inv1 : ran(inputbuffer) $\subseteq$ ran(sent)
inv2 : ran(outputbuffer) $\cap$ ran(inputbuffer) = $\varnothing$
inv3 : ran(inputbuffer) $\cap$ ran(chan) = $\varnothing$

The invariant expresses that each packet transiting in an input port (ip) has been sent by a source (s); inv2 demonstrates that each packet is transiting either in an output port (op) or an in input port (ip); and inv3 presents the fact that a packet is either in an input port or in a channel: the packet cannot be in an input port and a channel between two switches at the same time.

- The fifth refinement introduces the storage of packets in a switch: each output port of a switch can store a number of packets up to a limit (outputplaces) of three messages. Packets can be blocked in a switch, because of the "wait" or "occupation" signals from the neighbours. The event SWITCH_CONTROL is refined, and adds the fact that, following the transition of a packet from an input port of a switch (x) to an output port, if the switch (x) is not busy anymore, it sends a release signal to the previous switch linked to the input port. A new event RECEIVE_BUFFER_CREDIT models the receiving of a release signal by a switch (n).
- The last model xyM16 describes the architecture of the network (graph): the graph has a mesh topology (see Figure 11). A numerical limit (nsize) is introduced to bound the number of routers in the dimensions x and y of the network topology; the network will be a regular 2D-Mesh, with a size of (nsize $\times$ nsize); each switch is coupled with unique coordinates (x; y), with x 2 [0..nsize $-$ 1] and y 2 [0..nsize $-$ 1].

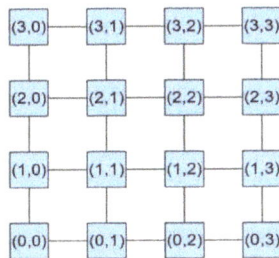

Figure 11. A regular Mesh with 2D-coordinates.

This coordinate system allows it to be more precise on the neighbours of each switch, as shown in Figure 11. This model also gives a fine-grained description of the structure of a switch (see Figure 12):

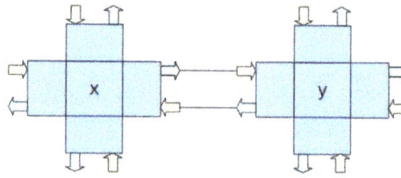

Figure 12. Switches: Structure and Links.

A switch generally has four output ports and four input ports (usually labelled N, S, E and W), used for communication with neighbours. However, two more cases are distinguished:

-   Boundary switches at the corners only have two output ports and two input ports (N-E, N-W, S-E, and S-W).
-   Other boundary switches have three output ports and three input ports (N-S-E, and N-S-W).

Moreover, this model also introduces the XY routing algorithm:

D: destination. Coordinates (Dx, Dy)
C: current node. Coordinates (Cx, Cy)
if (Cx>Dx) :
return W; (Case 1)
if (Cx<Dx) :
return E; (Case 2)
if ((Cx = Dx) ∨((Cx>Dx) ∧W is blocked) ∨
((Cx<Dx) ∧E is blocked)):
if (Cy <Dy):
return N; (Case 3)
if (Cy >Dy):
return S; (Case 4)

The cases of the XY routing algorithm are matched with refinements of event SWITCH_CONTROL:

-   SWITCH_CONTROL_LEFT models Case 1: A packet (p) is transmitted from an input port of a switch (x) to an output port, resulting in a neighbour (y) located at W. This event is triggered if the x-coordinate of the destination (d) (of the packet (p)) is inferior to the x-coordinate of the current node (x).
-   SWITCH_CONTROL_RIGHT models Case 2: A packet (p) is transmitted, from an input port of a switch (x), to an output port, leading to a neighbour (y), located at E. This event is triggered if the x-coordinate of the destination (d) (of the packet (p)) is superior to the x-coordinate of the current node (x).
-   SWITCH_CONTROL_UP models Case 3: A packet (p) is transmitted, from an input port of a switch (x), to an output port, leading to a neighbour (y), located at N. This event is triggered if the y-coordinate of the destination (d) (of the packet (p)) is superior to the y-coordinate of the current node (x), and either, if the x-coordinate of the destination (d) is equal to the x-coordinate of the current node (x), or if the packet (p) cannot transit along the x-axis.
-   SWITCH_CONTROL_DOWN models Case 4: A packet (p) is transmitted, from an input port of a switch (x), to an output port, leading to a neighbour (y), located at S. This event is triggered if the y-coordinate of the destination (d) (of the packet (p)) is inferior to the y-coordinate of the current node (x), and either, if the x-coordinate of the destination (d) is equal to the x-coordinate of the current node (x), or if the packet (p) cannot transit along the x-axis.

Table 1 gives the number of proofs obligations which are automatically discharged proving the checking of criteria for each considered level of abstraction and step by step. It can be noted that for context xyC15 and machine xyM14, there are more interactive proofs that automatic counterparts. This is explained by the fact that a majority of these interactive proofs are *quasi-automatic*: the proofs do not require significant efforts (no importing hypotheses, simplifying goals, etc.).

Table 1. Summary of proof obligations and discharged obligations.

Model	Total	Auto		Interactive	
xyC0	3	3	100%	0	0%
xyC1	6	6	100%	0	0%
xyC12	0	0	100%	0	0%
xyC13	0	0	100%	0	0%
xyC14	1	1	100%	0	0%
xyC15	5	0	0%	5	100%
xyM0	26	25	96.15%	1	3.85%
xyM1	38	28	73.68%	10	26.32%
xyM12	72	45	62.5%	27	37.5%
xyM13	74	37	50%	37	50%
xyM14	67	23	34.33%	44	65.67%
xyM15	24	14	58.33%	10	41.67%
xyM16	26	18	69.23%	8	30.77%
Total	342	200	58.48%	142	41.52%

2.4. Formal Devlopement of the Proposed Fault Tolerant Routing Algorithm and Discharge Obligations

In our case study, we consider an algorithm [1] that is a partially adaptive routing algorithm allowing both routing of messages in networks incorporating regions that are not necessarily rectangular, and all nodes, which are not completely blocked by faulty nodes. The considered routing scheme designed for a 2D mesh topology and based on the turn model and XY routing algorithm allows the handling of faulty nodes and regions of the chip [1]. More precisely, the switch uses a routing algorithm based on the classical XY algorithm that can be used initially in the network and which relies on the fact that the packets are routed according to X axis, and then to the Y axis of the array. If the routing packets encounter faulty nodes or regions that prevent them from going through XY paths, then the proposed routing algorithm allows circumvention of the faulty area(s) of the network. To achieve this sort of bypassing, modified routing rules are performed by the nodes surrounding network's faulty areas according to the strategy detailed below.

Definition 1. *In a 2D mesh, a node is called an **even** (respectively, **odd**) **node** if the sum of its coordinates (x and y dimensions) is an even (respectively, odd) number.*

The regular placement of even and odd nodes in the network is depicted in Figure 12. Each type of node is surrounded by elements of other types, even nodes by odds and vice versa. We distinguish two types of functioning of each node, *activated* and *deactivated* modes, which define *activated* and *deactivated* areas of a network as follows:

Definition 2. *One of the **activated area** of a network is a minimal rectangular area which envelops all faulty nodes or regions in the network.*

If a network does not have faulty nodes or regions, there is no activated area. Otherwise, a network can contain only one activated area. All nodes (except faulty nodes or regions) belonging to an activated area are activated. An activated area cannot have at the same time an odd (even) node at its top right-hand corner and an even (odd) node at its bottom left-hand corner.

Definition 3. *One of the **deactivated areas** of a network is the rest of the network which does not belong to the activated area.*

If a network does not have an activated area (no faulty nodes or regions), the whole network is deactivated. All nodes belonging to a deactivated area are deactivated. Figure 13 presents a network with an activated area formed around a faulty node. In this case, only the nodes wrapping the faulty node become activated. The nodes belonging to the rest of the network do not change their mode, and remain deactivated. A deactivated node routes a data packet according to the XY algorithm. Firstly, it routes along the X dimension and then along the Y dimension until the packet reaches its final destination. If a packet, before reaching its final destination, reaches the activated area's boundary, new routing rules are applied.

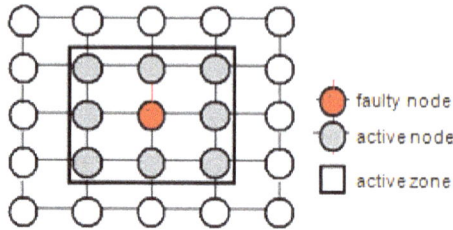

Figure 13. Example of an activated area.

Accurately, the activated nodes (boundary nodes of an activated zone) do not obey the same set of rules as deactivated nodes. These rules are presented in the following:

Rule 1. All packets in the active region can be routed only along the X-axis.
Rule 2. An activated node cannot route a packet from North to East side and vice versa.
Rule 3. An activated node cannot route a packet from South to West side and vice versa.
Rule 4. All activated nodes cannot route a packet from North to South and vice versa.

Each node can communicate directly with four neighbouring nodes on the network, which means that there are sixteen ($2^4 = 16$) possible pathways from one node in a mesh to other neighbouring nodes. To avoid routing loops, data packets cannot return to the sender node, hence the number of pathways left is twelve ($2^4 - 4 = 12$). To avoid livelock situation, it is necessary to eliminate the packets going in opposite direction of the authorized one, (Figure 14a) which leaves only eight possible routes. To route packets towards their final destination in an active zone, four paths are allowed and the others are blocked (see Figure 14b).

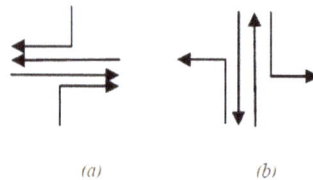

(a) (b)

Figure 14. Authorized (**a**); and Prohibited (**b**) turns and directions in activated nodes.

The network is scanned regularly in order to detect online the occurred faulty nodes. If a new faulty node or region is detected, the activated node's routing tables are updated with the new information about the positions of new faulty nodes or regions. More precisely, to achieve these routing decisions from the modified routing rule tables, each switch holds one input register in each of its ports (see Figure 1). Therefore, all packets pass through the input registers. Once the packets have arrived and are stored in the input registers, the routing logic block specifies the next direction (according to their authorized turns and directions) of the packets which are transmitted to the associated logic output port of the output direction. Consequently, the information contained in the routing tables help activated nodes to take routing decisions. For example, an activated area with the minimal number of active nodes, which forms a ring around faulty nodes, does not route the packets the same way as an activated area having more activated nodes. In the latter case, the packet not only uses the ring of activated nodes to get around the faulty nodes or regions, it can also use other activated nodes to route its packets. Before specifying and verifying the bypass algorithm, we propose a specification of conduct of event defining the bypass active zone around faulty nodes (knowing that there may be several faulty nodes in the network, thus more active zones can be constructed), it is proposed to use a colouring algorithm.

2.4.1. Vertex Colouring Algorithms

Symmetry breaking has always been a central problem in distributed systems. Several techniques have been developed for graph colouring algorithms including Maximal Independence Set (MIS). An algorithm for vertex colouring is a process of marking a graph [42] where the goal is to assign labels to the vertices of the graph. Labels are often treated as colours. Therefore, it is called coloured graph algorithm. Colouring/labelling is designed so that no two adjacent vertices share the same colour/label: an appropriate colouring of a graph G = V, E (with V the set of vertices of G and E the set of its edges), using a set of colours (N | COLOURS = 1 .. N) is a function f such that (F: V 3 COLOURS | f (i) = f 3 (j) where j, i ∈ E). The minimum for N is satisfied if it is called the chromatic number of G. These rules are generally applied to simple graphs (connected, reflexive, undirected, and unweighted). Several algorithms have been developed to colour a chart. As described in [43], the vertex colouring algorithms can be classified into two categories:

- Centralized algorithms [44,45]: The word "central" implies that there is at least one "administrator" who decides to graph colouring.
- Distributed algorithms [42–45]: These new algorithms involve all vertices of the graph that is coloured and the tops have their own "intelligence". Usually, they choose their own colours using random probabilities when they have chosen the same colour as their neighbours, and, when they have a good colour, in this case, they withdrew from the uncoloured curve [42–45]. In this work, we focus on the development of algorithms using distributed techniques. In fact, there is little or no verification of the accuracy of previous algorithms [42–45] considering some random numbers to define the process of secure coloration.

There are many practical applications of colourful graph algorithms that include:

- Planning graph colouring [46] can be used to control a set of nodes. Two nodes are considered adjacent when they may occur simultaneously. The aim is to prevent adjacent nodes occurring at the same time. However, in our case, there may be two nodes that have the same job but two test nodes cannot fix the failed node at the same time.
- Each correct node must be coloured in green; a correct node is a node that can send and receive packets.
- Each failed node must be coloured in red; a failed node is a node that cannot send or receive packets or one of the two.
- Each active node must be coloured in blue; an active node is a neighbouring node to a failed node.

2.4.2. Formal Specification of the System

Abstract level: We start with an abstract specification of the problem by defining the role of the network send and receive packets (See Figure 15). Thereby, two sets will be defined in this level; existing nodes (NODES) and packets (PACKETS) sent by a single source (src) and received by a single destination (dst). Sources are different from destinations. The following axioms are described in the context of the abstract level as follows:

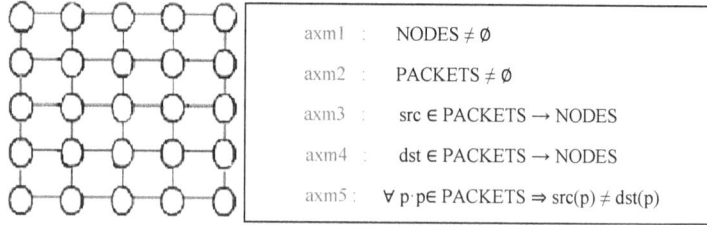

axm1	:	NODES $\neq \emptyset$
axm2	:	PACKETS $\neq \emptyset$
axm3	:	src $\in$ PACKETS $\rightarrow$ NODES
axm4	:	dst $\in$ PACKETS $\rightarrow$ NODES
axm5	:	$\forall$ p·p$\in$ PACKETS $\Rightarrow$ src(p) $\neq$ dst(p)

Figure 15. Abstract level.

We define the variable rcvd, which allows us to perform the SEND and RECEIVE actions. The following invariants are described in the abstract level of the machine as follows:

inv1 :	sent $\in$ NODES $\leftrightarrow$ PACKETS
inv2 :	rcvd $\in$ NODES $\leftrightarrow$ PACKETS
inv3 :	ran(rcvd) $\subseteq$ ran(sent)
inv4 :	$\forall$ s,p·s$\in$NODES $\wedge$ p $\in$ PACKETS $\wedge$ s$\mapsto$p $\in$ sent $\Rightarrow$ s = src(p)
inv5 :	$\forall$ d,p·d$\in$NODES $\wedge$ p $\in$ PACKETS $\wedge$ d$\mapsto$p $\in$ rcvd $\Rightarrow$ d = dst(p)
inv6 :	$\forall$ s1, s2, p·s1 $\in$ NODES $\wedge$s2 $\in$ NODES $\wedge$ p $\in$ PACKETS $\wedge$ s1$\mapsto$p $\in$ sent $\wedge$ s2$\mapsto$p $\in$ sent$\Rightarrow$ s1 = s2
inv7 :	$\forall$ d1, d2, p·d1 $\in$ NODES $\wedge$d2 $\in$ NODES $\wedge$ p $\in$ PACKETS $\wedge$ d1$\mapsto$p $\in$ rcvd $\wedge$ d2$\mapsto$p $\in$ rcvd$\Rightarrow$ d1 = d2

The initial values of variables are empty:

INITIALISATION $\triangleq$
act1 : sent := $\emptyset$
act2 : rcvd := $\emptyset$

The SEND event is action: act1: sen t$\mapsto$sent$\cup$\{s$\mapsto$p\} and RECEIVE event is action: act2: rcvd := rcvd $\cup$\{d$\mapsto$p\}

- **First refinement:** We assume that the graph is given a set of nodes. Next, we define a set of colours (Red_Colour, Green_Colour, and Blue_Colour), whose components are the colours selected by the nodes during the execution of the algorithms of graph colouring. We specify some properties of these constants GRAPH, Green_Colour, Red_Colour and Blue_Colour as follows:
- Axm5 axiom defines that all vertices belong to the GRAPH; they are not isolated.
- Axm6 axiom defines that the graph is irreflexive. The initial values of variables are empty:

axm1	: GRAPH ∈ NODES ↔ NODES
axm2	: GRAPH ≠ Ø
axm3	: COLOOR ≠ Ø
axm4	: ∀c·c∈COLOOR⇒ c=Red_Colour ∨ c=Green_Colour ∨ c=Blue_Colour
axm5	: ∀ n·n∈ NODES ⇒ n∈dom(GRAPH)
axm6	: NODES◁id ∩GRAPH =Ø
axm7	: GRAPH = GRAPH~
axm8	: ∀ s·s⊆ NODES ∧ s ≠ Ø ∧ GRAPH[s]⊆s⇒ NODES ⊆s

- Axm7 axiom expresses that the graph is symmetric.
- Axm8 axiom expresses that the graph is connected.

In this refined level abstract of the machine, we will consider that all nodes can send and receive packets, thus allowing them to be coloured in green (see Figure 16).

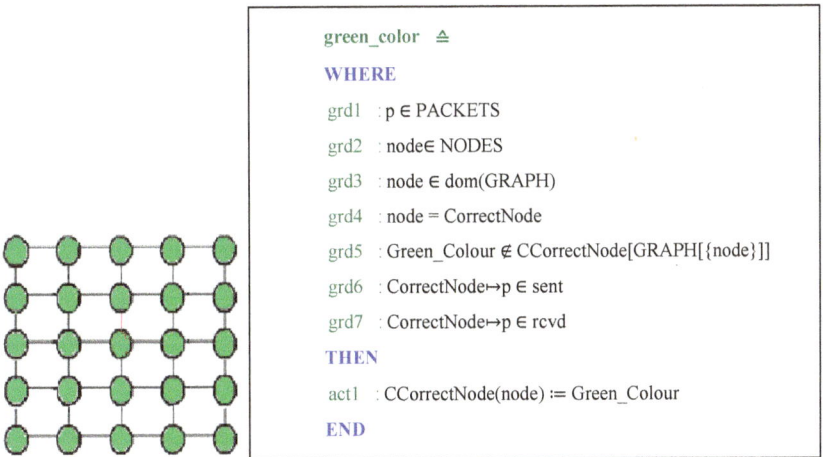

green_color ≙

WHERE

grd1 : p ∈ PACKETS

grd2 : node∈ NODES

grd3 : node ∈ dom(GRAPH)

grd4 : node = CorrectNode

grd5 : Green_Colour ∉ CCorrectNode[GRAPH[{node}]]

grd6 : CorrectNode↦p ∈ sent

grd7 : CorrectNode↦p ∈ rcvd

THEN

act1 : CCorrectNode(node) := Green_Colour

END

Figure 16. Colouring of correct nodes in the NoC with green.

The CCorrecteNode variable is defined with the following property:

inv1:ColouredCorrectNode∈ NODES ⇸ COLOUR

This property defines a node as a part of all coloured nodes. Therefore, the graph will be coloured green.

- **Second refinement:** The question turns now about the red colour and we need to find an inductive property which simulates the calculation of this function. Two variables will be added at this level; FaultyNode and CFaultyNode which are defined with the following properties where the FaultyNode is a failed node and coloured in red (see Figure 17):
- **Third refinement:** In this level, the calculation of the selection function of the active node (see Figure 18) is specified in a simple way to break the complexity of the role of this node:

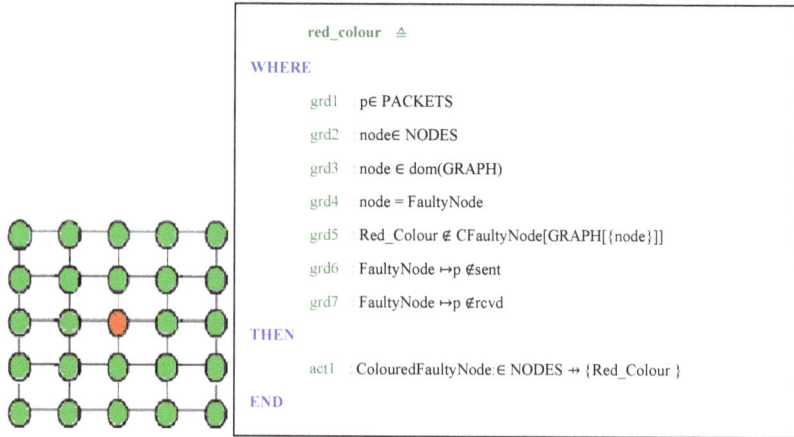

red_colour ≜

WHERE

    grd1   p∈ PACKETS

    grd2   node∈ NODES

    grd3   node ∈ dom(GRAPH)

    grd4   node = FaultyNode

    grd5   Red_Colour ∉ CFaultyNode[GRAPH[{node}]]

    grd6   FaultyNode ↦p ∉sent

    grd7   FaultyNode ↦p ∉rcvd

THEN

    act1   ColouredFaultyNode:∈ NODES ⇸ {Red_Colour }

END

Figure 17. Colouring of Faulty node in the NoC with red.

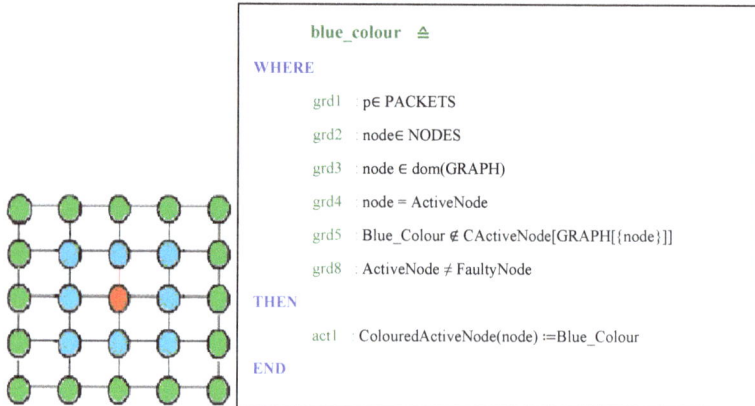

blue_colour ≜

WHERE

    grd1   p∈ PACKETS

    grd2   node∈ NODES

    grd3   node ∈ dom(GRAPH)

    grd4   node = ActiveNode

    grd5   Blue_Colour ∉ CActiveNode[GRAPH[{node}]]

    grd8   ActiveNode ≠ FaultyNode

THEN

    act1   ColouredActiveNode(node) :=Blue_Colour

END

Figure 18. Colouring of active node in the NoC with blue.

Table 2 gives the number of proofs obligations for the colouring algorithms of nodes which are automatically discharged proving the checking of criteria for considered abstraction level and step by step. These obligation results show that current development focuses on Vertex colouring algorithms, especially that they include possible errors in the choice of colours by the nodes. Thus, as the number of faulty nodes is not accurate according to network size and specification step, we help to prove the logical sequence of events.

Table 2. The proof obligations for the colouring algorithm of nodes.

Element Name	Total	Auto	Manuel	Reviewed	Undischarged
ColourActiveZone	27	25	2	0	0
Test C00	1	1	0	0	0
Test C01	0	0	0	0	0
Test M00	16	14	2	0	0
Test M01	3	3	0	0	0
Test M02	3	3	0	0	0
Test M03	4	4	0	0	0

2.4.3. Formal Development of the Bypassing Routing

This section presents the proposed formal development of the fault tolerant routing scheme of the considered NoC architecture. This proven formalism is based on refinement, which allows breaking the operation complexity of the routing algorithm and performing this formalization with different levels of abstraction carried out step-by step. Figure 18 presents the step-by-step modelling of the proposed fault tolerant routing scheme (see Figure 19).

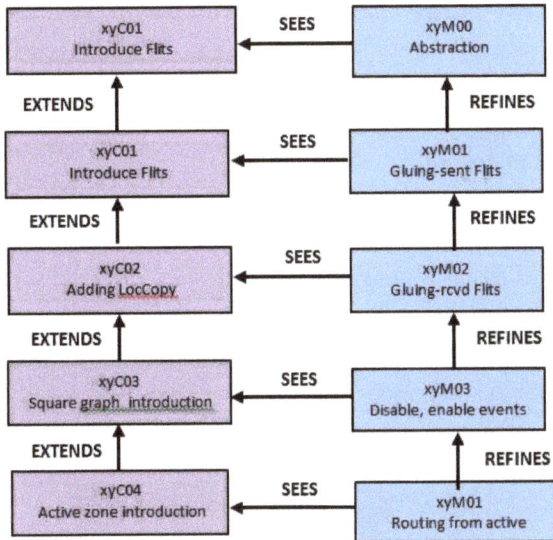

Figure 19. Step-by-step Modelling of fault tolerant routing algorithm suitable for NoC.

a. Abstract specification level: *xyC00*

The abstract level defines the role of the network to send an infinite number of messages which are packetized and encapsulate (*Flitization*) into sequence of packets from a source (s) to a destination (d) (see Figure 20).

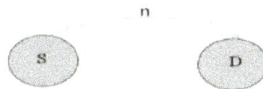

Figure 20. Abstract level.

A set of switches (NODES), a set of packets (PACKETS), a function src, associating packets (p) and their sources, a function dst, coupling packets (p) and their destinations, are defined in context xyC00. The machine xyM00 uses (sees) the contents of context xyC00, and describes an abstract view of the service provided by the NoC. An event SEND presents the sending of a packet (p), by its source (s), to a switch destination (d). An event RECEIVE depicts the receiving of a sent packet (p) by its destination (d).

The first model xyC00 is an abstract description, which specifies the packet nodes, sources (src), destinations (dst) of each packet (p). The following axioms describe how the source (src), and destination (dst) have been defined for a package (p). Each packet has a single source and single destination that are different:

axm3 : src ∈ PACKETS → NODES
axm4 : dst ∈ PACKETS → NODES
axm5 : ∀ p · p ∈ PACKETS ⇒ src(p) ≠ dst(p)

The machine xyM00 specifies the sending (sent) and receiving (rcvd) packets (p) in the abstract way. Thereby, the send is set to the action:

$$sent := sent \cup \{s \mapsto p\}$$

The reception is set with the following action:

$$rcvd := rcvd \cup \{d \mapsto p\}$$

b. The first refinement: cutting packets on flits (Flitization)

The machine xyM01 refines xyM00 and introduces scutting packets on flits.

xyC01: FLITS is a new set introduced by this context by cutting each packet on flits (axm1), and the flits of each packet are different from those of other packets (axm2).

axm1 : flits ∈ PACKETS → ℙ1(FLITS)
axm2 : ∀ p1, p2 · p1 ∈ PACKETS ∧ p2 ∈ PACKETS ∧ p1 ≠ p2 ⇒ flits(p1) ∩ flits(p2) = ∅

xyM01: Instead of sending (act1) a whole package, this sends packet flits (see Figure 21) and can be received only when all the flits that were sent up (grd6);

Figure 21. Sends a packet as flits.

The SEND_FLIT event is defined at this level by the action:

WHERE
grd6 : f ∉ f_sent[{s}]
THEN
act1 : f_sent := f_sent ∪ {s ↦ f}

c. The second refinement: adding LocCopy variable

xyC02: The copy of the package in node (axm7) is in the original sources (axm8) and in the sources of these packages. *Theorem* (axm10) states that the local copy is originally in one place on the network.

axm7 : InitLocCopy ∈ NODES ↔ PACKETS
axm8 : ∀ p · p ∈ PACKETS ⇒ src(p) ↦ p ∈ InitLocCopy
axm9 : ∀ n, p · n ∈ NODES ∧ p ∈ PACKETS ∧ n ↦ p ∈ InitLocCopy ⇒ n = src(p)
axm10 : ∀ n1, n2, p · n1 ∈ NODES ∧ n2 ∈ NODES ∧ p ∈ PACKETS ∧ n1 ↦ p ∈InitLocCopy∧
n2 ↦ p ∈InitLocCopy ⇒ n1 = n2

xyM02: In this machine, refinement adds RECEIVE_FLIT to indicate when we can get a package (view the context xyC01).

WHERE
grd7 : flits(p) ⊆ f_sent[{s}]
THEN
act1 : f_rcvd := f_rcvd ∪ {d ↦ f}

d. Third refinement: Faulty node

xyM03: It is a refinement of the behaviour of a node if it is broken and/or when it returns to normal as expressed in both disable and enable events. This level also allows us to create the variable locCopy (see Figure 22, case (02)) to ensure flit sends of a packet without loss.

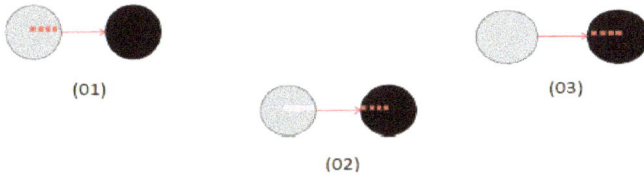

(01)

(02)

(03)

Figure 22. Routing flits keeping the local copy.

The following invariants are used to show how the model has been specified:

> inv1 : gr ⊆ NODES × NODES
> inv2 : str ∈ NODES ↔ FLITS
> inv3 : ran(str) ⊆ ran(f_sent)
> inv4 : ∀f· f ∉ ran (f_rcvd) ∧ f ∈ ran(f_sent) ⇒ f ∈ ran(str)
> inv5 : locCopy ∈ NODES ↔ PACKETS
> inv9 : gr = gr~
> inv10 : gr ⊆ g

The invariant *inv1* expresses that graph (gr) is set of NODES related between them in the current network. In *inv2*, str is a variable which contains the current position of a flit in the network. *inv3 is a* flit circulating in the network was sent by a source. *inv4* is a flit sent but not received and starved in the network. In *inv5*, locCopy is a packet in node. *inv9 indicates* that graph (gr) is symmetric while in *inv10* graph gr is always an amendment contained in the original graph g. Those are important invariants which represented in our graph.

Disable: a node n becomes faulty/off. It can no longer receive or route messages of its neighbours. The new graph new_gr will be the current without bidirectional links between the node n and its neighbours (grd4). grd6 expresses that the new_gr is symmetric and not-empty (grd6), and the action updating of the current graph with the new graph.

> **WHERE**
> grd1 : n ∈ NODES
> grd2 : n ∈dom(gr)
> grd3 : n ew_gr⊆ NODES × NODES
> grd4 : new_gr = gr \ (({n} × gr[{n}]) ∪ (gr[{n}] × {n}))
> grd5 : new_gr~ = new_gr
> grd6 : new_gr ≠ ∅
> **THEN**
> act1 : gr :=new_gr
> **END**

Enable: A node n becomes active. It can receive and route messages from its neighbours when n is no longer part of the current graph. It gives n in the current graph with bidirectional links with his former neighbours.

```
WHERE
    grd1 : n ∈ NODES
    grd2 : n ∉ dom(gr)
    grd3 : n ∈ dom(g)
THEN
    act1 : gr := gr ∪ (({n}×g[{n}])∪(g[{n}]×{n}))
END
```

e. Fourth refinement:

xyC03: The environment has allowed us to break the xyC04 context to simplify the proof with the RODIN tool.

- Extension of the definition of the initial graph by adding the concept of square graph. **g** is the initial graph in which all nodes are connected together since their coordinates (prj1(c), prj2(c)) are neighbours. This is a graph that is part of the eligible graphs.

```
∀ n1, n2, c1, c2 ·n1 ↦ c1 ∈posnd∧ n2 ↦ c2 ∈ posnd∧ n1 ≠ n2 ∧
(
  (prj1(c1) = prj1(c2)−1 ∧ prj2(c1) = prj2(c2)) ∨
  (prj2(c1) = prj2(c2)−1 ∧ prj1(c1) = prj1(c2)) ∨
  (prj1(c1) = prj1(c2)+1 ∧ prj2(c1) = prj2(c2)) ∨
  (prj2(c1) = prj2(c2)+1 ∧ prj1(c1) = prj1(c2))
)
⇒ n1 ↦ n2 ∈ g)
```

- Let x be a node x with coordinates cx (posnd(x)). If there is another node with coordinates cd as the axis of the coordinate but which smaller then it can be moved on the x-axis to the left from x.

```
∀ x, cx · x ∈ NODES ∧ cx = posnd(x) ∧
(∃ d, cd · d ∈ NODES ∧ cd = posnd(d) ∧ prj1(cx) > prj1(cd))
⇒
(prj1(cx)−1 ↦ prj2(cx)) ∈ ran(posnd)
```

- Let x be a node with coordinates cx. If there is a node with coordinates cd as the axis of the coordinate but which are larger than that of the x-axis, then it can be moved to the right from x.

```
∀ x, cx · x ∈ NODES ∧ cx = posnd(x) ∧
(∃ d, cd · d ∈ NODES ∧ cd = posnd(d) ∧ prj1(cx) < prj1(cd))
⇒
(prj1(cx)+1 ↦ prj2(cx)) ∈ ran(posnd)
```

- g is an initial graph in which all nodes are interconnected in pairs of adjacent coordinates of x and y. This is a graph that is part of the eligible graphs.

```
(∀ n1, n2, c1, c2 ·n1 ↦ c1 ∈posnd∧ n2 ↦ c2 ∈posnd∧ n1 ≠ n2 ∧ n1 ↦ n2 ∈ g
⇒ (
  (prj1(c1) = prj1(c2)−1 ∧ prj2(c1) = prj2(c2)) ∨
  (prj2(c1) = prj2(c2)−1 ∧ prj1(c1) = prj1(c2)) ∨
  (prj1(c1) = prj1(c2)+1 ∧ prj2(c1) = prj2(c2)) ∨
  (prj2(c1) = prj2(c2)+1 ∧ prj1(c1) = prj1(c2))
)
)
```

- We give the "neighbours" (ind_nbg) diagonal "distance 1" node:

$$\boxed{\text{ind\_nbg} \in \text{NODES} \rightarrow \mathbb{P}(\text{NODES})}$$

- If a node n is part of the faulty area and a node fails new_f_nodef_node, the set_of_f_nodes set is a direct neighbour or diagonal n. Thus, it is part of failing zone around new_f_node.

$$\boxed{\forall \, n, r, nod, nd \cdot n \mapsto r \in dom(zn) \wedge nod \in r \wedge nd \in zn(n \mapsto r) \wedge (nod \mapsto nd \in g \vee nd \in ind\_nbg(nod)) \Rightarrow nod \in zn(n \mapsto r)}$$

- *Theorem:* If a node fails f_nodeset_of_f_nodes, all nodes are direct neighbours or diagonal new_f_node. Then, it is part of the defective area new_f_node.

$$\boxed{\forall \, n, r, nod \cdot n \mapsto r \in dom(zn) \wedge nod \in r \wedge (n \mapsto nod \in g \vee nod \in ind\_nbg(n)) \Rightarrow nod \in zn(n \mapsto r)}$$

xyC04: Introduces the operators for calculating the active zone surrounding a node near a faulty zone. The step of the construction rectangular shape of an active zone za is as follows:

- Xmin(b) contains the minimum X coordinate found in b:

$$\boxed{\forall \, a, b \cdot b \subseteq \text{NODES} \wedge b \neq \emptyset \wedge a \in b \Rightarrow (\exists \, c \cdot c \in b \wedge prj1(posnd(c)) \leq prj1(posnd(a)) \wedge Xmin(b) = prj1(posnd(c)))}$$

- XMax(b) contains the maximum X coordinate found in b:

$$\boxed{\forall \, a, b \cdot b \subseteq \text{NODES} \wedge b \neq \emptyset \wedge a \in b \Rightarrow (\exists \, c \cdot c \in b \wedge prj1(posnd(c)) \geq prj1(posnd(a)) \wedge Xmax(b) = prj1(posnd(c)))}$$

- Ymin(b) contains the minimum Y coordinate found in b:

$$\boxed{\forall \, a, b \cdot b \subseteq \text{NODES} \wedge b \neq \emptyset \wedge a \in b \Rightarrow (\exists \, c \cdot c \in b \wedge prj2(posnd(c)) \leq prj2(posnd(a)) \wedge Ymin(b) = prj2(posnd(c)))}$$

- YMax(b) contains the maximum Y coordinate found in b:

$$\boxed{\forall \, a, b \cdot b \subseteq \text{NODES} \wedge b \neq \emptyset \wedge a \in b \Rightarrow (\exists \, c \cdot c \in b \wedge prj2(posnd(c)) \geq prj2(posnd(a)) \wedge Ymax(b) = prj2(posnd(c)))}$$

- Lim Xmin(a) contains less than the X coordinate of the bounding rectangle and has Xmin(a) -1 within the graph g:

$$\boxed{\forall \, a \cdot a \subseteq \text{NODES} \wedge a \neq \emptyset \wedge Xmin(a) - 1 > 0 \Rightarrow LimXmin(a) = Xmin(a) - 1}$$

- Lim max (a) contains the coordinate of the upper X bounding rectangle and has Xmax (a) if one within the graph g:

$$\boxed{\forall \, a \cdot a \subseteq \text{NODES} \wedge a \neq \emptyset \wedge Xmax(a) + 1 < nsize - 1 \Rightarrow LimXmax(a) = Xmax(a) + 1}$$

- Lim Ymin (a) contains the Y coordinate of the bounding rectangle and has lower Xmin (a) -1 within the graph g:

$$\boxed{\forall \, a \cdot a \subseteq \text{NODES} \wedge a \neq \emptyset \wedge Ymin(a) - 1 > 0 \Rightarrow LimYmin(a) = Ymin(a) - 1}$$

- Lim Ymax (a) contains the coordinate of upper rectangle encompassing a Y. Ymax (a) if one within the graph g:

$$\boxed{\forall \, a \cdot a \subseteq \text{NODES} \wedge a \neq \emptyset \wedge Ymax(a) + 1 < nsize - 1 \Rightarrow LimYmax(a) = Ymax(a) + 1}$$

- The rectangle given by za(a) and including a contains n nodes whose coordinates (x, y) are defined as:

LimXmin(a) $\leq$ x $\leq$ LimXmax(a) and LimYmin(a) $\leq$ y $\leq$ LimYmax(a)

$$\boxed{\begin{aligned} &\text{axm24}: \forall \, a \cdot a \subseteq \text{NODES} \wedge a \neq \emptyset \\ &\Rightarrow za(a) = \{nd \mid (prj1(posnd(nd)) \geq LimXmin(a) \wedge prj1(posnd(nd)) \leq LimXmax(a) \wedge prj2(posnd(nd)) \geq LimYmin(a) \wedge \\ &\quad prj2(posnd(nd)) \leq LimYmax(a))\} \end{aligned}}$$

We build the rectangle of the active area after forcing the evidence and limit the active node in NoC inside the network size limit. Thus, we add a rule to the firewall rule sets as follows:

Rules 5: The active zone has the option to occasionally disable unauthorized cases (see Figure 14b) for routing flits of a packet that has no possibility to arrive at the destination.
xyM04: This machine contains a refinement of two events:

- The routing flits in different directions depending on the destination. If after a node (s) is transmitted flit (f) to the node (y), (x) still has flits of (f), the local copy (LocCopy) does not change and x no longer has flits of (f). The local copy (LocCopy) of packet (p) changes from x to y. This is expressed in the following warning:

flits(p) $\not\subseteq$ (str[{y}] $\cup$ {f}) $\Rightarrow$ newLocCopy = locCopy
flits(p) $\subseteq$ (str[{y}] $\cup$ {f}) $\Rightarrow$ newLocCopy = (locCopy\ {x $\mapsto$ p}) $\cup$ {y $\mapsto$ p}

The cases of the XY routing algorithm are matched with refinements of event FORWARD:

- **Case 01:** (FORWAD-W) It forwards a flit (f) of a pack (P) to a neighbouring node (y) located W. This event is raised if x-coordinated destination (d) (the flit (f)) is inferior to x-coordinated the current node (s).
- **Case 02:** (FORWARD-E) It forwards a flit (f) of a pack (P) to a neighbouring node (y) located E. This event is raised if x-coordinated destination (d) (the flit (f)) is superior to x-ordinated the current node (s).
- **Case 03:** (FORWARD-N) It forwards a flit (f) of a pack (P) to a neighbouring node (y) located in N. This event is triggered if y-coordinated destinations (d) (the flit (f)) is superior to y-coordinated current (x) and x-coordinated the destination node (d) is equal to x-ordinated the current node (s), or if conflicts cannot pass along the x-axis.
- **Case 04:** (FORWARD-S) It forwards a flit (f) of a pack (P) to a neighbouring node (y) situated S. This event is triggered if y-coordinated destination (d) (the flit (f)) is less than y-coordinated to the current (x) and x-ordinated the destination node (d) is equal to x-ordinated the current node (s), or if conflicts cannot pass along the x-axis.

- The delivery of a flit (f) of the packet (p) from the node (s) in an active zone.

- **FORWARD_AUTH1:** Can forward a flit (f) a packet (p) to a neighbouring node (y) to the east of the current node (x) knowing that the source node (b) in the south compared to the current node (x) and headed to a destination (d) to the east. With RODIN, this case is expressed as follows:

grd14 : b $\in$ NODES
grd15 : d $\in$ NODES
grd16 : x $\neq$ d
grd17 : x $\in$ z_a
grd18 : prj1(posnd(x))< prj1(posnd(d))
grd19 : prj2(posnd(y)) = prj2(posnd(x))
grd20 : prj1(posnd(y)) = prj1(posnd(x))+1
grd21 : prj2(posnd(b))< prj2(posnd(x))
grd22 : prj1(posnd(x)) = prj1(posnd(b))
grd23 : b=src(p)

- **FORWARD_AUTH2:** Can forward a flit (f) and a packet (p) to a neighbouring node (y) to the right of the current (x) node knowing the source node (b) to the West relative to the current node (x) and headed to a destination (d) to the east. With RODIN, this case is expressed as as follows:

$$grd20 : prj1(posnd(y)) = prj1(posnd(x))+1$$
$$grd21 : prj2(posnd(b))= prj2(posnd(x))$$
$$grd22 : prj1(posnd(b)) = prj1(posnd(x)) -1$$

- **FORWARD_AUTH3:** Can forward a flit (f) and a packet (p) to a neighbouring node (y) to the West of the current (x) node knowing the source node (b) is located in relation to node current (x) and headed to a destination (d) to the west. With RODIN, this case is expressed as follows:

$$grd18 : prj1(posnd(x))> prj1(posnd(d))$$
$$grd19 : prj2(posnd(y)) = prj2(posnd(x))$$
$$grd20 : prj1(posnd(y)) = prj1(posnd(x))-1$$
$$grd21 : prj2(posnd(b))= prj2(posnd(x))$$
$$grd22 : prj1(posnd(b)) = prj1(posnd(x)) +1$$

- **FORWARD_AUTH4:** Can forward a flit (f) and a packet (p) to a neighbouring node (y) to the West of the current (x) node knowing the source node (b) to the east from the node current (x) and headed to a destination (d) to the west. With RODIN, this case is expressed as follows:

$$grd18 : prj1(posnd(x))> prj1(posnd(d))$$
$$grd19 : prj2(posnd(y)) = prj2(posnd(x))$$
$$grd20 : prj1(posnd(y)) = prj1(posnd(x))-1$$
$$grd21 : prj2(posnd(b))> prj2(posnd(x))$$
$$grd22 : prj1(posnd(b)) = prj1(posnd(x))$$

Table 3 gives the number of proofs obligations for the refinements development of the bypassing routing which are automatically discharged proving the checking of criteria for each considered levels of abstraction and step by step. The proposed incremental formal verification of a fault tolerant routing scheme of the NoC-switch strategy has been performed using RODIN environment. The number of proof obligations measures the complexity of the development that are automatically/manually discharged. Note that, in the case of M03 Machine, there is more evidence than interactive machines, which explains that the goal was simplified (semi-automatic). By cons in the context C03, we needed more assumptions in order to find more evidence that a automatically discharged.

Table 3. Results of proof obligations for the bypassing routing models.

Element Name	Total	Auto	Manuel	Reviewed	Undischarged
ActiveZone	27	25	2	0	0
xyC00	1	1	0	0	0
xyC01	0	0	0	0	0
xyC02	16	14	2	0	0
xyC03	3	3	0	0	0
xyC04	3	3	0	0	0
xyM00	0	0	0	0	0
xyM01	0	0	0	0	0
xyM02	13	9	4	0	0
xyM03	41	33	8	0	0
xyM04	22	10	12	0	0

3. Discussion

We have given a formal verification of a bypassing routing algorithm for the design of reliable NoC by using the Event-B methodology. The main originality of the proposed routing algorithm based on the bypass rules is that the routing is performed in the activated area depending on the number of faulty nodes or regions where, in the case of a rectangular region (or a group of faults forming a rectangular shape), the activated nodes form a ring around the region and the routing in activated area is reduced to route along the formed ring. Therefore, a formal proof on the bypassing operations inside NoC and the dependability problems are related to the choice of path routing of data packets and strategies imposed for diversion in the case of detected faulty routers. The formalization process is based on an incremental and validated correct-by-construction development of the adaptive NoC architecture. We have considered the event B refinement as stepwise validation. Indeed, the Event B modelling language can express safety properties, which are invariants, theorems or safety properties in a machine corresponding to the system. Thereby, Event B allows a progressive development of models through refinements by considering two main structures available in Event B which are Contexts expressing static information about the model and Machines expressing dynamic information about the model, invariants, safety properties, and events. The mere usage/running of provers (provided by the RODIN platform) allowed us to discharge these obligations (see Tables 2 and 3). These proof operators can be used to improve the NoC architecture specified using Event-B models through refined models while to get out all the bugs during the creation of hardware architecture which cannot found with simulations or analytic methods. Indeed, contrary to verification by simulation only, our work provides a framework for developing the Network-on-Chip Architecture based XY algorithm. Thus, this demonstration is comprehensive in order to detect problems even for configurations that designers have not yet considered. The purpose is to develop a dependable routing algorithm using essential safety properties together with a formal proof that asserts its correctness. This will allow proving the functionality of the associated hardware architecture by replacing the time consuming simulations in the design flow by a formal proof method.

4. Conclusions and Future Work

In this paper, we have proposed a formal method using Event-B to specify, verify and prove the behaviour of fault tolerant routing scheme suitable for design reliable or adaptive NoC architectures. More precisely, we have considered the correct-by-construction paradigm for specifying the behaviour hardware systems. The correct-by-construction paradigm offers an alternative approach to prove and derive correct systems and architectures, through the reconstruction of a target system using stepwise refinement and validated methodological techniques. Our goal is to complement the time consuming simulations in the design flow with a formal proof method. This formal verification [28] can either compare the computer data that will enable the realization of a future NoC Switch with its specification, or verify that the circuit obeys certain constraints characterizing its proper functioning. The novelty of our proposed design approach for the NoC routing associated with error detectors is that it can prove the dependability of the routing of the data packets to get around the faulty switches or regions (zone bypass decision) which is adapted for the design of fault tolerant NoCs. We have formalized and proven that our proposed algorithm allows the routing of packets in the networks incorporating faulty nodes and regions. Our approach relies on an incremental development of routing operations of the Network-on-Chip Architecture, using the Event B formalism. The formalization of the architecture is presented from an abstract level to a more concrete level in a hierarchical way. We have proved our incremental formal verification of our proposed fault tolerant routing scheme of the NoC witch strategy with environmental RODIN, where the number of proof obligations measures the complexity of the development, which have been automatically/manually discharged. This approach will develop the general process of modelling with Event-B which makes the notion of refinement intuitive.

A framework for developing NoC Architectures and the XY routing algorithm using essential safety properties together with a formal proof that asserts its correctness has been developed. As a part of our future efforts, we have considered the translation of the most concrete (detailed and close to algorithmic form) model into an intermediate language, from which hardware description (e.g., in *VHDL*) can be extracted. Moreover, it is noted that the first levels of the Event-B design of the NoC Architecture express general cases of routing methodologies and fall in the interesting domain of reusable and generic refinement based structures. We plan to investigate further this domain of generic and reusable proof-based models.

Systems **2017**, *5*, 17

Author Contributions: Hayat Daoud and Mikael Heil performed the experiments and have contributed to wrote the paper. Camel Tanougast conceived the experiments and wrote the paper. Mostefa Belarbi analyzed the results. Camille Diou contributed to the analysis of RODIN tool and the NoC architecure as well as to wrote the paper.

Conflicts of Interest: The authors declare no conflict of interest.

References

1. Killian, C.; Tanougast, C.; Monteiro, F.; Dandache, A. Smart Reliable Network-on-Chip. *IEEE Trans. Very Large Scale Integr. Syst.* **2014**, *22*, 242–255. [CrossRef]
2. Leonidas, T.; Sere, K.; Plosila, J. Modeling Communication in Multi–Processor Systems–on–Chip Using Modular Connectors. Innovations in Embedded and Real-Time Systems Engineering for Communication. *IGI Glob.* **2012**, 219–240. [CrossRef]
3. Guang, L.; Plosila, J.; Isoaho, J.; Tenhunen, H. Hierarchical Agent Monitored Parallel On-Chip System: A Novel Design Paradigm and Its Formal Specification. In *Innovations in Embedded and Real-Time Systems Engineering for Communication*; IGI Publishing: Hershey, PA, USA, 2010; pp. 86–105.
4. Ostroumov, S.; Tsiopoulos, L.; Plosila, J.; Sere, K. Formal approach to agent-based dynamic reconfiguration in Networks-On-Chip. *J. Syst. Archit.* **2013**, *59*, 709–728. [CrossRef]
5. Verbeek, F.; Schmaltz, J. Easy Formal Specification and Validation of Unbounded Networks-on-Chips Architectures. *ACM Trans. Design Autom. Electron. Syst.* **2012**, *17*. [CrossRef]
6. Borrione, D.; Helmy, A.; Pierre, L.; Schmaltz, J. A Formal Approach to the Verification of Networks on Chip. *EURASIP J. Embed. Syst.* **2009**, *2009*, 548324. [CrossRef]
7. Aydi, Y.; Tligue, R.; Elleuch, M.; Abid, M.; Dekeyser, J. A Multi Level Functional Verification of Multistage Interconnection Network for MPSOC. In Proceedings of the 16th IEEE International Conference on Electronics, Circuits, and Systems, Yasmine Hammamet, Tunisia, 13–16 December 2009; pp. 439–442. [CrossRef]
8. Van den Broek, T.; Schmaltz, J. Towards A Formally Verified Network-on-Chip. In Proceedings of the 2009 Formal Methods in Computer-Aided Design, (FMCAD 2009), Austin, TX, USA, 15–18 November 2009; pp. 184–187.
9. Schmaltz, J.; Borrione, D. A functional formalization of on chip communications. *Form. Asp. Comput.* **2008**, *20*, 241–258. [CrossRef]
10. Voros, J.; Snook, C.; Hallerstede, S.; Masselos, K. Embedded System Design Using Formal Model Refinement: An Approach Based on the Combined Use of UML and the B Language. *Design Autom. Embed. Syst.* **2004**, *9*, 67–99. [CrossRef]
11. Laibinis, L.; Troubitsyna, E.; Leppänen, S. Service-Oriented Development of Fault Tolerant Communicating Systems: Refinement Approach. *Int. J. Embed. Real-Time Commun. Syst.* **2010**, *1*, 61–85. [CrossRef]
12. Verbeek, F.; Schmaltz, J. Formal Specification of Networks-on-Chips:Deadlock and Evacuation. In Proceedings of the Design, Automation Test Europe Conference & Exhibition, Dresden, Germany, 8–12 March 2010; pp. 1701–1706.
13. Helmy, A.; Pierre, L.; Jantsch, A. Theorem Proving Techniques for the Formal Verification of NoC Communications with Non-minimal Adaptive Routing. In Proceedings of the IEEE 13th International Symposium on Design and Diagnostics of Electronic Circuits and Systems, Vienna, Austria, 14–16 April 2010; pp. 221–224.
14. Yang, S.-A.; Baras, J.S. Correctness Proof for a Dynamic Adaptive Routing Algorithm for Mobile Ad-hoc Networks. In *IFAC Workshop—Modeling and Analysis of Logic Controlled Dynamic Systems*; Elsevier: New York, NY, USA, 2003; pp. 1–10.
15. Wu, J. A fault-tolerant and deadlock-free routing protocol in 2D meshes based on odd-even turn model. *IEEE Trans. Comput.* **2003**, *52*, 1154–1169.
16. Park, D.; Nicopoulos, C.; Kim, J.; Vijaykrishnan, N.; Das, C. Exploring fault-tolerant network-on-chip architectures. In Proceedings of the International Conference on Dependable Systems and Networks (DSN 2006), Philadelphia, PA, USA, 25–28 June 2006; pp. 93–104.

17. Cansell, D.; Tanougast, C.; Beviller, Y. Integration of the proof process in the design of microelectronic architecture for bitrate measurement instrumentation of transport stream program MPEG-2 DVB-T. In Proceedings of the IEEE International Workshop on Rapid System Prototyping, Geneva, Switzerland, 28–30 June 2004; pp. 157–163.

18. Leavens, G.T.; Abrial, J.-R.; Batory, D.S.; Butler, M.J.; Coglio, A.; Fisler, K.; Hehner, E.C.R.; Jones, C.B.; Miller, D.; Jones, S.L.P.; et al. Roadmap for enhanced languages and methods to aid verification. In *Proceedings of the 5th International Conference on Generative Programming and Component Engineering, Portland, OR, USA, 22–26 October 2006;* Jarzabek, S., Schmidt, D.C., Veldhuizen, T.L., Eds.; ACM: New York, NY, USA, 2006; pp. 221–236.

19. Abrial, J.-R.; Cansell, D.; Mery, D. A mechanically proved and incremental development of IEEE 1394 tree identify protocol. *Form. Asp. Comput.* **2003**, *14*, 215–227. [CrossRef]

20. Clarke, E.M.; Grumberg, O.; Jha, S. Verifying parameterzed networks. *ACM Trans. Program. Lang. Syst.* **1997**, *19*, 726–750. [CrossRef]

21. Bharadwaj, R.; Felty, A.; Stomp, F. Formalizing Inductive Proofs of Network Algorithms. In Proceedings of the 1995 Asian Computing Science Conference, Springer-Verlag, London, UK, 11–13 December 1995.

22. Curzon, P. Experiences formally verifying a network component. In Proceedings of the IEEE Conference on Computer Assurance, Gaithersburg, MD, USA, 27 June–1 July 1994.

23. Gordon, M.; Melham, T. *Introduction to HOL: A Theorem Proving Environment for Higher Order Logic*; Cambridge University Press: Cambridge, UK, 1993.

24. Chenard, J.S.; Bourduas, S.; Azuelos, N.; Boul'e, M.; Zilic, Z. Hardware Assertion Checkers in On-line Detection of Network-on-Chip Faults. In Proceedings of the Desig Automation and Test in Europe, Worshop on Diagnostic Services in Networks-on-Chips, Nice, France, 16–20 April 2007.

25. IEEE. *IEEE Standard for Property Specification Language (PSL)*; IEEE STD 1850. Available online: http:// ieeexplore.ieee.org/document/1524461/ (accessed on 20 February 2017).

26. Boule, M.; Zilic, Z. Automata-based assertion-checker synthesis of PSL properties. *ACM Trans. Des Autom. Electron. Syst.* **2008**, *13*, 1–21. [CrossRef]

27. Clarke, E.M., Jr.; Grumberg, O.; Peled, D.A. *Model Cheking*; The MIT Press: Cambridge, MA, USA; London, UK, 1999; p. 6.

28. Ogras, U.; Hu, J.; Marculescu, R. Key Research Problems in NoC Design: A Holistic Perspective. In Proceedings of the International Conference on Hardware/Software Codesign and System Synthesis (CODES+ISSS'2005), New York, NY, USA, 18–21 September 2005; pp. 69–74.

29. Goossens, K. Formal Methods for Networks on Chips. In Proceedings of the Fifth International Conference on Application of Concurrency to System Design (ACSD'05), Saint Malo, France, 7–9 June 2005; IEEE Computer Society: Washington, DC, USA, 2005; pp. 188–189.

30. Pande, P.; De Micheli, G.; Grecu, C.; Ivanov, A.; Saleh, R. Design, Synthesis, and Test of Networks on Chips. *IEEE Design Test Comput.* **2005**, *22*, 404–413. [CrossRef]

31. Bjerregaard, T.; Mahadevan, S. A survey of research and practices of Network-on-chip. *ACM Comput. Surv.* **2006**, *38*, 1–51. [CrossRef]

32. Jantsch, A. Models of Computation for Networks on Chip. In Proceedings of the Sixth International Conference on Application of Concurrency to System Design (ACSD'06), Turku, Finland, 28–30 June 2006; IEEE Computer Society: Washington, DC, USA, 2006; pp. 165–178.

33. Nielsen, S.F.; Sparso, J. Analysis of low-power SoC interconnection networks. In Proceedings of the IEEE 19th Norchip Conference, Kista, Sweden, 12–13 November 2001; pp. 77–86.

34. Grecu, C.; Ivanov, A.; Saleh, R.; Sogomonyan, E.; Pande, P. On-line Fault Detection and Location for NoC Interconnects. In Proceedings of the International On-Line Testing Symposium (IOLTS'06), Lake of Como, Italy, 10–12 July 2006; pp. 1–8.

35. Murali, S.; De Micheli, G.; Benini, L.; Theocharides, T.; Vijaykrishnan, N.; Irwin, M.J. Analysis of Error Recovery Schemes for Networks on Chips. *IEEE Design Test Comput.* **2005**, *22*, 434–442. [CrossRef]

36. Schafer, M.; Hollstein, T.; Zimmer, H.; Glesner, M. Deadlock-free routing and Component placement for irregular mesh-based networks-on-chip. In Proceedings of the 2005 IEEE/ACM International Conference on Computer-Aided Design (ICCAD'05), San Jose, CA, USA, 6–10 November 2005; IEEE Computer Society: Washington, DC, USA, 2005; pp. 238–245.

37. Gebremichael, B.; Vaandrager, F.W.; Zhang, M.; Goossens, K.; Rijpkema, E.; Radulescu, A. Deadlock Prevention in the Æthereal protocol. In Proceedings of the 13th IFIP WG 10.5 Advanced Research Working Conference on Correct Hardware Design and Verification Methods (CHARME'05), Saarbrücken, Germany, 3–6 October 2005.

38. Bendisposto, J.; Leuschel, M.; Ligot, O.; Samia, M. La validation de modèles Event-B avec le plug-in ProB pour RODIN. *TSI* **2008**, *27*, 1065–1084. [CrossRef]

39. Sayar, I.; Bhiri, M.-T. From an abstract specification in event-B toward an UML/OCL model. In Proceedings of the 2nd FME Workshop on Formal Methods in Software Engineering (FormaliSE 2014), Hyderabad, India, 3 June 2014; pp. 17–23.

40. Jastram, M. (Ed.) *RODIN User's Handbook*; Sponsored by the Deploy Project, 2012. Available online: https://www3.hhu.de/stups/handbook/rodin/current/pdf/rodin-doc.pdf (accessed on 20 February 2017).

41. Andriamiarina, M.B.; Daoud, H.; Belarbi, M.; Méry, D.; Tanougast, C. Formal verification of fault tolerant NoC-based architecture. In Proceedings of the First International Workshop on Mathematics and Computer Science (IWMCS 2012), Tiaret, Algeria, 16–17 December 2012.

42. Métivier, Y.; Robson, J.M.; Saheb-Djahromi, N.; Zemmari, A. Brief Annoucement: Analysis of an Optimal Bit Complexity Randomised Distributed Vertex Colouring Algorithm (Extended Abstract). In Proceedings of 13th International Conference on Principles of Distributed Systems (OPODIS 2009), Nîmes, France, 15–18 December 2009; pp. 359–364.

43. Duffy, K.; O'Connell, N.; Sapozhnikov, A. Complexity analysis of a decentralised graph colouring algorithm. *Inf. Process. Lett.* **2008**, *107*, 60–63. [CrossRef]

44. Nickerson, B.R. Graph colouring register allocation for processors with multi-register operands. In *Proceedings of the ACM SIGPLAN 1990 Conference on Programming Language Design and Implementation (PLDI'90), White Plains, NY, USA, 20–22 June 1990*; ACM: New York, NY, USA; pp. 40–52.

45. Schneider, J.; Wattenhofer, R. A new technique for distributed symmetry breaking. In *Proceedings of the 29th ACM SIGACT-SIGOPS Symposium on Principles of Distributed Computing (PODC'10), Zurich, Switzerland, 25–28 July 2010*; ACM: New York, NY, USA; pp. 257–266.

46. Malkawi, M.; Hassan, M.A.-H.; Hassan, O.A.-H. A new exam scheduling algorithm using graph coloring. *Int. Arab J. Inf. Technol.* **2008**, *5*, 80–86.

systems MDPI

Article

Building the Observer into the System: Toward a Realistic Description of Human Interaction with the World

Chris Fields

243 West Spain St., Sonoma, CA 95476, USA; fieldsres@gmail.com; Tel.: +1-707-980-5091

Academic Editors: Gianfranco Minati, Eliano Pessa and Ignazio Licata
Received: 26 July 2016; Accepted: 24 October 2016; Published: 28 October 2016

Abstract: Human beings do not observe the world from the outside, but rather are fully embedded in it. The sciences, however, often give the observer both a "god's eye" perspective and substantial *a priori* knowledge. Motivated by W. Ross Ashby's statement, "the theory of the Black Box is merely the theory of real objects or systems, when close attention is given to the question, relating object and observer, about what information comes from the object, and how it is obtained" (*Introduction to Cybernetics*, 1956, p. 110), I develop here an alternate picture of the world as a black box to which the observer is coupled. Within this framework I prove purely-classical analogs of the "no-go" theorems of quantum theory. Focussing on the question of identifying macroscopic objects, such as laboratory apparatus or even other observers, I show that the standard quantum formalism of superposition is required to adequately represent the classical information that an observer can obtain. I relate these results to supporting considerations from evolutionary biology, cognitive and developmental psychology, and artificial intelligence.

Keywords: black box; classicality; environment as witness; Landauer's principle; pragmatic information; quantum Darwinism; separability; superposition; time

---

The theory of the Black Box is merely the theory of real objects or systems, when close attention is given to the question, relating object and observer, about what information comes from the object, and how it is obtained.

— W. Ross Ashby, 1956 ([1], p. 110)

1. Introduction

Modern science is built on two far-reaching ideas: that there is a way that the world works, and that we human beings are not separate from but are rather part of the world. The first idea matured in the 17th and 18th centuries, reaching its full expression—but by no means widespread acceptance—in Laplace's concept of a mechanical, fully-deterministic universe. The maturation of the second idea did not begin until the birth of modern biology and psychology in the mid-19th century, and it is not yet complete. While both biology and psychology firmly place human beings and other organisms within the world, and while the enormous progress made in these disciplines during the past 50 years allows the construction of increasingly-sophisticated models of human beings and other organisms as systems that acquire information from their environments and use that information to act on their environments, "the observer" and her alter ego, "the experimenter" still stand outside of our most basic physical theories.

The fundamental fact required to incorporate the observer/experimenter (hereafter, in keeping with tradition, simply "the observer") into physical theory has, however, been recognized for well over a century. It was first formally elucidated by Boltzmann in the 1880s: reducing uncertainty by acquiring information requires the expenditure of energy. The minimal energetic cost of acquiring and recording one bit, i.e., the answer to one yes/no question, is known: it is $ln2\,kT$, where k is Boltzmann's constant and T is temperature [2–4]. However, k is small, $k \sim 1.38 \times 10^{-23}$ J/K in SI units, so at physiological temperature, $T = 310$ K, the minimal per-bit energetic cost $ln2\,kT \sim 3 \times 10^{-21}$ J is vastly smaller than typical macroscopic energies. This tiny value for the per-bit energetic cost of observation justified the traditional ideal, implicit in classical physics, of the "detached" observer whose observational activities had no impact on the observed world.

When, almost 50 years after Boltzmann's discovery, the early development of quantum theory introduced a formalism in which observation *did* impact the world—collapsing wave functions and even determining whether entities would behave as waves or particles—this violation of the ideal of detachment quickly assumed metaphysical proportions. From the remarkable Solvay Conference of 1927 [5] onwards, debates about the physical meaning of the quantum formalism have remained largely metaphysical (for synoptic reviews, see [6,7]; for recent surveys of interpretative positions, see [8–10]). Physicists regularly deplore this situation, with Fuchs, for example, likening interpretative stances to religions [11] and Cabello labelling the entire interpretative landscape a "map of madness" [12]. Nonetheless, substantial numbers of physicists—perhaps most famously, Bell [13]—maintain that the observer and the process of observation can play no role in any theory, and in particular no role in any *physical* theory, describing a world that worked in whatever way it worked long before humans existed and will continue to work in that way long after humans are gone. Even physicists who embrace some form of what Fuchs [14] has called, following Wheeler [15], "participatory realism" may insist, as Fuchs himself has done, that physics can provide no *theory* of the observer [11]. If physics can offer no theory of the observer, however, the observer still stands outside of physics, whether as a "participant" or not.

We *know*, however, what observers do. Observers acquire information and, in their complementary role as actors on their environments, they create information. Biology and psychology tell us, in ever increasing detail, *how* they acquire information and *how* they act to create more of it. Hence observation, together with its complement, experimentation, is a process of information exchange between an observer and that observer's environment. Both the observer and the observer's environment are parts of the world, and if the "environment" is expanded to become the observer's complement, the two together constitute the world. Boltzmann's insight then tells us that any exchange of information between observer and environment involves an exchange of energy, and hence a *physical interaction*, between observer and environment. This simple consideration renders the detached observer a mere approximation, one that may be expected sometimes—and possibly always—to fail.

That the idealization of the detached observer must be rejected even on classical-physics grounds is, of course, not a new idea: it is a cornerstone of second-order cybernetics. von Uexküll [16], von Foerster [17], Kampis [18], Koenderink [19] and many others have argued for it explicitly. Rössler's concept of endophysics [20] is based on it. The critical question in any theoretical framework that rejects the detached observer is that of how the "epistemic cut" [21] separating the observer from the observed environment is defined, and in particular, whether it is defined in a way that truly places the observer within the world. As Kampis puts it, "what endophysics aspires for is the study of the encompassing big black box of which the observer and his epistemology are a part...an enclosed observer is bound to the same laws as those of the system observed" ([18], p. 265). As von Neumann recognized [21], the observer being "bound by the same laws" is assured if the position of the cut is arbitrary (*cf*. Pattee, "The cut itself is an epistemic necessity, not an ontological condition" [22], p. 13). If the observer is "special" in some way, so that the observer-environment boundary cannot be "erased" within the theory without affecting the behavior of the world—as it appears to be the case, for example, in the endophysical framework of Kauffman and Gare [23]—then the observer remains in an important sense outside of the theory.

Classical cybernetics provides us with a simple formal model of the information exchange between the observer and the observed environment: the theory of the Black Box. My goal in this paper is to take seriously both the Black Box model and the requirement that the epistemic cut be arbitrary and to see where these assumptions jointly lead. What is offered here is, therefore, not a new model of observation, but rather an exploration in some depth of an old model of observation conjoined with an even-older general constraint on model construction. After reviewing the Black Box model and the associated epistemic cut in the next section, I show that conjoining the Black Box model with an arbitrarily-movable cut leads to "no-go" theorems, i.e., theorems limiting the inferences that may be drawn from either theoretical considerations or observational outcomes, that are classical analogs of no-go theorems familiar from quantum information theory. I then show that if a classical black-box world is assumed, by hypothesis, to comprise a collection of bounded, causally-independent "objects", finite observations can only identify superpositions of such objects. The classical Black Box model with an arbitrary epistemic cut thus requires us to view even a classical world in quantum-theoretic terms. The traditional "classical worldview" of bounded, independently-observable and independently-manipulable objects only emerges when we ignore the constraints of the Black Box model and assume that observers have a priori knowledge and local causal power that the Black Box model—and indeed, the formalisms of both quantum and classical physics—tell us that observers cannot obtain. To place this result in a larger context, I briefly discuss ideas and observations from evolutionary biology, cognitive and developmental psychology, computer science and artificial intelligence that point toward this same conclusion. I close by suggesting that if observers and observation are placed firmly within the world, its theory of the world is the only theory of the observer that physics needs. The observer only steps out of physical theory—and indeed, out of the physical world—when physics contradicts itself, granting to the observer knowledge and causal power that physics itself tells us no worldly observer can have.

2. The Black Box Model

The Black Box model is a formal representation of the interaction between an observer and the system being observed. This section reviews the informal notion of a "system" with particular emphasis on the assumption of separability. It then presents the Black Box model and provides reasons for believing that it accurately describes the interactions between observers and physical systems. It briefly discusses the relation between the sense of observer dependence entailed by the Black Box model and the informal notion of "subjectivity" with which it is sometimes associated.

2.1. Systems and Separability

The terms 'universe' and 'world' are here used interchangeably to denote the maximal closed system containing all possible observers and all possible observed systems as components. Here 'closed' has the usual meaning of not interacting with any other system. As observation requires interaction, any closed system is unobservable; the universe, as a closed system, is therefore unobservable. Setting special relativity aside for simplicity, the 'observed world' or 'environment' of any observer can be regarded as the complement of that observer within the world. In this case, given any arbitrary choice of observer, that observer together with that observer's environment jointly compose the world (taking special relativity into account converts this complement into a light cone).

The terms 'system' and 'physical system' are here used interchangeably to denote a component of the world, *not* a mathematical or other formal model of such a component as is sometimes intended by other authors. A "system" in the latter, formal-model sense can be stipulated. A system in the sense employed here can only be observed and/or acted upon, after which a model of that system may be constructed or stipulated. The term 'component' is here used informally; it can be made precise by stipulating a physical model of the universe which is then decomposed. If the universe is represented by a classical configuration (or phase) space, decomposition is implemented by the Cartesian product operator $\times$, while if the universe is represented by a Hilbert space, decomposition is implemented by the tensor product operator $\otimes$. In either case, the factors into which the universe is decomposed represent systems as defined here. The Black Box model is a model of the interaction between observers and systems, i.e., a model of the interaction between two components of the world, an observer (which is always itself a system) and a system (which may itself also qualify as an observer).

The usage employed here may be compared to that recently employed by Kitto [24]. It is consistent with the idea that a system is "a set of entities that are interacting via a set of relationships" ([24], p. 542), provided that the entities and relationships referred to are components and characteristics, respectively, of the world, as opposed to models, approximations, or simplifications of such components or characteristics. The current usage differs, however, from that of Kitto in that it *does not* assume that a system is "separate from its surroundings" ([24], p. 541). In particular, it does not assume that a system is *separable* from its surroundings as this term is employed in physics, i.e., it does not assume that a system occupies or can be assigned a state independently of the state occupied by or assigned to its surroundings. As discussed in detail below, this latter provision allows the observer-system boundary to be moved arbitrarily without altering the composition or the behavior of the world, i.e., of the composite system comprising the observer plus the observed system.

Separability is an intuitively appealing and quite natural assumption; hence rejecting it from the outset can be regarded as radical. Separability is, however, despite its intuitive appeal an *assumption*, indeed (as will be shown below) a very strong one. While failures of separability are typically identified (by definition) with entanglement in the quantum theory literature, moreover, they can arise in multiple formal and interpretative contexts. Tipler, for example, has shown that requiring classical dynamics to be strictly deterministic reproduces the formalism of unitary quantum theory and hence entails non-separability [25]; indeed he shows that the "quantum potential" required for strict classical determinism is identical to the "guiding field" of Bohmian quantum mechanics [26]. While Bohm interpreted this guiding field as non-local, however, Tipler interprets it as strictly local. In a similar vein, Rosen [27] formalizes the notion of a constraint hierarchy in classical complex systems proposed by Polanyi [28], showing that it can be represented as an arbitrarily extended hierarchy of potential functions. Rosen concludes, "a complex system requires an *infinite* mathematical object for its description ... it is quite clear that there is no such thing as a set of *states*, assignable to such a system once and for all" ([27], p. 190, emphasis in original). Here again we have a failure of separability in a *classical* context, one to which the quantum-theoretic concept of entanglement would ordinarily be thought irrelevant.

Separability is closely associated with contextuality, i.e., with the potential dependence of system behavior on how, in what setting, or with what preparations measurements are made. Kitto [24] distinguishes three kinds of contextuality: dependencies of system components on each other, dependencies of system behavior on how it is measured, and dependencies of system behavior on the environmental setting. The distinctions between these types of contextuality depend, however, on the assumption of separability. If the components of a separable system are non-separable, the first kind of contextuality appears as Kitto describes ([24], p. 551); similarly if a system **S** is not separable from the apparatus employed to measure its state, but the joint system **S**$'$ comprising **S** plus the apparatus is separable from its environment, the second kind of contextuality appears. If, however, the assumption of separability is dropped altogether, all three kinds of contextuality can appear and they cannot be distinguished. As shown in detail in §3 below, if a system cannot be bounded, its components cannot be identified; *ipso facto* they cannot be distinguished from components of the apparatus, the environment, or for that matter, the observer. Here the distinction between the *identification* of components—something that must be accomplished by observation—and the *stipulation* of components within an hypothesized model (and hence the fundamental distinction between systems and models) is clearly essential.

By not assuming separability as an a priori condition on systems, the present analysis permits arbitrary contextuality. As shown in §3 below, an arbitrarily-movable epistemic cut renders the possibility of such contextuality inescapable. Potential effects of contextuality may in practice be *ignored*, i.e., simply assumed not to exist as they often are, but the consequences of doing so cannot be predicted and may be severe.

2.2. Definition of the Black Box

A *black box* as described by classical cybernetics [1] is a system to which observers have only exterior access. They can form hypotheses about what is going on inside the box, they can formulate and test by further observation models of what is going on inside the box, but they cannot disassemble the box, examine its interior workings, and determine by such observations the dynamics that the box executes.

An observer interacts with a black box by sending it strings of bits; traditionally this interaction is considered to involve turning dials, pushing buttons, and so forth. The black box, in turn, sends the observer strings of bits, e.g., by moving pointers on dials or changing the numbers in a digital display. The observer-box interaction is considered to take place in episodes, i.e., the observer does something to the box, waits for a certain period for the box to respond, and then does something else. The box may respond by doing nothing. The observer may also do nothing, in hopes of observing spontaneous behavior on the part of the box.

The observer is assumed to be a finite system with finite energy resources. The finite energetic cost of encoding each bit limits the observer to sending only a finite number of bits to the box, and to recording only a finite number of bits from the box, in each of a finite number of episodes of interaction. The lengths of these "input" and "output" bit strings can be considered the "resolution" of the inputs to and outputs from the box. For simplicity, the input and output resolutions are considered to be identical. In this case, the information exchange between observer and box can be considered a classical information channel [29] with a fixed capacity equal to the resolution. Again for simplicity, this classical information channel can be considered to be noise-free. The consequences of relaxing the restriction to finite communication through this classical channel is discussed in Section 3.3 below.

Nothing prevents the observer from sending the same input to the box multiple times; similarly, nothing prevents the box from sending the same output to the observer multiple times. If the channel capacity and hence the resolution of the inputs and outputs is small and the number of episodes of interaction is large, such repetitive behavior is inevitable.

The observer is assumed to be capable of observing her own inputs to the box, and to have a memory on which both the input sent and the output received in each episode are both written. Because the number of interaction episodes is finite, this memory is finite. It is assumed for simplicity that the observer's memory is sufficient to record all of the interaction episodes; this assumption is discussed further in Section 3.3 below. The observer can be considered to consult the memory in order to frame hypotheses about the box's behavior and to freely choose, based on these hypotheses, what input to send to the box in the next episode. The meaning of "freedom" in this context is also discussed in Section 3.3 below.

These descriptions, characterizations and assumptions can be formalized with the following:

Definition 1. *A black box is a system about which no observer can have more (non-hypothetical) information than is contained in a finite list of finite-length bit strings representing observed input-output transitions.*

The interaction between the observer and the box can be represented as in Figure 1a. The observer collects finite data about the behavior of the box "from the outside" and uses these data—the only data that are available—to understand and frame hypotheses about the box's internal structure and dynamics and to use these hypotheses to predict the output that will be produced by the box in response to each input.

2.3. The Black Box as a Model of Physical Systems

As indicated by the opening quotation, Ashby considered all "real objects or systems" to be black boxes, or at least to become black boxes as their structures and behaviors were probed at finer and finer resolutions. Opening the case of a laptop computer—or increasingly, even the engine compartment of an automobile—reveals within merely an arrangement of fully-encapsulated parts, "components" with internal structures hidden from the user. Such components are black boxes by design, meant to be installed within the larger system without modification and to be fully interchangable with replacements having the same input-output behavior without affecting the performance of the larger system. Further investigation of such components reveals that they too contain components. Even when a component is easily recognizable and "simple"—perhaps it is a machine screw or a wire—it is still possible to ask what internal structure explains its properties or behavior. However, at some point such questioning stops, leaving whatever components that remain black boxes.

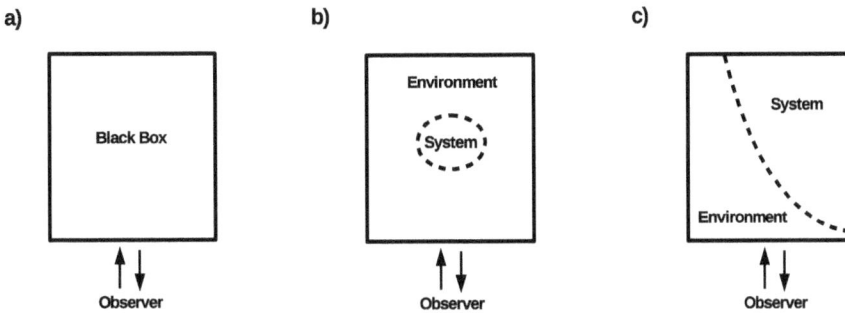

Figure 1. The Black Box model and two related concepts. (**a**) An observer interacts with a black box by exchanging information with the box through a classical channel of fixed, finite capacity; (**b**) The "environment as witness" formulation of decoherence permits observers to access information about a physical system only via the mediation of a surrounding environment; (**c**) Decompositional equivalence requires an observer's interactions with a composite system to be independent of the placement of subsystem boundaries within that system.

Probing any physical system deeply enough reveals structure and behavior only adequately describable using quantum theory. It is here that the black-box nature of physical systems becomes explicit. As Bohr repeatedly emphasized, in making measurements of the behavior of a quantum system one is inevitably forced to make use of macroscopic instruments. One must interact, in other words, with a macroscopic composite system comprising the quantum system of interest together with the apparatus used to measure its behavior. No independent access to the quantum system is possible. This lack of independent access renders the quantum system a black box. In addition, the physical coupling of the quantum system of interest to the apparatus blurs the boundary between the two, a blurring that is formalized as entanglement by the unitary time evolution of the Schrödinger equation. As Bohr put it, "it is no longer possible sharply to distinguish between the autonomous behavior of a physical object and its inevitable interaction with other bodies serving as measuring instruments, the direct consideration of which is excluded by the very nature of the concept of observation in itself" ([30], p. 290).

It is important here to emphasize that characterizing a quantum system—or any system—as a black box is making a statement about *observational access* to that system. It is straightforward to write down precise formal models of many quantum systems in either mixed or pure states and to claim, on the basis of such a model, that a system is well understood. Observing a model, however, is not the same thing as observing a system. Arbitrarily precise models can be constructed for any black box, and they can accurately predict some finite number of observations or not. The existence of such a model, however, has no bearing on whether the system being modelled is a black box. A system is a black box if and only if it satisfies the definition given above, i.e., if and only if it is a system about which no observer can have more (non-hypothetical) information than is contained in a finite list of finite-length bit strings representing observed input-output transitions. The statements of Ashby and

Bohr quoted above are statements about observational access, not statements about the possibility of model construction.

In the Copenhagen interpretation of Bohr and Heisenberg that became the standby of "textbook" quantum theory (e.g., [31]; see also [32]), it is the amplification of signals indicating the presence of quantum phenomena by a directly coupled macroscopic apparatus that converts the fragile information contained in quantum states into the robust, recordable classical information of observational outcomes. The more modern, though by now almost 50 years old, version of this view is the theory of environmental decoherence [33–40], in which exposure to and hence entanglement with any macroscopic environment, intentional or not, converts quantum information into classical information. Ollivier, Poulin and Zurek introduced the "environment as witness" formulation of environmental decoherence [41,42], later elaborated as the theory of "quantum Darwinism" [43–47] or "quantum state information broadcasting" [48–51] in recognition of the fact that observers typically obtain information about physical systems by interacting with the surrounding environment, typically the surrounding ambient photon field. This more recent formulation of environmental decoherence makes it explicit that observers are *not directly coupled* to the systems that their observational outcomes purport to describe, but are rather directly coupled to an intervening environment—effectively a very large apparatus—comprising vast numbers of unobserved degrees of freedom.

Central to the environment as witness formulation and quantum Darwinism is the requirement that environmental encodings of quantum states be *redundant* in the sense that many observers at different locations can each obtain the same information about the system of interest without disturbing either the interaction of the system with its environment or the information accessed by the other observers (cf. [42], Figure 1 and surrounding discussion). This redundancy requirement prevents observers from accessing information about the system by any means *other than* through the environment, as any such access would disturb the system and hence destroy the redundancy of the encoding [52,53]. The redundancy requirement, in other words, makes the environment a black box, a container that prevents direct observational access to any systems contained in its interior (Figure 1b). If observations of quantum systems satisfy the constraints imposed by the environment as witness formulation of decoherence, therefore, they can be characterized as interactions with a black box.

The theoretical framework of environmental decoherence assumes "no-collapse" or unitary quantum theory, i.e., quantum theory in which there is no *physical* process of quantum state collapse and hence no *physical* violation of unitarity [54]. Unitary quantum theory applies, however, only to closed systems. As first explicitly recognized by Everett [55], this restriction of unitary evolution to closed systems has two important consequences: (1) the universe, being a closed system, must evolve unitarily; and (2) the universe can contain no closed systems, as the absence of interactions between such systems and their surroundings would permit violations of unitarity. The environmental decoherence framework is explicitly designed to cope with the openness of all possible systems to interaction—and hence to observation—that is implied by point # 2. Schlosshauer, for example, remarks when introducing the decoherence framework that "over the past three or so decades it has been slowly realized that the isolated-system assumption—which, as we have described above, had proved so useful in classical physics and had simply been taken over to quantum physics—had in fact been the crucial obstacle to an understanding of the quantum-to-classical transition" ([40], p. 3–4).

If unitary quantum theory is correct, then there are no *physically* classical systems, only apparently or "for all practical purposes" (FAPP; terminology due to Bell [13]) classical systems. Alternative theories in which unitarity is physically violated have been proposed, and are sometimes (incorrectly) regarded as "interpretations" of quantum theory as opposed to alternative theories; examples include the theories of Ghirardi, Rimini and Weber [56], Penrose [57] or Weinberg [58]. In such theories, there are physically classical systems. However, these theories have at present no empirical support (e.g., [54,59,60]) and are increasingly challenged by theoretical approaches such as holography that invoke quantum theory at very large scales (e.g., [61–63]) and by derivations of quantum theory from purely information-theoretic foundations (see Section 3.4 below). This situation presents at least a *prima facie* challenge to any theoretical approach that assumes or implies the existence of physically-classical systems. Kitto, for example, remarks that in at least some cases "measurement merely records reality; it does not influence what is found during that measurement" while cautioning that "such assumptions, while sometimes correct, can be markedly dangerous" ([24], p. 547). If unitary quantum theory is correct, such assumptions maybe *useful* in some circumstances, but they are never correct in the strong sense of accurately describing reality. If the universe indeed exhibits large-scale entanglement as unitary quantum theory would imply, regarding "the notion of separability to be key to the definition of a system" ([24], p. 542) results in a universe without systems.

2.4. The Epistemic Cut and Decompositional Equivalence

The universality of the Black Box model for physical systems is, however, also supported by theoretical considerations far more fundamental than decoherence. As noted earlier, both classical and quantum physics represent the states of physical systems as vectors in state spaces, the real configuration (or phase) space in classical physics and the complex Hilbert space in quantum physics. The product operators that implement state-space factorization and its inverse, state-space combination are in each case associative, i.e., $A \times B \times C = (A \times B) \times C = A \times (B \times C)$ in classical physics and $A \otimes B \otimes C = (A \otimes B) \otimes C = A \otimes (B \otimes C)$ in quantum physics. Both theories, moreover, represent interactions between systems by linear operators, specifically Hamiltonians. These operators are additive, i.e., $H_{(A \times B)} = H_A + H_B + H_{AB}$ in classical physics and $H_{(A \otimes B)} = H_A + H_B + H_{AB}$ in quantum physics, where in each case the final term H_{AB} represents the $A - B$ interaction. In both theories, therefore, *decomposing a system into two or more components makes no difference to the composite system's structure or behavior.* The epistemic cut between components is, in this case, purely epistemic as Pattee [22] requires; where it is placed makes no difference to the composite system. An observer interacting with the composite system cannot tell, either by characterizing its state space or by measuring its dynamics, that it has been decomposed into components. Both the boundaries between the internal components and the interactions defined at these internal boundaries are invisible to an outside observer. The composite system is, from the outside observer's perspective, a black box within which its components are fully contained (Figure 1c). In describing quantum Bayesianism, Fuchs employs phenomenological language to make this point, referring to the "hermetically sealed" character of a black-box world as "interiority" and attributing an inviolable interiority to all systems, large and small ([11], p. 20).

The invisibility of internal component boundaries to outside observers, previously termed *decompositional equivalence* [64], has significant consequences for decoherence and particularly for quantum Darwinism [65–67]. If the classical information acquired by outside observers of a composite system does not depend on the placement of subsystem boundaries within that system, then it cannot depend on interactions defined at those boundaries. Observers of an environment cannot, therefore, determine the boundaries of systems embedded within that environment given only classical information encoded by that environment. Any such encoding is arbitrarily ambiguous about the boundaries of embedded systems. Multiple observers can interact with the same, redundant encoding, hypothesize radically different system-environment boundaries, and infer radically different system states as a result. They can agree completely about the observational outcomes they have obtained, but disagree about both the system and the system-environment interaction that produced them. This principled ambiguity about the states of internal systems extends even to whether they are coherent, i.e., entangled, or environmentally decohered, a phenomenon termed "entanglement relativity" [68–75].

The goal of observation is to determine the state space and the dynamics, i.e., the self-Hamiltonian, of the system being observed. The classical Black Box model, the environment as witness formulation of decoherence, and the formal constraints imposed on either quantum or classical physics by decompositional equivalence all locate the physical interaction that implements observation at a single boundary: the boundary between the observer and the observed composite environment, i.e., between the observer and the entire rest of that observer's observable universe. The physical interaction at this boundary implements the classical information channel between the observer and the composite environment that is that observer's observed world. Only information that flows through this classical channel contributes to observational outcomes. The observer can hypothesize, on the basis of these observational outcomes, the existence of internal boundaries within the composite environment and hence internal structure and interactions within the observed world. These hypotheses can only be tested, however, by examining their consequences for the observed world as a composite whole, because the observed world as a whole is the only system with which the observer directly interacts.

2.5. Observer Dependence and "Subjectivity"

A Black Box model of the observer-environment interaction entails a precise sense of observer dependence: each decomposition of the universe into an observer and that observer's environment is unique, so each observer observes a unique environment. It thus extends the observer-relativity of quantum states common to Everett's "relative state" interpretation [55], Rovelli's "relational" interpretation [76] and Fuch's quantum Bayesianism [11] to an observer-relativity of quantum *systems* (cf. Zanardi [68] where observer-relative systems are already taken for granted and Dugić and Jeknić-Dugić [70,71] where they are discussed explicitly). Moreover, each observer's observational capabilities—in quantum-theoretic language, the observables that each observer is capable of deploying—are exactly specified by the observer-environment interaction Hamiltonian. Decompositional equivalence entails that this sense of observer dependence is shared by classical and quantum physics. Many authors from von Uexküll [16] onwards have emphasized that observation is, by its very nature, observer-dependent in this way.

The sense of observer dependence implied by the Black Box model, and hence the model itself, is often criticized as subjectivist or even solipsist. Fuchs responds to the charge of solipsism with characteristic vigor: "No agent, no outcome for sure, but that's not solipsism: For, no system, no outcome either!" ([11], p. 19, Figure 5 caption). What the observer is *capable of* observing is determined by the observer-environment interaction (i.e., the Hamiltonian H_{OE}), and in the limit of a very large environment, primarily by the internal dynamics of the observer (i.e., by H_O). However, what the observer actually observes at any given instant—the output delivered by the black box in response to any given input—is determined by the environment (i.e., by H_E). This is "subjectivity" in the sense that an observer can only observe what she is capable of observing, but it is not "subjectivity" in the (typically pejorative) sense of an observer that observes only what she expects to observe or, worse yet, only what she wants to observe. As Moore [77] so clearly emphasized, the essence of the Black Box model is the ever-present possibility of surprise.

"Subjectivity" in the sense entailed by the Black Box model is, moreover, consistent with FAPP intersubjectivity and hence FAPP objectivity. Two very similar observers embedded in a very large world in such a way as to make their respective environments and hence their respective observer-environment interactions very similar can be expected to record very similar observational outcomes. If the bandwidths of their respective information channels and hence their effective measurement resolutions are small, their outcomes as recorded may even be identical. Observer dependence does not mean that agreement about outcomes is impossible. It rather means, as discussed in more detail in §3 and §4 below, that such agreement can carry no *ontological* weight. Nothing the observers can do can guarantee that their observations are caused by the same system; indeed the Black box model entails that their observations are caused by different systems, by their own unique and distinct, albeit very similar, environments. Each observer has, by definition, her own personal *umwelt* [16], and every observation she makes is constrained by it.

3. No-Go Theorems for Black Boxes

"No-go" theorems restrict the inferences that observers can make from either theory or observations. If the Black Box model indeed provides a general description of observational interactions with physical systems, it should provide no-go theorems that impose the same restrictions on inferences from theory or observations that the familiar no-go theorems of quantum information theory impose. This section shows that this is indeed the case.

3.1. Moore's Theorem (1956)

Edward Moore proved, in 1956, the fundamental no-go theorem for observers of black boxes ([77], Theorem 2):

Theorem 1 (Moore, 1956). *Finite observations of a black box are insufficient to determine its machine table.*

The "machine table" of a black box is the list of all *possible* input-output transitions of the black box. The finite list of observed input-output transitions accumulated by an observer of some black box **B** is at least a partial machine table of **B**. Moore's theorem shows that no such partial machine table can be demonstrated to be complete. Hence it prohibits the inference of a *complete* machine table from any finite list of observed input-output transitions, and hence from any model based on such a list, no matter how long this list may be.

Proof (Moore, 1956). Given any partial machine table T obtained by finite observation of some black box **B**, it is possible to construct arbitrarily many hypothetical machine tables that contain T. These constructed machine tables correspond to black boxes with larger state spaces than the minimal black box $\mathbf{B}_T$ required to generate T. Because any of these larger black boxes could have generated T, nothing about T is sufficient to determine that the observed black box **B** is $\mathbf{B}_T$, i.e., nothing about T is sufficient to determine that T is the complete machine table of **B**. □

Moore's theorem shows that the very next output from any black box can be a complete surprise. Any black box could, for example, contain a clock that counts to some large number, after which it enables the execution of a qualitatively different dynamics than that executed previously; alarm clocks, time bombs, commonplace computer programs and many biological systems are obvious examples. A significant special case to which this applies is that of "tests for contextuality" such as discussed by Kitto [24]; no finite sequence of observations during which a system appears to behave in a non-contextual way provides any guarantee that the system will not behave in a contextual way in the future. One can appeal to Occam's razor and hypothesize that the behavior that has already been observed is the only behavior of which a black box is capable, but Moore's theorem reminds us that this is always merely an hypothesis. The very next observation could prove it wrong.

Moore's theorem gives hypotheses about the behavior of a black box, whether framed as input-output "laws" or as proposed internal mechanisms, exactly the status that Popper [78] gives scientific hypotheses in general: they are candidates for empirical falsification. They are at best satisficing, not optimal. To borrow Bell's [13] somewhat pejorative term, they may be true FAPP, but they can never be considered true *simpliciter*. They cannot, therefore, support the weight of metaphysical or ontological claims. As noted above, ontological claims about *physical* boundaries and hence physical structure inside a black box fall immediately afoul of decompositional equivalence. Moore's theorem generalizes this prohibition of ontology to internal structures described in any language. Explicitly hypothetical claims—models—of course suffer no such prohibition. Provisional or FAPP "knowledge" of the workings of a black box—the kind of knowledge represented by not-yet-disconfirmed models, is allowed by Moore as it is by Popper; it is certain or god's-eye knowledge that is prohibited.

3.2. A Black Box Cannot Be Bounded

Moore's theorem provides the basis from which to prove the principle result of this paper: that a black box cannot be bounded. The informal "picture" of a black box found in the classical cybernetics literature (e.g., [1,77]) or in Figure 1 places the observer outside the box. The box is, in the canonical example, a device captured from an enemy and given to an engineer to investigate, or in a design example, a device about which only the input-output interface is known. In either case, the box is *bounded*, clearly separate from and causally independent of the other items in the laboratory or workshop. The following theorem shows that this traditional, intuitively-appealing picture is seriously misleading.

Theorem 2 (no-boundary). *If the observable world contains a black box, it is a black box.*

To prove this theorem, it is useful to introduce the following:

Definition 2. *Let* **B** *be a black box and* ξ *be an arbitrarily-chosen observable degree of freedom. The degree of freedom* ξ *is separable from* **B** *if and only every state of* **B** *is independent of the state of* ξ.

This is the definition of separability familiar from physics (cf. Section 2.1), phrased from the perspective of **B**. The proof is now simple: Moore's Theorem shows that knowing that ξ is separable from **B** is knowing more than can be known.

Proof. If the observable world contains a black box **B**, but is not itself a black box, it must contain at least one observable degree of freedom ξ that is separable from **B**. If this is the case, all states of **B** and therefore the complete machine table of **B** must be independent of the state of ξ. This, however, is by Moore's theorem information about **B** that cannot be obtained by finite observation, contradicting the assumption that **B** is a black box. □

This no-boundary theorem confirms in general—taking it as obvious that the world contains at least one black box—the conclusion reached above from the observer as witness formulation and from decompositional equivalence: each observer obtains information only at her boundary with the surrounding "environment", i.e., the rest of her observable universe. This environment is a black box. The systems it contains, if any, are not accessible for direct observation.

The no-boundary theorem appears radical, but it is not. Indeed it rests not on a strong underlying assumption, but on the absence of one: the proof would not go through if it were permitted to assume, a priori, that systems are separable. In this case **B** could be assumed, a priori, to be separable from its environment and hence from ξ. The traditional picture of a black box as a device resting on an engineer's workbench implicitly makes this assumption; the engineer in this traditional picture does not need to worry about the box being dynamically coupled to his heartbeat or to the phases of the Moon. Removing the assumption of separability turns these worries, however far-fetched they may be, into real possibilities. If they are to be removed from consideration definitively, not just FAPP, they must be removed by observational evidence. Moore's Theorem shows that definitive observational evidence of independence cannot be obtained by finite means.

The no-boundary theorem moves each observer into an all-encompassing black box, the rest of that observer's observable universe (Figure 2). It denies observers independent access to "systems" embedded within their observable universes; the physical interaction by which an observer obtains information is defined at the observer—observable universe boundary, and all information obtained by the observer must cross this boundary. The "source" of the information within the box cannot be identified, as doing so would require access to the interior of the box, access that no observer can have.

An immediate consequence of the no-boundary theorem is that the composite observer-box system is, from the observer's point of view, a closed system. The observer cannot, in particular, detect anything else that interacts with the box; the box is all she sees. The closure of the composite observer-box system is only a fact FAPP, but it is, for the observer, a conclusion that no observation can falsify.

The no-boundary theorem does not, of course, limit the observer's ability to *hypothesize* the existence of distinct "systems" within the black box that constitutes her observable universe. Nor does it imply that such hypotheses cannot be useful FAPP for choosing what inputs to give the box or for predicting what outputs to expect from the box. It merely establishes the Popperian point that while such hypotheses can be supported by observation, they can never be confirmed by observation. They cannot, therefore, support ontological claims. In particular, they cannot be regarded as "self-evident truths" and accorded the status of axioms. "Systems exist" is sometimes regarded as an axiom of quantum theory [39,79]. As discussed in more detail below, the apparent existence of distinct physical systems should instead be regarded as an explanadum of quantum theory.

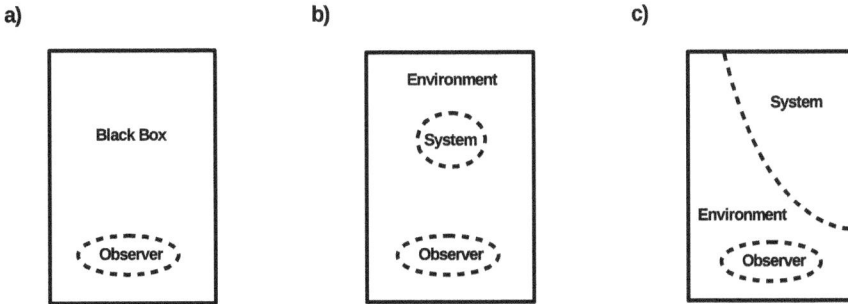

Figure 2. The observer within the black box. (**a**) The observer is embedded within "the rest of the observable universe", which by Theorem 2 is a black box; (**b**) The observer is embedded within the environment postulated by the environment as witness formulation of decoherence, from which any information about systems embedded within that environment must be obtained; (**c**) The observer is embedded within an observable universe that satisfies decompositional equivalence. Subsystem boundaries (other than the observer's) within that universe have no effect on the information that the observer can obtain.

3.3. Corollaries

The no-boundary theorem shows that separability between any two observable degrees of freedom, i.e., between any two components of the state(s) of one or more systems, is at best a FAPP assumption. Separability is, however, the fundamental enabling assumption of the "classical worldview" of independently-observable, independently-manipulable physical objects. Einstein put it this way: "Without such an assumption of the mutually independent existence (the "being-thus") of spatially distant things, an assumption which originates in everyday thought, physical thought in the sense familiar to us would not be possible". (quoted in [79], p. 6). Separability underpins the concept of statistical independence and, therefore, Bell's [80] inequality. If the observable universe is a black box, the idea that any two observable degrees of freedom would satisfy Bell's inequality is at best a FAPP assumption. There is no reason for it to be true, and no justifications beyond FAPP utility and convenience for assuming that it is true. The no-boundary theorem tells us, indeed, to *expect* Bell's inequality to be violated, the surprise elicited in fact by experimental confirmations of its violation by Aspect, Dalibard and Roger [81] and now many others notwithstanding. It is important to emphasize that this expectation follows purely and solely from *classical* cybernetics. No "quantum" assumptions are needed. If "physical thought in the sense familiar to us" is not possible without separability, then it is not possible even in *classical* physics.

As shown previously [82], the restrictions on inferences from observational outcomes imposed by Bell's theorem [80], the Kochen-Specker theorem [83] and the no-cloning theorem [84] can also be derived from Moore's theorem; they therefore characterize any observations of a black box. The no-boundary theorem shows that the restrictions imposed by these theorems characterize observations across the board. In every case, these restrictions follow from the inability of observers to identify the particular collection of degrees of freedom within the black box under observation—the particular embedded "system"—that produced a given observational outcome. If the sources of observational outcomes cannot be singled out, i.e., if separability fails, then outcomes that appear to come from two distinct systems cannot be assumed to be independent (Bell's theorem), outcomes that appear to come from a single system cannot be assumed to be independent of the order in which they are obtained (Kochen-Specker), and identical outcomes do not indicate the presence of identical systems (no cloning). This last result is simply a special case of Moore's theorem: no two finite sequences of observational outcomes obtained from two black boxes can demonstrate that the two black boxes have the *same* machine table, i.e., no two such sequences of observational outcomes can demonstrate that two black boxes are clones. This and the other results are, once again, consequences of classical cybernetics alone. That they were discovered in the context of quantum systems, and that they are still often regarded as "intrinsically quantum" (e.g., [85] where Bell's theorem is regarded as "intrinsically quantum" although the Kochen-Specker and no-cloning theorems are not), indicate the depth with which the view that the observable universe cannot be a black box pervades "classical" thinking, i.e., the extent to which the decompositional equivalence of the classical formalism is ignored.

Several additional corollaries follow immediately from the no-boundary theorem.

Corollary 1 (no-communication). *Two or more observers cannot employ their shared environment as a classical communication channel.*

Proof. By the no-boundary theorem, the environment shared by two or more observers is a black box. No observer of this black box can determine the internal source or causal history of any output received from the box. Therefore, no observer can distinguish states of the box that result from inputs from other observers from states of the box that have no dependence on inputs from the other observers. □

It cannot, in other words, be determined by finite observation that an output from a black box is a "message" from another observer. Indeed from the point of view of any one observer, all other observers are inside that observer's observable universe and hence inside a black box (Figure 3).

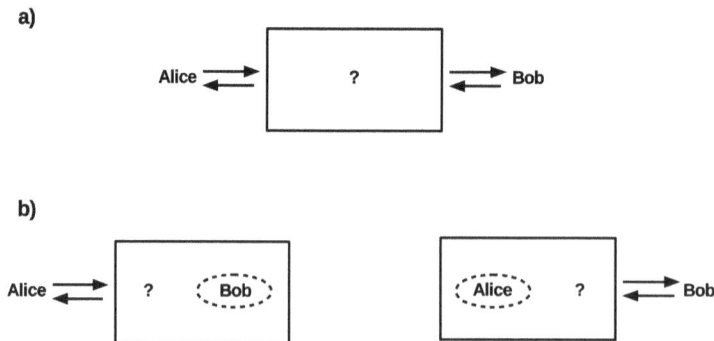

Figure 3. (**a**) Two observers, Alice and Bob, interact with a black box; (**b**) It is equivalent to regard Alice as interacting with a black box that contains Bob, and Bob as interacting with a black box that contains Alice. It is clear in this latter picture that Alice cannot determine by observation that an output from her box is a message from Bob, nor can Bob determine by observation that an output from his box is a message from Alice.

The no-communication corollary is analogous to the no-communication theorem for shared quantum systems [86], which shows that the local measurement outcomes obtained from a quantum system by one observer cannot depend on the activities of another observer, located somewhere else, who also interacts with the system. This latter result is generally interpreted as a limitation on *instantaneous* information transfer between observers via the shared quantum system, i.e., as showing that entanglement cannot be used for super-luminal communication. However, spacelike-separated observers can obtain no direct observational evidence that they are manipulating the same system; it is for this reason that local observations must be supplemented with classical communication via a separate classical channel in communication protocols (local observations, classical communication or "LOCC" protocols [87]) that depend on shared quantum systems and hence entanglement as a resource. If *only* a quantum system is shared, classical communication is impossible and no information can be exchanged, i.e., entanglement ceases to be a resource.

If observers cannot exchange classical information through a shared black box, they clearly cannot exchange physical systems either. Physical systems such as clocks, meter sticks and gyroscopes are, however, the *reference frames* with respect to which measurements are made [88]. Hence we have:

Corollary 2 (no-external-reference). *Observers have no access to external reference frames.*

Proof. By the no-boundary theorem, any external reference frame is inside a black box and therefore inaccessible. □

In particular, observers cannot specifically and exclusively consult external clocks, spatial reference frames, or items of laboratory apparatus, as all of these are internal to the black box with which each observer interacts. Information from such devices can only be obtained if it is "packaged" into an output from the black box. By the definition of a black box, this packaging process precludes observational access to the source(s) of the packaged information.

The no-external-reference corollary seems radical, but in fact has a simple interpretation: the reference frames employed by observers to make sense of their observations must be internal. Without an internal representation of time, for example, the motion of a clock hand has no temporal meaning; without an internal memory and an internal means of comparing current and remembered events, even the motion itself is undetectable. Observers, in other words, cannot be "blank slates" that just record data if they are to obtain what Roederer has called "pragmatic information", information that enables doing something [89,90]. An observer's internal reference frames determine what kinds of outputs from a black box that observer can observe. They allow the *interpretation* of observed outputs in terms of hypotheses about the box's states and dynamics. Without such internal reference frames, the observer cannot do science, indeed the observer cannot *do* anything at all. It was assumed earlier that observers could choose what to do next; this assumption can now be re-stated as the assumption that all observers of interest have internal reference frames, including in particular an internal time reference frame, that enable them to obtain and act on pragmatic information (cf. [64]).

If observers have internal structure, it is natural to ask if that internal structure is accessible to other observers. The no-boundary theorem restricts this question to one case only: the case in which the "other observer" in question is the black box with which the observer of interest interacts. In this case, however, the answer has already been decided: the black box obtains only a finite sequence of finite bits strings from the observer. The situation shown in Figure 1a is, therefore, completely symmetrical: the black box can "know" no more about the observer than the observer can know about the black box. Hence we have:

Corollary 3 (observer-box equivalence). *Observers are black boxes and vice-versa.*

Proof. Consider an observer (Alice) interacting with a black box **B** that contains another observer (Bob) as shown in Figure 3b. Because Bob's presence within **B** is undetectable by Alice, this situation is completely general. By the no-boundary theorem, the outputs from **B** cannot depend on Bob's boundary, therefore it can expand arbitrarily within **B** (Figure 4). In the limit, Bob and **B** are identical. □

This observer-box equivalence corollary extends Ashby's remark about "real objects or systems" to include observers, an extension that comports well with Ashby's lifelong commitment to understanding biological intelligence in terms of adaptive information processing [91]. It removes, in particular, any justification for assuming that observers are made out of a different kind of "stuff" or have an intrinsically different structure from the environments in which they are embedded. It allows the boundary around an observer embedded in a black box to be erased; i.e., it recognizes that a black box containing an observer satisfies decompositional equivalence.

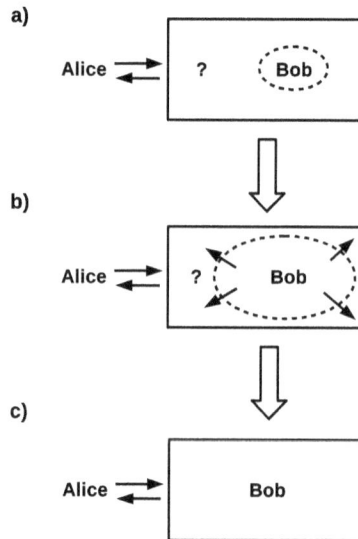

Figure 4. (a) Alice interacts with a black box containing Bob, as in Figure 3b; (b) Bob's boundary within the box can expand arbitrarily without affecting the box's interaction with Alice; (c) In the limit, Bob and the box are identical.

The observer-box equivalence corollary allows us to re-conceptualize the observer-box interaction as an interaction between two observers or, equivalently, between two black boxes (Figure 5). The observers/boxes together comprise a closed system, a "universe" in which each observer/box is the other observer/box's "environment" or observable universe. Classical information exchanged by the observers flows through an inter-observer boundary S, which plays the role of a classical information channel. As this boundary is an oriented surface separating the observers, both the information sent and the information received by each observer can be viewed as "written on" the side of this surface facing that observer. Sending and receiving are symmetric, so each observer "sees" exactly the same information on their own side of the surface; they differ only in the labelling of what bits have been "sent" and what bits have been "received". As there is no process by which to "unsend" or "unreceive" a message, this "writing" on S is permanent. Indeed the "memories" of both Alice and Bob can, without loss of generality, be regarded as implemented by their shared boundary.

If two observers/boxes, e.g., Alice and Bob in Figure 5, are allowed to interact indefinitely, the amount of information encoded on the inter-observer boundary S will increase, limited only by the assumed-finite energy resources of the observers. Let $E_{Max}(\mathbf{O})$ be the maximum energy available to an observer $\mathbf{O}$ for information exchange.

Definition 3. *Let $\mathbf{A}$ and $\mathbf{B}$ be interacting black boxes that together comprise a closed system, and suppose they are allowed to interact indefinitely. The observable machine tables of $\mathbf{A}$ and $\mathbf{B}$ are then the exactly complementary sequences of inputs and outputs exchanged by $\mathbf{A}$ and $\mathbf{B}$ prior to the smaller of $E_{Max}(\mathbf{A})$ and $E_{Max}(\mathbf{B})$ being exhausted.*

The observable machine table of a black box can clearly be its complete machine table if its number of states is small. If energy resources are unlimited, the observable machine table of a finite black box is the complete machine table. If both black boxes are infinite and can access infinite energy, completeness is approached only as a limit.

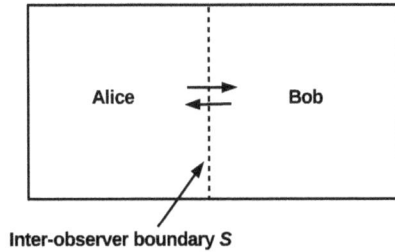

Figure 5. Alice and Bob exchange classical information through an inter-observer boundary *S*.

Corollary 4 (holographic-encoding). *If a black box is a component of a closed system, its observable machine table can be written on its boundary.*

Proof. Let **B** be the black box in question. If **B** is itself a closed system, it has no boundary but also no inputs or outputs, so it has no observable machine table. If **B** together with one other black box compose a closed system, the result follows by definition of the observable machine table. If **B** interacts with more than one other black box, the no-boundary theorem allows their union to be considered a single black box, so the previous case applies. □

The holographic-encoding corollary provides an alternative way of thinking about the "inaccessible interior" of a black box: to say that the interior of a black box is inaccessible is just to say that all available information about the box can be "read off" from its exterior. Physical systems for which this is true are obviously familiar: they are black holes. In the case of a black hole, the information written on the horizon *S* is proportional to its mass [92]. In the case of a black box, the information written on the horizon *S* is proportional to its available energy (and hence its mass) or to its behavioral and hence computational complexity, whichever is less. The connection between mass, horizon-encoded information and computational complexity has been noted previously in the case of black holes [63,93]; it is interesting to see it re-appear here in the purely-classical case of black boxes. While in the black-hole case the computational complexity in question is the computational complexity faced by an observer stationed near the horizon who must analyze Hawking radiation from the black hole before it evaporates, in the black box case the computational complexity in question is the computational complexity of the procedure with which the box chooses outputs in response to inputs. It can be conjectured that the two cases are in fact fully analogous: that the computational complexity of an observer's analysis of Hawking radiation mirrors the computational complexity of the black hole's "choice" of what Hawking radiation to emit.

The holographic-encoding corollary holds for black boxes, or by the observer-box equivalence corollary, observers that are components of closed systems. As noted earlier, a system being closed can only be a fact FAPP. All systems *appear*, however, to be closed to observers embedded in them. Hence all observers appear, to themselves, to have their observable machine tables encoded as memories on their boundaries.

It was remarked in the Introduction that modern science assumes that "there is a way that the world works". This assumption is naturally interpreted, and historically has been interpreted, as an assumption of determinism. Standard quantum theory with no *physical* "collapse" process is fully unitary and therefore strictly deterministic. As noted earlier, Tipler has shown that the "quantum potential" that the Schrödinger equation adds to the classical Hamilton-Jacobi equation is exactly the potential required to make classical physics strictly deterministic [25]. Let us assume, therefore, that the dynamics of any closed system is strictly deterministic. Even in this case, the following holds:

Corollary 5 (free-will). *In a closed system comprising two interacting black boxes, neither box determines the other's behavior.*

Proof. For one box to determine the other's behavior, it must have information specifying the other's state and dynamics. However, such information cannot be obtained for any black box. □

This free-will corollary is an analog for black boxes of the Conway-Kochen free-will theorem, which states that the behavior of a quantum system cannot be determined by the information available in its own past lightcone [94]. While quantum theory can be globally deterministic, no *local* information can be determinative for any system. It is in this sense of *local* non-determinism that any observer/box can be said to "freely choose" its next output (cf. [95]). Note that this freedom also applies to each observer's own local information: as each observer's memory contains only its own observable machine table, no observer "knows" its own current state or dynamics and hence no observer can locally determine its own next output. Observers are free and appear autonomous, but they are not autonomous in the literal sense of locally self-determining; local non-determinism includes local non-*self*-determinism. Local non-determinism prevents strict "objective" autonomy for any bounded system, including any observer, but requires apparent autonomy for all observed systems, including all observers. Local non-determinism can be stated in terms of decompositional equivalence: any system's action on its observed world, including any observer's action on her observed world, remains the same when the boundary around the system/observer is erased, rendering the action no longer *the system's/observer's action on the observed world* but rather *the entire world's action on itself.* Within a black-box based theory, these two actions are indistinguishable.

3.4. What Is Physics about?

The no-go results demonstrated here, whether in their current classical cybernetic form or in their more traditional quantum-information form, naturally raise the question of what physics is about. This question has traditionally been debated in realist-versus-instrumentalist terms, with the added twist of realists about the "quantum world" typically being instrumentalists about the "classical world" and vice-versa (e.g., [6] or for a more recent survey of competing positions, [12]). The no-boundary theorem and the no-communication corollary add a further issue to this debate: that of observations being intrinsically *personal*. This individuality and principled non-redundancy of observations has previously been emphasized by Fuchs as a central feature of quantum Bayesianism [11]. If communications from other observers are regarded as "observations" like any others and accorded no special, epistemically-privileged status, the usual notion of objectivity based on inter-subjective agreement breaks down. It becomes impossible for the agreeing agents to establish unambiguously what they are agreeing about (cf. [66]). They can agree FAPP, but they can do no better than this.

The concept of classical information—information that can be irreversibly encoded as a bit string on some physical medium—is taken as a primitive in classical cybernetics and in the current discussion. Bits are discrete "quanta" of information; hence classical information is itself "quantized". As noted above, any observer able to obtain classical information and use it to make predictions must have an internal time reference frame, i.e., an internal clock. No generality is lost in assuming that the observer's internal clock "ticks" whenever an outcome is received from the system being observed; the unit time interval of the observer's internal clock can, therefore, be set to $\Delta t = 1$. As the observer must expend a minimum energy $E_{Min} = ln2\, kT$ to record each bit, the observer's minimal action per recorded outcome is $E_{Min}\Delta t$. Hence *the observer's action is quantized* even in classical physics. Boltzmann's recognition that obtaining information requires expending energy is thus the fundamental idea needed to develop quantum theory. The quantization $E_{Min}\Delta t = Constant$ fixes the value of the minimal action; at human physiological temperature, $T = 310$ K, for example, and $\Delta t \sim 200$ ns, the response time of rhodopsin to light at physiological temperature [96], $E_{Min}\Delta t \sim 6.0 \times 10^{-34}$ J · s, a value remarkably close to Planck's $h \sim 6.6 \times 10^{-34}$ J · s.

As Jennings and Leifer have shown [85], the quantization of the observer's action is sufficient to understand the non-commutativity of position and momentum measurements and the position-momentum uncertainty principle, as well as no-cloning, Kochen-Specker contextuality, and other "quantum" effects. While Jennings and Leifer do not ask what additional assumptions are needed to obtain all of quantum theory within a classical setting, others have done so. As Tipler shows, requiring time evolution to be strictly deterministic is sufficient [25]; Bohmian mechanics provides a similar demonstration from a different starting point, that of preserving quantum behavior while assuming an ontology of classical "particles" [26]. Motivated in part by Wheeler's provocative "it from bit" proposal [97], a number of information-theoretic [98–103] or more generally, operational [104–106] axiomatizations of quantum theory have been proposed. Within these formulations, "physical" interactions become transformations of bits, typically with qubits as "inside the box" intermediaries. Hence physics in these formulations is "about" computation, typically quantum computation. Reviewing "device-independent" operational methods originating in quantum cryptography, which disallow assumptions about the degrees of freedom or dynamics of the systems from which observational outcomes are obtained, Grinbaum goes even farther, concluding that physics is "about languages" ([107], p. 14), specifically, languages as formal entities without imposed semantics.

Physics being about information, computation or languages is a far cry from the more traditional, and far more common, view that physics is about the everyday objects of ordinary human experience, even if its theories tend toward the abstruse and postulate non-ordinary objects that are not directly observed. *Effectively* embedding the human observer into the theory requires explaining, within the theory, why the everyday observations made by human observers can be described so readily in terms of bounded, causally-independent objects. The next section makes a step in this direction by showing that "objects" can be considered to be collections of outcome values that are held fixed while making measurements. "Objects" are thus internal reference frames. It is then shown in Section 5 that objects construed in this way are the natural external referents of the "object tokens" that have been proposed as components of human episodic memories [108,109].

4. "Objects" within a Black Box

The world may be a black box, but when human observers look at it, they see discrete objects. The "environment" in which these objects are embedded is effectively invisible, a condition that has obtained, cosmologically, since photons decoupled from matter [110]. Hence while human observers physically interact with the environment, predominately the ambient photon field, they *see* objects. From the present perspective, the question of interest is how to describe these objects in a way that is consistent with the Black Box model.

4.1. Object Identification in Practice

The almost-automatic human assumption that distinct objects exist in an observer-independent "objective" way within the world diverts attention away from a question that is critical for understanding observation: how does an observer *identify* the object being observed? How, for example, does one identify a familiar object like a coffee cup? Such objects do not appear to us in isolation; they must be picked out from scenes that contain many other objects. The only incoming information available for doing so is that contained in the observational outcomes obtained from the scene as a whole.

As will be discussed further in Section 5 below, answering this system-identification question is non-trivial [111]. For the present purposes, however, the most important part of the answer is that observers must dedicate some of the observational outcomes that they associate with an object to its identification, leaving the rest to indicate its state. A coffee cup, for example, may be identified by its fixed size, shape and color, while its location and whether it is filled with coffee are allowed to vary, i.e., to be state variables. What is critical is that *not all observables associated with an object can* simultaneously *be state variables*; some must be set aside for object identification. Coffee cups are in this sense like electrons: some observable properties must be regarded as definitional and hence invariant in order for system identification to succeed.

Recognizing that system identification requires a designated collection of observables that are fixed and invariant destroys a key component of the "classical worldview". Ollivier, Poulin and Zurek, for example, operationally define *objectivity* thus ([42], p. 3):

> "A property of a physical system is *objective* when it is:
>
> 1.   simultaneously accessible to many observers,
> 2.   who are able to find out what it is without prior knowledge about the system of interest, and
> 3.   who can arrive at a consensus about it without prior agreement."

This definition clearly applies to no systems at all, as observers "without prior knowledge about the system" would have no designated observables with which to identify it and observers "without prior agreement" about a designated collection of identifying observables would have no basis for any subsequent agreement. The no-external-reference corollary shows that the reference frames required to characterize system states must be included within the "prior knowledge" of observers; inter-observer agreement about reference frames must, similarly, be included among the "prior agreements" between observers. Shared conceptual schemes and shared category hierarchies are examples of such prior agreements. As discussed in more detail below, the collection of observable properties that are regarded as defining an object can be considered a reference frame for characterizing that object: without agreement about which observables constitute a reference frame for identifying an object, multiple observers cannot agree unambiguously about object identification. As noted earlier, a critically important special case is the collection of reference frames that observers use to identify each other, including their prior agreements about how they will communicate.

4.2. Superpositions Encode the Unresolvable Ambiguity of Object Identification

The role of observational outcomes in identifying objects can be illustrated with a simple thought experiment. Consider an observer equipped with a single reference frame, a vertically-oriented meter stick, who is embedded in a world containing multiple objects, some of which have some linear dimension equal, within the resolution of the meter stick, to 1 m while others do not (Figure 6). The meter stick registers a "1" outcome if it comes into contact with an object having a vertical dimension of 1 m and registers a "0" outcome if it comes into contact with an object having a different vertical dimension; it registers no outcome if it is not in contact with an object. An unseen mechanism occasionally puts one of the objects in contact with the meter stick. The observer receives no information other than the "1" and "0" outcomes from the meter stick; in particular, the observer is unable to observe the size, shape, orientation, or anything else about the objects other than the vertical linear dimension reported by the meter stick. The world of this observer is clearly a black box.

With what system does the observer interact when an outcome value of "1" is received? The *observed state* of that system is known: it is whatever state yields an outcome value of "1." Call this state "$|1\rangle$" and call its complement, the state that yields an outcome value of "0", "$|0\rangle$." What, however, is the system that occupies these states? While the no-boundary theorem gives the precise answer that the "system" is the black-box world as a whole, the human tendency to associate distinct observational outcomes with distinct systems suggests an answer postulating two systems, an "object" S_1 perpetually in, and hence identified by $|1\rangle$ and a second "object" S_2 perpetually in and identified by $|0\rangle$. These two alternatives are operationally equivalent: interacting with a single system that can occupy two states is indistinguishable from interacting with two systems, each of which permanently occupies a single state.

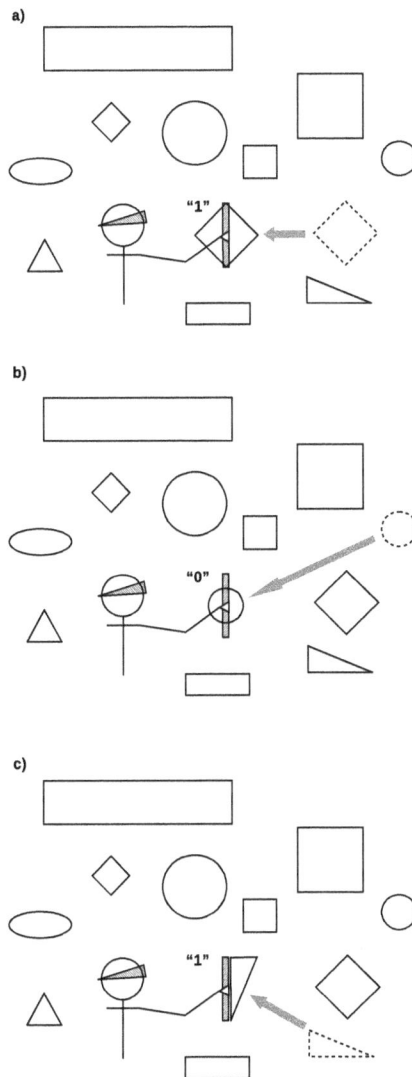

Figure 6. (a) An observer equipped with a meter stick and embedded in a world of multiple objects. The meter stick reports an outcome of "1" whenever an object with a linear dimension of 1 m is placed in contact with it. The observer has no ability to observe any features of the objects other than the linear dimension reported by the meter stick; (b) The meter stick reports an outcome of "0" whenever an object with a linear dimension other than 1 m is placed in contact with it; (c) A mechanism that places successive objects in contact with the meter stick produces a stream of binary outcome values.

From the "god's eye" perspective of Figure 6, one can associate multiple distinct objects distinguished by the unobserved and hence "hidden" variable of geometric shape with both S_1 and S_2. Providing the observer in Figure 6 with a shape detector would allow a choice between identifying objects by size and treating shape as a state variable or identifying objects by shape and treating size as a state variable. These descriptions are, once again, operationally equivalent, and both are equivalent to a description in which there is only one object—the black box—and both size and shape are state variables. The only basis for predictions in any of these descriptions is the sequence of bit strings observed so far, a sequence that provides no information beyond a lower bound on complexity about the mechanism that selects objects to put in contact with the meter stick. In particular, the observations provide no basis for deciding whether the mechanism selects "objects" for presentation at random or by following a deterministic algorithm.

It is natural for an observer who knows no more about what is being observed than the observational outcomes obtained thus far to represent the source of these outcomes, whatever it is, by a linear superposition of the observed outcomes. As distinct outcomes can indicate either distinct states or distinct systems, this representation can be interpreted as either a superposition of states or as a superposition of systems; the two interpretations are entirely equivalent [67]. While the former is by far the more familiar, the latter is implicit in Feynman diagrams and is even more evident in more abstract representations, such as the amplituhedron [112], in which particles and even spacetime are elided altogether.

4.3. Objects as Internal Reference Frames

It is increasingly recognized that the reference frames employed to make physical measurements are not mere abstracta but rather are physical objects [88]. Successfully employing a LOCC protocol with entangled photons, for example, requires sharing not just a *description* of the measurement conditions at the two spatially-separated sites, but also a physical orientation reference frame such as the Earth's gravitational field. The physicality of reference frames forces the physicality of instrument calibration procedures to be taken explicitly into account; when this is done, the standard "quantum" uncertainty relations emerge even for purely classical measurements [113]. The complementarity between system-identifying observables and state observables makes it clear why this must be the case across the board: one or the other set of observables must be measured first, and the finite time required for measurement is time during which the unmeasured observables are free to undergo unpredictable dynamic evolution.

The no-external-reference corollary shows, however, that reference frames cannot be *external* physical objects; they must be internal to the observer. Even an external calibration standard requires identification by observation: from the observer's perspective, it is merely a collection of observational outcomes. Calibration itself is a process of comparing observational outcomes that identify and characterize the state of the "standard" to observational outcomes that identify and characterize the state of the "instrument" being calibrated. This comparison is an operation on bit strings carried out by the observer; it is a computation. It is essential to the meaning of this computation that the outcomes that identify both the calibration standard and the instrument remain invariant during the calibration process. As with the fixed size, shape and color of a coffee cup that enable reaching for it, grasping it, and filling it with coffee, it is the invariance of outcomes that identify objects with respect to outcomes that indicate changes of state that allow state changes to be recognized as such.

If all objects, including objects that function as calibration standards and reference frames, are identified by sets of observational outcomes that remain invariant while other relevant observational outcomes change, it is natural to consider these sets of invariant outcomes to be the *internal reference frames* that the no-external-reference corollary requires. These internal reference frames are information-bearing structures—effectively, data structures—that encode the observer's expectations about the objects present "within" the black box under observation. In Bayesian predictive-coding [114] terms, they are prior probability distributions. Without such priors, the observer cannot identify objects, and hence cannot recognize state changes in objects.

There is, however, no need to postulate the real existence of external objects at all; the no-boundary theorem guarantees that their existence cannot be revealed by observation. Any observer's observational outcomes are generated by that observer's impenetrable black box. The observer encoding a set of internal reference frames that identify objects is all that is required to explain the *appearance*, to the observer, of objects within the box (von Foerster [17] refers to such internal reference frames as "eigenforms"; see also [115,116]). Hence the observer encoding a set of internal reference frames is all that physics needs postulate. Internal reference frames are necessary for object perception in a black-box world; they are also sufficient.

5. The Black Box as a Cross-Disciplinary Paradigm

The assumption that the sources, within the world, of observational outcomes can be spatially localized and bounded is central to the classical worldview. It goes hand-in-hand with the assumption that specific states of spatially localized, bounded systems can be "prepared" by local manipulations. The classical cybernetics of the mid-20th century was by no means the first or the only intellectual movement to question these assumptions. Philosophers have criticized them at least since Heraclitus. The 20th century saw, however, an explosion of criticism of these assumptions that ranged across scientific disciplines. As science requires replicable experiments performed on shareable apparatus and reported on shareable media, this burst of criticism of the assumptions on which these conditions rest indicate that the logic of science itself is at most paraconsistent [117].

5.1. Formal Semantics, Device Independence and Virtual Machines

By the 1930s, Wittgenstein's essentially 19th-century conception of propositions in natural or formal languages "picturing" unique relations between independently-existing objects [118] had been largely replaced, in no small part due to Wittgenstein's own efforts, by a radically different conception of languages and of symbol systems generally as supporting arbitrarily many distinct interpretations. This reached maturity in Tarski's development of formal model theory as a general approach to semantics [119]; however, it is evident in the Turing machine [120] and in the work of Gödel, Church and others. Via the later work of Quine and others, the disconnection of semantics from syntax imposed by formal model theory informs all recent thinking about natural languages. The implications of this break for ontology have not gone unnoticed. Quine, for example, points out that with this disconnection, "physical objects are postulated entities which round out, and simplify our account of the flux of experience ... the conceptual scheme of physical objects is a convenient myth, simpler than the literal truth" ([121], p. 32). Physical objects are, in other words, for Quine just *semantic interpretations* of the observational outcomes that constitute "the flux of experience." Experience itself is experience of a black box.

The most powerful practical impact of the idea that symbols could support arbitrarily many distinct interpretations was in computer science; indeed it can be argued that the disconnection of syntax and hence implementation from semantics is the foundational idea of computer science. It enables the two defining characteristics of post-1950s computing: device independence (the same software can run on arbitrarily many distinct hardware platforms) and virtual machines (a single hardware system, appropriately programmed, can exhibit the input-output behaviors of arbitrarily many distinct devices). These two concepts are closely coupled: virtual machines, in the form of interpreters and compilers, enable device independence, while device independence is what makes virtual machines so useful and valuable [122,123]. Without these concepts, we would still be programming special-purpose computers by re-arranging their hardware. With them, we are able to regard a device that fits comfortably in a shirt pocket as performing the specialized functions of hundreds to thousands of different tools.

Despite its everyday familiarity within computer science, the virtual machine concept has yet to penetrate the disciplinary discourse of physics. "Device-independent methods," however, are virtual machine methods. An experimenter who "prepares" the state of a quantum system is in fact preparing the state of a virtual machine, often literally in a computer-controlled laboratory. "Experimental data" are, similarly, outputs of virtual machines, again often literally. From the experimenter's point of view, these virtual machines are black boxes, in many cases black boxes that are literally functionally interchangeable, and in practice regularly interchanged, with other black boxes that offer the same functionality FAPP. What goes on in the interiors of these black boxes is unknown, and in many cases arguably unknowable, even in principle, by human beings [124]. To do an experiment is, therefore, to undertake an extended exercise in *semantic interpretation* of observational outcomes obtained from unknown and (at least FAPP but very possibly in principle) unknowable sources. While some "physical reality" or other must be postulated for the exercise to make sense, the observational outcomes obtained leave this reality arbitrarily underspecified. All of physics, in other words, is device-independent by its very nature. Experimentally-accessible "physical systems" are bits of semantics, or as Grinbaum [107] puts it, bits of language.

5.2. Cognitive Science and AI

The 1960s "cognitive revolution" in psychology reintroduced the idea that understanding what is going on inside the subject's—i.e., the observer's—head is important for understanding the subject's behavior. Mere stimulus-response laws are not sufficient [125]. Cognitive subjects are, however, notoriously black boxes: the observational tools available to experimental psychology and neuroscience do not directly reveal either the contents of cognition or the processes that manipulate those contents. In particular, they do not reveal the relationship between cognitive states and states of the external world. The question of whether this semantic relationship can be considered fixed by causal processes and hence unique underlies the "symbol grounding problem" (SGP) [126,127] in cognitive science and AI. As "grounding" a symbol in the external world requires identifying the object, event, or class of objects or events to which it refers, the *external* SGP is equivalent to the problem of object or system identification [128]. Responses to the SGP range from treating the observer's world as a black box, at least methodologically [129], to considering essentially arbitrary components of the state of the world as components of the observer's cognitive state [130]. The former response replicates the conceptual structure of Figure 5; the latter reinforces the arbitrariness with which the observer-world boundary is drawn.

While symbol grounding can be viewed as a purely theoretical or even philosophical problem in cognitive science, it becomes a practical problem in AI, particularly for "embodied" and "situated" systems such as mobile robots [131]. In "open" task environments in which not all objects can be identified using a priori knowledge, object identification criteria and hence the "grounds" for object-specifying symbols must be learned. How such criteria can be learned, even just FAPP, is a central problem in both developmental robotics [132] and developmental psychology [111]. *Internally* grounding object symbols in object-appropriate, context-relative perceptual-motor associations provides a FAPP solution to the SGP; this solution effectively replicates the embedding relationship between object tokens [108] and event files [133] or episodic memories [134] in theories of human object recognition [109]. The external or ontological version of the SGP is widely regarded as both unsolvable and irrelevant to practical robotics [135]. Rejecting the external SGP as a problem for robotics is considering the world in which a robot behaves to be a black box.

5.3. Evolution and Development

Cognitive scientists who reject a black-box view of the observer-world interaction often appeal to biological evolution as a mechanism for enforcing a causal connection between world states and cognitive states. Mark, Marion and Hoffman have, however, shown using evolutionary game simulations that agents responsive only to fitness payoffs out-compete agents responsive to the actual distribution of resources in a simulated world [136]. This result motivates the "interface theory of perception" (ITP), which conceptualizes perceptual systems as providing finite-resolution information about the fitness consequences of actions in a black-box world [137]. Within ITP, the semantic relation between the agent's internal representations and the world states or structures to which fitness consequences are associated is essentially arbitrary, as expected within a model-theoretic or virtual-machine formulation of the semantics. The principle of "conscious realism" of Hoffman and

Prakash [138], which treats an agent's world as another agent and hence treats agent-world games as two-agent games, replicates the symmetric structure of Figure 5 within ITP.

While ITP is explicit in conceptualizing perception and hence cognition as interaction with a black box, it is not alone in making this assumption. The Bayesian predictive coding framework [114] provides an organizing principle for not only neurocognitive architecture [139], but for the evolution and development of complex biological structures in general [140]. As Friston et al. point out, the predictive coding framework "only make(s) one assumption; namely, that the world can be described as a random dynamical system" ([140], p. 9). The world is, in other words, from the observer's perspective a black box. The boundary of the box in this formalism is the "Markov blanket" of the observer; the observer effectively perceives and acts on only the inside surface of this blanket [141], making it an "interface" in the sense employed by ITP. The interactions of the observer with the blanket are highly dependent on the observer's Bayesian priors, i.e., on the observer's conceptual and categorical structures; hence they display significant contextuality of the second kind identified by Kitto [24]. Theoretical approaches to conditions such as autism within the predictive coding framework (e.g., [142]) emphasize this contextuality.

6. Conclusions

As shown here, the over 60-year-old concept of the Black Box provides a powerful basis for conceptualizing observer-system interactions. While traditional discussions of the Black Box placed the observer outside the box, the no-boundary theorem moves the observer inside the box. Moving the observer inside the box, and then recognizing that the observer-box boundary is itself arbitrarily movable, leads immediately to no-go theorems of the type familiar from quantum information theory. The observer-box equivalence corollary then shows that observers are themselves black boxes, while the holographic-encoding corollary shows that both observers and their observed worlds are analogous, from an information-theoretic perspective, to black holes. Physics, therefore, already has a detailed and well established theory of the observer: the theory of the black hole. Conceptualizing the observer as a black hole forces observer-observer interactions to be regarded as physical interactions occurring at a defined interface between otherwise mutually-inaccessible systems. It forces us to acknowledge that we do not know what Schrödinger saw when he opened his infamous box and looked for his cat. We only know what he *reported* seeing. Indeed we each only know *our own interpretation*, based on our own physical interaction with his physically-tokened report, of what he reported seeing. As Fuchs has pointed out, with this level of epistemic personalization the usual "measurement problem" of quantum theory simply disappears [11]. It is no longer unclear why observers report classical outcomes: only classical outcomes—outcomes encodable by classical bit strings—are recordable in a thermodynamically-irreversible way, and hence only classical outcomes are reportable. It is no longer unclear how the measurement basis is chosen: observers deploy the observables they are able to deploy, represented in the measurement bases they are able to deploy. The traditional measurement problem is replaced, however, by an even deeper problem: a deep and in-principle unresolvable ambiguity about both the ontological structure and the state of the world. This deep ambiguity is not a "quantum" phenomenon. It follows solely from classical physics, the classical physics of the Black Box. Had Moore's theorem been proved 20 years before the development of quantum theory instead of 20 years after, this deeper problem may have been considered the "measurement problem" all along.

Taking the Black Box seriously as a model of observation not only clarifies the relation between classical and quantum physics; it also clarifies the relation between physics and computer science, cognitive science and even biology. It forces physics to view semantics in model-theoretic terms, and thereby enables physics to view physical systems in virtual-machine terms. While this view is implicit in computational conceptualizations of physical dynamics, it typically only appears explicitly in speculations about simulated worlds (e.g., [143]). The simulation assumption is, however, not required for the conclusion that any observer's observed world is a virtual world. This conclusion requires only that the world itself, the world in which the observer is embedded, is a black box.

Acknowledgments: This work was supported in part by The Federico and Elvia Faggin Foundation. Conversations with Eric Dietrich, Federico Faggin, Don Hoffman, Ken Krechmer, Mike Levin, Chetan Prakash and Manish Singh, as well as the comments and criticisms of two anonymous referees have helped to clarify the ideas in this paper.

Conflicts of Interest: The author declares no conflict of interest. The funding sponsor had no role in the design of the study; in the collection, analyses, or interpretation of data; in the writing of the manuscript, or in the decision to publish the results.

Abbreviations

The following abbreviations are used in this manuscript:

AI	Artificial intelligence
FAPP	For all practical purposes
ITP	Interface theory of perception
LOCC	Local observations, classical communication
SGP	Symbol grounding problem

References

1. Ashby, W.R. *Introduction to Cybernetics*; Chapman and Hall: London, UK, 1956.
2. Landauer, R. Irreversibility and heat generation in the computing process. *IBM J. Res. Dev.* **1961**, *5*, 183–195.
3. Landauer, R. Information is a physical entity. *Phys. A* **1999**, *263*, 63–67.
4. Bennett, C.H. Notes on Landauer's Principle, reversible computation, and Maxwell's Demon. *Stud. Hist. Philos. Mod. Phys.* **2003**, *34*, 501–510.
5. Bacciagalupi, G.; Valenti, A. *Quantum Theory at the Crossroads: Reconsidering the 1927 Solvay Conference*; Cambridge University Press: Cambridge, UK, 2006.
6. Landsman, N.P. Between classical and quantum. In *Handbook of the Philosophy of Science: Philosophy of Physics*; Butterfield, J., Earman, J., Eds.; Elsevier: Amsterdam, The Netherlands, 2007; pp. 417–553.
7. Wallace, D. Philosophy of quantum mechanics. In *The Ashgate Companion to Contemporary Philosophy of Physics*; Rickles, D., Ed.; Ashgate: Aldershot, UK, 2008; pp. 16–98.
8. Norsen, T.; Nelson, S. Yet another snapshot of foundational attitudes toward quantum mechanics. 2013, preprint arXiv:1306.4646v2.
9. Schlosshauer, M.; Kofler, J.; Zeilinger, A. A snapshot of foundational attitudes toward quantum mechanics. *Stud. Hist. Philos. Mod. Phys.* **2013**, *44*, 222–223.
10. Sommer, C. Another survey of foundational attitudes towards quantum mechanics. 2013, preprint arXiv:1303.2719v1.
11. Fuchs, C. QBism, the perimeter of Quantum Bayesianism. 2010, preprint arxiv:1003.5201v1.
12. Cabello, A. Interpretations of quantum theory: A map of madness. 2015, preprint arXiv:1509.04711v1.
13. Bell, J.S. Against measurement. *Phys. World* **1990**, *3*, 33–41.
14. Fuchs, C. On participatory realism. 2016, preprint arxiv:1601.04360v1.
15. Wheeler, J.A. Law without law. In *Quantum Theory and Measurement*; Wheeler, J.A., Zurek, W.H., Eds.; Princeton University Press: Princeton, NJ, USA, 1983; pp. 182–213.
16. Von Uexküll, J. A stroll through the worlds of animals and men. In *Instinctive Behavior*; Schiller, C., Ed.; van Nostrand Reinhold: New York, NY, USA, 1957; pp. 5–80.
17. Von Foerster, H. Objects: Tokens for (eigen-) behaviors. *ASC Cybern. Forum* **1976**, *8*, 91–96.
18. Kampis, G. Explicit epistemology. *Revue de la Pensee d'Aujourd'hui* **1996**, *24*, 264–275.
19. Koenderink, J. The all-seeing eye. *Perception* **2014**, *43*, 1–6.
20. Rössler, O.E. Endophysics. In *Real Brains—Artificial Minds*; Casti, J., Karlquist, A., Eds.; North-Holland: New York, NY, USA, 1987; pp. 25–46.
21. Von Neumann, J. *The Mathematical Foundations of Quantum Mechanics*; Princeton University Press: Princeton, NJ, USA, 1955.
22. Pattee, H.H. The physics of symbols: Bridging the epistemic cut. *Biosystems* **2001**, *60*, 5–21.
23. Kauffman, S.A.; Gare, A. Beyond Descartes and Newton: Recovering life and humanity. *Prog. Biophys. Mol. Biol.* **2015**, *119*, 219–244.

24. Kitto, K. A contextualised general systems theory. *Systems* **2014**, *2*, 541–565.
25. Tipler, F.J. Quantum nonlocality does not exist. *Proc. Natl. Acad. Sci. USA* **2014**, *111*, 11281–11286.
26. Bohm, D.; Hiley, B.J.; Kaloyerou, P.N. An ontological basis for the quantum theory. *Phys. Rep.* **1987**, *144*, 321–375.
27. Rosen, R. On information and complexity. In *Complexity, Language, and Life: Mathematical Approaches*; Casti, J.L., Karlqvist, A., Eds.; Springer: Berlin, Germany, 1986; pp. 174–196.
28. Polanyi, M. Life's irreducible structure. *Science* **1968**, *160*, 1308–1312.
29. Shannon, C.E. A mathematical theory of communication. *Bell Syst. Tech. J.* **1948**, *27*, 379–423.
30. Bohr, N. Causality and complementarity. *Philos. Sci.* **1937**, *4*, 289–298.
31. Landau, L.D.; Lifshitz, E.M. *Quantum Mechanics: Non-Relativistic Theory*; Pergamon: Oxford, UK, 1977.
32. Mermin, D. What's wrong with this pillow? *Phys. Today* **1989**, *42*, 9–11.
33. Zeh, D. On the interpretation of measurement in quantum theory. *Found. Phys.* **1970**, *1*, 69–76.
34. Zeh, D. Toward a quantum theory of observation. *Found. Phys.* **1973**, *3*, 109–116.
35. Zurek, W.H. Pointer basis of the quantum apparatus: Into what mixture does the wave packet collapse? *Phys. Rev. D* **1981**, *24*, 1516–1525.
36. Zurek, W.H. Environment-induced superselection rules. *Phys. Rev. D* **1982**, *26*, 1862–1880.
37. Joos, E.; Zeh, D. The emergence of classical properties through interaction with the environment. *Z. Phys. B Condens. Matter* **1985**, *59*, 223–243.
38. Zurek, W.H. Decoherence, einselection and the existential interpretation (the rough guide). *Philos. Trans. R. Soc. A* **1998**, *356*, 1793–1821.
39. Zurek, W.H. Decoherence, einselection, and the quantum origins of the classical. *Rev. Mod. Phys.* **2003**, *75*, 715–775.
40. Schlosshauer, M. *Decoherence and the Quantum to Classical Transition*; Springer: Berlin, Germany, 2007.
41. Ollivier, H.; Poulin, D.; Zurek, W.H. Objective properties from subjective quantum states: Environment as a witness. *Phys. Rev. Lett.* **2004**, *93*, 220401.
42. Ollivier, H.; Poulin, D.; Zurek, W.H. Environment as a witness: Selective proliferation of information and emergence of objectivity in a quantum universe. *Phys. Rev. A* **2005**, *72*, 042113.
43. Blume-Kohout, R.; Zurek, W.H. Quantum Darwinism: Entanglement, branches, and the emergent classicality of redundantly stored quantum information. *Phys. Rev. A* **2006**, *73*, 062310.
44. Zurek, W.H. Quantum Darwinism. *Nat. Phys.* **2009**, *5*, 181–188.
45. Riedel, C.J.; Zurek, W.H. Quantum Darwinism in an everyday environment: Huge redundancy in scattered photons. *Phys. Rev. Lett.* **2010**, *105*, 020404.
46. Riedel, C.J.; Zurek, W.H.; Zwolak, M. The rise and fall of redundancy in decoherence and quantum Darwinism. *New J. Phys.* **2012**, *14*, 083010.
47. Zwolak, M.; Riedel, C.J.; Zurek, W.H. Amplification, redundancy, and quantum Chernoff information. *Phys. Rev. Lett.* **2014**, *112*, 140406.
48. Chiribella, G.; D'Ariano, G.M. Quantum information becomes classical when distributed to many users. *Phys. Rev. Lett.* **2006**, *97*, 250503.
49. Korbicz, J.K.; Horodecki, P.; Horodecki, R. Objectivity in a noisy photonic environment through quantum state information broadcasting. *Phys. Rev. Lett.* **2014**, *112*, 120402.
50. Horodecki, R.; Korbicz, J.K.; Horodecki, P. Quantum origins of objectivity. *Phys. Rev. A* **2015**, *91*, 032122.
51. Brandão, F.G.S.L.; Piani, M.; Horodecki, P. Generic emergence of classical features in quantum Darwinism. *Nat. Commun.* **2015**, *6*, 7908.
52. Fields, C. Quantum Darwinism requires an extra-theoretical assumption of encoding redundancy. *Int. J. Theor. Phys.* **2010**, *49*, 2523–2527.
53. Fields, C. Classical system boundaries cannot be determined within quantum Darwinism. *Phys. Essays* **2011**, *24*, 518–522.
54. Schlosshauer, M. Experimental motivation and empirical consistency of minimal no-collapse quantum mechanics. *Ann. Phys.* **2006**, *321*, 112–149.
55. Everett, H., III. "Relative state" formulation of quantum mechanics. *Rev. Mod. Phys.* **1957**, *29*, 454–462.
56. Ghirardi, G.C.; Rimini, A.; Weber, T. Unified dynamics for microscopic and macroscopic systems. *Phys. Rev. D* **1986**, *34*, 470–491.
57. Penrose, R. On gravity's role in quantum state reduction. *Gen. Relativ. Gravit.* **1996**, *28*, 581–600.

58. Weinberg, S. Collapse of the state vector. *Phys. Rev. A* **2012**, *85*, 062116.
59. Jordan, T.F. Fundamental significance of tests that quantum dynamics is linear. *Phys. Rev. A* **2010**, *82*, 032103.
60. Wiseman, H. Quantum physics: Death by experiment for local realism. *Nature* **2015**, *526*, 649–650.
61. Swingle, B. Entanglement renormalization and holography. *Phys. Rev. D* **2012**, *86*, 065007.
62. Saini, A.; Stojkovic, D. Radiation from a collapsing object is manifestly unitary. *Phys. Rev. Lett.* **2015**, *114*, 111301.
63. Susskind, L. Computational complexity and black hole horizons. *Fortschr. Phys.* **2016**, *64*, 24–43.
64. Fields, C. If physics is an information science, what is an observer? *Information* **2012**, *3*, 92–123.
65. Fields, C. A model-theoretic interpretation of environment-induced superselection. *Int. J. Gen. Syst.* **2012**, *41*, 847–859.
66. Fields, C. On the Ollivier-Poulin-Zurek definition of objectivity. *Axiomathes* **2014**, *24*, 137–156.
67. Fields, C. Decompositional equivalence: A fundamental symmetry underlying quantum theory. *Axiomathes* **2016**, *26*, 279–311.
68. Zanardi, P. Virtual quantum subsystems. *Phys. Rev. Lett.* **2001**, *87*, 077901.
69. Zanardi, P.; Lidar, D.A.; Lloyd, S. Quantum tensor product structures are observable-induced. *Phys. Rev. Lett.* **2004**, *92*, 060402.
70. Dugić, M.; Jeknić, J. What is "system": Some decoherence-theory arguments. *Int. J. Theor. Phys.* **2006**, *45*, 2249–2259.
71. Dugić, M.; Jeknić-Dugić, J. What is "system": The information-theoretic arguments. *Int. J. Theor. Phys.* **2008**, *47*, 805–813.
72. De la Torre, A.C.; Goyeneche, D.; Leitao, L. Entanglement for all quantum states. *Eur. J. Phys.* **2010**, *31*, 325–332.
73. Harshman, N.L.; Ranade, K.S. Observables can be tailored to change the entanglement of any pure state. *Phys. Rev. A* **2011**, *84*, 012303.
74. Thirring, W.; Bertlmann, R.A.; Köhler, P.; Narnhofer, H. Entanglement or separability: The choice of how to factorize the algebra of a density matrix. *Eur. Phys. J. D* **2011**, *64*, 181–196.
75. Dugić, M.; Jeknić-Dugić, J. Parallel decoherence in composite quantum systems. *Pramana* **2012**, *79*, 199–209.
76. Rovelli, C. Relational quantum mechanics. *Int. J. Theor. Phys.* **1996**, *35*, 1637–1678.
77. Moore, E.F. Gedankenexperiments on sequential machines. In *Autonoma Studies*; Shannon, C.W., McCarthy, J., Eds.; Princeton University Press: Princeton, NJ, USA, 1956; pp. 129–155.
78. Popper, K. *Conjectures and Refutations: The Growth of Scientific Knowledge*; Routledge & Kegan Paul: London, UK, 1963.
79. Fuchs, C.A.; Stacey, B.C. Some negative remarks on operational approaches to quantum theory. 2014, preprint arxiv:1401.7254v1.
80. Bell, J.S. On the Einstein-Podolsky-Rosen paradox. *Physics* **1964**, *1*, 195–200.
81. Aspect, A.; Dalibard, J.; Roger, G. Experimental test of Bell's inequalities using time-varying analyzers. *Phys. Rev. Lett.* **1982**, *49*, 1804–1807.
82. Fields, C. Bell's theorem from Moore's theorem. *Int. J. Gen. Syst.* **2013**, *42*, 376–385.
83. Kochen, S.; Specker, E.P. The problem of hidden variables in quantum mechanics. *J. Math. Mech.* **1967**, *17*, 59–87.
84. Wootters, W.K.; Zurek, W.H. A single quantum cannot be cloned. *Nature* **1982**, *299*, 802–803.
85. Jennings, D.; Leifer, M. No return to classical reality. *Contemp. Phys.* **2016**, *57*, 60–82.
86. Peres, A.; Terno, D.R. Quantum information and relativity theory. *Rev. Mod. Phys.* **2004**, *76*, 93–123.
87. Chitambar, E.; Leung, D.; Mančinska, L.; Ozols, M.; Winter, A. Everything you always wanted to know about LOCC (but were afraid to ask). *Commun. Math. Phys.* **2014**, *328*, 303–326.
88. Bartlett, S.D.; Rudolph, T.; Spekkens, R.W. Reference frames, superselection rules, and quantum information. *Rev. Mod. Phys.* **2007**, *79*, 555–609.
89. Roederer, J.G. *Information and Its Role in Nature*; Springer: Berlin, Germany, 2005.
90. Roederer, J.G. Pragmatic information in biology and physics. *Philos. Trans. R. Soc. A* **2016**, *374*, 20150152.
91. Asaro, P.M. From mechanisms of adaptation to intelligence amplifiers: The philosophy of W. Ross Ashby. In *The Mechanical Mind in History*; Husbands, P., Holland, O., Wheeler, M., Eds.; MIT/Bradford: Cambridge, MA, USA, 2008.
92. Bekenstein, J.D. Black holes and entropy. *Phys. Rev. D* **1973**, *7*, 2333–2346.

93. Harlow, D.; Hayden, P. Quantum computing vs. firewalls. *J. High Energy Phys.* **2013**, *2013*, 85.
94. Conway, J.; Kochen, S. The free will theorem. *Found. Phys.* **2006**, *36*, 1441–1473.
95. Fields, C. A whole box of Pandoras: Systems, boundaries and free will in quantum theory. *J. Exp. Theor. Artif. Intell.* **2013**, *25*, 291–302.
96. Wang, Q.; Schoenlein, R.W.; Peteanu, L.A.; Mathies, R.A.; Shank, C.V. Vibrationally coherent photochemistry in the femtosecond primary event of vision. *Science* **1994**, *266*, 422–424.
97. Wheeler, J.A. Information, physics, quantum: The search for links. In *Complexity, Entropy, and the Physics of Information*; Zurek, W.H., Ed.; Westview: Boulder, CO, USA, 1990; pp. 3–28.
98. Clifton, R.; Bub, J.; Halvorson, H. Characterizing quantum theory in terms of information-theoretic constraints. *Found. Phys.* **2003**, *33*, 1561–1591.
99. D'Ariano, G.M. Physics as Information Processing. In Proceedings of the Physics Education Research Conference, Omaha, NE, USA, 3–4 August 2011; Volume 1327, pp. 7–18.
100. Chiribella, G.; D'Ariano, G.M.; Perinotti, P. Informational derivation of quantum theory. *Phys. Rev. A* **2011**, *84*, 012311.
101. Chiribella, G.; D'Ariano, G.M.; Perinotti, P. Quantum theory, namely the pure and reversible theory of information. *Entropy* **2012**, *14*, 1877–1893.
102. Hardy, L. Reconstructing quantum theory. *Fund. Theor. Phys.* **2015**, *181*, 223–248.
103. Müller, M.P.; Masanes, L. Information-theoretic postulates for quantum theory. *Fund. Theor. Phys.* **2015**, *181*, 139–170.
104. Knuth, K.H. Information-based physics: An observer-centric foundation. *Contemp. Phys.* **2014**, *55*, 12–32.
105. Deutsch, D.; Marletto, C. Constructor theory of information. *Proc. R. Soc. A* **2015**, *471*, 20140540.
106. Höhn, P.A.; Wever, C.S.P. Quantum theory from questions. 2015, preprint arxiv:1511.01120v2.
107. Grinbaum, A. How device-independent approaches change the meaning of physical theory. 2015, preprint arxiv:1512.01035v3.
108. Zimmer, H.D.; Ecker, U.K.D. Remembering perceptual features unequally bound in object and episodic tokens: Neural mechanisms and their electrophysiological correlates. *Neurosci. Biobehav. Rev.* **2010**, *34*, 1066–1079.
109. Fields, C. The very same thing: Extending the object token concept to incorporate causal constraints on individual identity. *Adv. Cognit. Psychol.* **2012**, *8*, 234–247.
110. Harrison, E.R. Standard model of the early universe. *Annu. Rev. Astron. Astrophys.* **1973**, *11*, 155–186.
111. Fields, C. Visual re-identification of individual objects: A core problem for organisms and AI. *Cognit. Process.* **2016**, *17*, 1–13.
112. Arkani-Hamed, N.; Trnka, J. The amplituhedron. *J. High-Energy Phys.* **2014**, *10*, 030.
113. Krechmer, K. Relational measurements and uncertainty. *Measurement* **2016**, doi:10.1016/j.measurement.2016.06.058.
114. Friston, K. The free-energy principle: A unified brain theory? *Nat. Rev. Neurosci.* **2010**, *11*, 127–138.
115. Kauffman, L. Eigenforms—Objects as tokens for eigenbehaviors. *Cybern. Hum. Knowing* **2003**, *10*, 73–90.
116. Kauffman, L. EigenForm. *Kybernetes* **2005**, *34*, 129–150.
117. Dietrich, E.; Fields, C. Science generates limit paradoxes. *Axiomathes* **2015**, *25*, 409–432.
118. Wittgenstein, L. *Tractatus Logico-Philosophicus*; Kegan Paul, Trench, Trubner: London, UK, 1922.
119. Tarski, A. The semantic conception of truth and the foundations of semantics. *Philos. Phenomenol. Res.* **1944**, *4*, 341–376.
120. Turing, A.R. On computable numbers, with an application to the *Entscheidungsproblem*. *Proc. Lond. Math. Soc.* **1936**, *442*, 230–265.
121. Quine, W.V.O. On what there is. *Rev. Metaphys.* **1948**, *2*, 21–38.
122. Tanenbaum, A.S. *Structured Computer Organization*; Prentice Hall: Upper Saddle River, NJ, USA, 1976.
123. Smith, J.E.; Nair, R. The architecture of virtual machines. *IEEE Comput.* **2005**, *38*, 32–38.
124. Partridge, D. *The Seductive Computer*; Springer: London, UK, 2010.
125. Chomsky, N. Review of B. F. Skinner, *Verbal Behavior. Language* **1959**, *35*, 26–58.
126. Harnad, S. The symbol grounding problem. *Phys. D* **1990**, *42*, 335–346.
127. Taddeo, M.; Floridi, L. Solving the symbol grounding problem: A critical review of fifteen years of research. *J. Exp. Theor. Artif. Intell.* **2005**, *17*, 419–445.
128. Fields, C. Equivalence of the symbol grounding and quantum system identification problems. *Information* **2014**, *5*, 172–189.

129. Fodor, J.A. Methodological solipsism considered as a research strategy in cognitive science. *Behav. Brain Sci.* **1980**, *3*, 63–73.

130. Clark, A; Chalmers, D. The extended mind. *Analysis* **1998**, *58*, 7–19.

131. Anderson, M.L. Embodied cognition: A field guide. *Artif. Intell.* **2003**, *149*, 91–130.

132. Cangelosi, A.; Schlesinger, M. *Developmental Robotics: From Babies to Robots*; MIT Press: Cambridge, MA, USA, 2015.

133. Hommel, B. Event files: Feature binding in and across perception and action. *Trends Cognit. Sci.* **2004**, *8*, 494–500.

134. Eichenbaum, H.; Yonelinas, A.R.; Ranganath, C. The medial temporal lobe and recognition memory. *Annu. Rev. Neurosci.* **2007**, *30*, 123–152.

135. Cubek, R.; Ertel, W.; Palm, G. A critical review on the symbol grounding problem as an issue of autonomous agents. *Lect. Notes Comput. Sci.* **2015**, *9324*, 256–263.

136. Mark, J.T.; Marion, B.B.; Hoffman, D.D. Natural selection and veridical perceptions. *J. Theor. Biol.* **2010**, *266*, 504–515.

137. Hoffman, D.D.; Singh, M.; Prakash, C. The interface theory of perception. *Psychon. Bull. Rev.* **2015**, *22*, 1480–1506.

138. Hoffman, D.D.; Prakash, C. Objects of consciousness. *Front. Psychol.* **2014**, *5*, 577.

139. Friston, K.J. Functional and effective connectivity: A review. *Brain Connect.* **2011**, *1*, 13–36.

140. Friston, K.; Levin, M.; Sengupta, B.; Pezzulo, G. Knowing one's place: A free-energy approach to pattern regulation. *J. R. Soc. Interface* **2015**, *12*, 20141383.

141. Friston, K. Life as we know it. *J. R. Soc. Interface* **2013**, *10*, 20130475.

142. Lawson, R.P.; Rees, G.; Friston, K.J. An aberrant precision account of autism. *Front. Hum. Neurosci.* **2014**, *8*, 302.

143. Bostom, N. Are we living in a computer simulation? *Philos. Quart.* **2003**, *53*, 243–255.

systems

MDPI

Concept Paper
Transdisciplinarity Needs Systemism

Wolfgang Hofkirchner

Bertalanffy Center for the Study of Systems Science; Paulanergasse 13, 1040 Vienna, Austria;
wolfgang.hofkirchner@bcsss.org

Academic Editors: Gianfranco Minati, Eliano Pessa and Ignazio Licata
Received: 1 November 2016; Accepted: 10 February 2017; Published: 16 February 2017

Abstract: The main message of this paper is that systemism is best suited for transdisciplinary studies. A description of disciplinary sciences, transdisciplinary sciences and systems sciences is given, along with their different definitions of aims, scope and tools. The rationale for transdisciplinarity is global challenges, which are complex. The rationale for systemism is the concretization of understanding complexity. Drawing upon Ludwig von Bertalanffy's intention of a General System Theory, three items deserve attention—the world-view of a synergistic systems technology, the world picture of an emergentist systems theory, and the way of thinking of an integrationist systems method.

Keywords: systems thinking; systems science; systems practice; way of thinking; world picture; world-view; integrationism; emergentism; synergism

---

1. Introduction

Since we live in an age of global challenges, the responsibility of the members of the scientific community is a topical issue. The trend towards transdisciplinary studies can be seen in that context. Many researchers devoted to transdisciplinarity use systems terms but may fail in a sound usage of the terms. The further development of systems thinking, systems sciences and systems practice should address that demand. The paper at hand offers three clear-cut specifications of the systems paradigm that might support the transdisciplinary impetus.

2. Aims, Scope, and Tools of Science

From a philosophy of science perspective, sciences can be classified according to three dimensions of knowledge: the technological (praxiological), the theoretical (ontological), and the methodological (epistemological) dimension.

Technology is objectivated knowledge about objects for objectives. Theory is objectivated knowledge about objects. Methodology is knowledge of how to objectivate knowledge, that is, to process it in a social procedure such that it is not only subjective but might be used as an objective basis for knowledge about objects, which is required for fulfilling objectives.

Technology incorporates the aims of scientific studies; it directs theory towards practical application. Applications intervene in the real world so as to help solve problems. Problems stand at the beginning of any science because they form ends for any science. Problems are, in the last resort, social. Sciences provide the means to reach a goal, given a point of departure.

Theory embraces the scope of scientific studies; it gives deep insights in the functioning of the real world—insights that can be functionalised for the solution of problems, by informing the practice about the way to a goal from a point of departure.

Methodology provides the tools of scientific studies; it is a framework through which understanding of the functioning of the real world can be generated to serve its function during problem-solving.

3. Transcending the Disciplines

Scientific disciplines are determined by specific aims, by a specific scope and by specific tools. The objective is a determinate problem solution, the object of study is a determinate piece of reality, and the objectivation is guided by a determinate mixture of methods.

However, given the rise of complex problems, monodisciplinary approaches do not fit the situation any more. Multi-, inter- and transdisciplinary approaches are needed. Transdisciplinarity has been gaining considerable attention since. It differs from monodisciplinarity in aims, scope, and tools.

As to the aims, disciplines shall be transcended by the inclusion of stakeholders through participation in the processes of research and development as well as through diffusion of innovations, which allows them to co-determine what shall be regarded as a problem and what shall be regarded as a solution. By doing so, technological knowledge shall be constructed for solving problems that are complex.

As to the scope, disciplines shall be transcended by the inclusion of interdependencies between factors across space (long-range effects), time (long-term effects), and matter (side effects) in the focus of the study. By doing so, theoretical knowledge shall be enabled to depict a bigger picture than mere isolated pieces of reality to underpin complex problem-solving.

As to the tools, disciplines shall be transcended by the inclusion of a common code that shall perform the translation of concepts of one domain to those of other domains. By doing so, methodological knowledge shall orient towards the identification of similarities across domains to gain a deeper understanding of complex problems.

4. The Systems Paradigm

So far, mainstream disciplines have been regarding systems practice, systems sciences, systems thinking—subsumed here under the term "systemism" (Mario Bunge)—as a determinate approach that aims at managing, intervening in, transforming, engineering or designing systems (systems technologies), which presupposes real-world systems, their dynamics and structures, or their elements as in the scope of sciences (systems theories), which, in turn, presupposes tools that frame empirical data as systems phenomena (systems methods). As such, systemism might or might not be the means of choice.

Figure 1 gives a sketch of systemism that is inspired by Robert Rosen's conception of the modelling relation. His drawing shows the natural system, that is, the object to be modelled, on the left side and the formal system, that is, the model, on the right side. From the left side to the right side, there is an encoding arrow, and from the right back to the left, a decoding arrow. After Rosen, the model is an attempt to formalise causations in the object to be modelled as inferences, whereas the encoding and decoding processes cannot be formalised themselves but make modelling "as much an *art*, as it is a *methodology*" [1] (p. 54). In Figure 1 the natural system is specified as a real-world system of any kind, and the formal system as those sciences that follow a systems approach. Rosen's encoding process is turned into a framing process and his decoding process into a designing process, both of which involve creative acts. Inside the systems, sciences layers are added to represent aims, scope and tools. The arrows between those layers do not represent entailment operations but operations that involve creative acts as well in order to accomplish the jumps between systems methods, theories and technologies.

However, systemism is not merely an option that can be taken up according to the researchers' predilection. It is far more a well-founded approach best suited for cases in which transdisciplinarity is needed. This holds not only for particular projects. It has the potential to give the whole edifice of sciences a new shape.

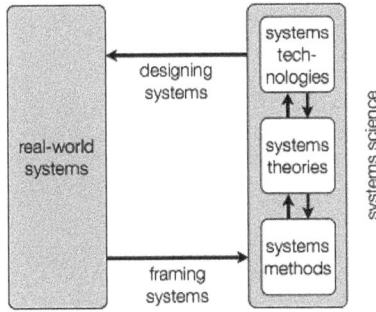

Figure 1. Modelling Systems.

The customary description of sciences involves (see Figure 2)

(a)   philosophy on the uppermost general level;

(b)   formal sciences, real-world sciences, and applied sciences on a next level and

(c)   subgroups of the former—as logic and mathematics in the case of formal sciences, as natural sciences such as physics, etc., and social and human sciences such as sociology, etc., in the case of real-world sciences, and as engineering science (that develops technological innovations) such as computer science, etc., and management science and science of arts (that develop social innovations) in the case of applied sciences—and

(d)   their sub-disciplines and sub-sub-disciplines, and so forth, on ever more specific levels.

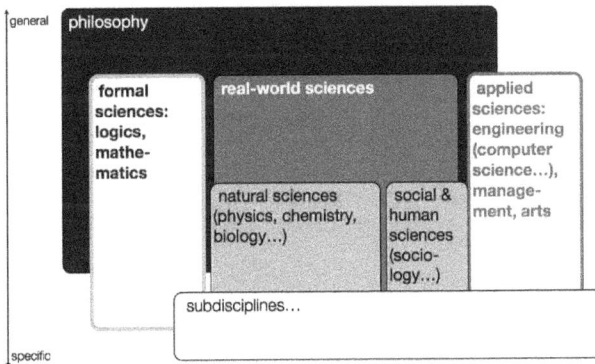

Figure 2. The Edifice of Sciences from a positivistic perspective.

Those sciences are imagined to have well-defined boundaries and to interact at best without undergoing fundamental changes themselves. Thus is the image of positivist sciences.

Systemism comes with a new understanding of sciences (see Figure 3). The systemist conception of sciences assumes semi-permeable boundaries and upward and downward interactions across the levels between all sciences. This is substantiated through defining any science in the context of discovering, describing and dealing with an overall systemic interconnectedness. Accordingly, formal sciences become part of a systems methodology that embraces formal and non-formal methods to understand systems features; real-world sciences turn into a science of different real-world systems that can be categorised as material systems, living (material) systems and social (living) systems, and applied sciences turn out to be a science of the artificial design of those systems. Any science is open to further specifications of subsystems, sub-subsystems and so on.

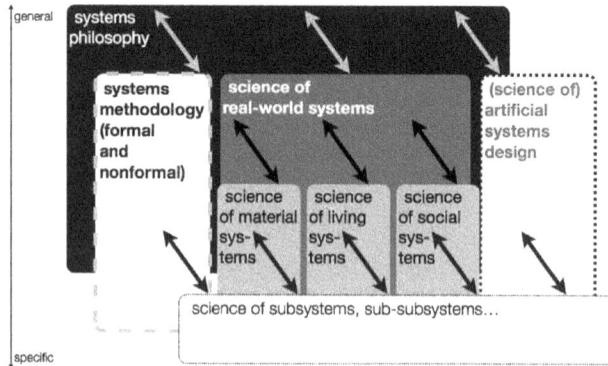

Figure 3. The Edifice of Sciences from a systems perspective.

Thus a lively body of scientific knowledge can be considered as growing and developing along the feed-forward and feedback cycles in the direction of new generalisations (upward) and new specifications (downward), while consistency is taken care of and the silos of the disciplines in the positivist perspective are broken up for the benefit of a convergent whole of science.

Systemism has been effecting a paradigm shift that can transform positivist disciplines into parts of an overarching transdisciplinary endeavour. Of course, there have been drawbacks: the resistance of the positivist science establishment has proven strong; systemism has itself branched into a plethora of different schools. Though systems terms have flooded the disciplines, the meaning of the terms is heterogeneous. So the paradigm shift is far from being completed.

The need for science and technology responses to global challenges in an intelligent way has been supporting the attraction of transdisciplinarity among the scientific community.

Drawing on Ludwig von Bertalanffy's intention to form a General System Theory [2], as well as several other compatible insights revolving around issues such as complexity, emergence, and self-organisation, can help sharpen transdisciplinary efforts. In particular, it is these features that play an innovative part in supporting the transdisciplinary agenda: systems technologies can be characterised by a new world view, systems theories by a new world picture, and systems methods by a new way of thinking.

The discussion of these issues here will follow the systems modelling processes sketched in Figure 1 and start at the bottom level with framing systems.

4.1. A New, Systemic Way of Thinking for Transdisciplinarity: Integrationism

Complex problems need an epistemological approach that does justice to the complexity of reality from which systems phenomena emanate. In many cases, if not in any case, an assumption has to be made about which is the interrelation of phenomena of different degrees of complexity: how does the lower-complexity phenomenon relate to the higher-complexity phenomenon (and vice versa)?

This is a question of the way of thinking. There are, in principle, three (or four) possibilities [3] (see Table 1).

Table 1. Ways of thinking.

		Complexities	Identity and Differences
universalism	reductionism	levelling down higher complexity	identity at the cost of differences (uniformity not diversity)
	projectionism	levelling up lower complexity	identity for the benefit of one difference (uniformity not diversity)
particularism: disjunctionism		segregating degrees of complexity	any difference at the cost of identity (singularity or duality/plurality not unity)
systemic framing: integrationism		conjoining complexity degrees of several levels through integration and differentiation	identity and differences united (dialectic of unity through diversity, including as less unity as necessary and as much diversity as possible)

First, there is a universalist way of thinking that gives priority to uniformity over diversity. It comes in two varieties:

(a)   the levelling down of phenomena of higher complexity to phenomena of lower complexity; identity of the phenomena is established at the cost of differences; this is known as reductionism;

(b)   the levelling up of phenomena of lower complexity to phenomena of higher complexity; identity of the phenomena is established for the benefit of one difference; this is called "projectionism"; higher complexity is erroneously conceptualised at a level where it does not exist.

The second way of thinking is a particularist one. Priority is given to the singularity of a difference or the plurality of all differences over unity. The disjoining of phenomena of different degrees of complexity establishes the identity of a particular difference or identities of any difference, thus an equivalence of differences—indifference—at the cost of an identity common to the phenomena. That is called "disjunctionism".

A third option is that way of thinking that inheres in systemism. It negates universalism and particularism as well and interrelates phenomena to each other through integration and differentiation of their complexity degrees. The union of identity and differences yields unity through diversity. That is the meaning of integrationism here. That is, the phenomenon with a lower degree of complexity shares with the phenomenon with a higher degree of complexity at least one property, which makes both of them, to a certain extent, identical, but the latter phenomenon is in the exclusive possession of at least another property, which makes it, to a certain extent, distinct from the former. So both phenomena are identical and different at the same time.

The method of transdisciplinarity can take advantage of bringing this new systems method to bear: framing the phenomena through the equilibration of integration and differentiation during the processes of conceptualisation in order to rule out reductionist, projectionist, and disjunctionist ways of thinking.

Example. Let us take the relationship of social science and engineering science as an example for how to transcend the borders of both disciplines by making use of a systemic framing and transform their relationship into a true transdisciplinary one (see Table 2).

Table 2. Ways of thinking in society and technology.

		Complexities	Identity and Differences
technomorphism		man is deemed a machine	the conception of the artificial is sufficient for comprehending social forms
anthropo-socio-morphism		the machine is deemed man-like	the conception of social forms is necessary for comprehending the artificial
hetero-morphism	socio-centrism	man is deemed exclusive	social forms are construed as distinctive over and against the artificial
	techno-centrism	the machine is deemed exclusive	the artificial is construed as distinctive over and against social forms
	relativism	man and machine are deemed apart	both social and artificial forms are construed side by side
systemic framing: techno-social systems		man and machine are deemed nested	the understandings of social and artificial forms join into the understanding of techno-social systems

In order to combine social science with engineering science, representatives of the latter might be inclined to reduce that which is human to that which is engineerable: man is deemed a machine. Operation Research, Cybernetics, Robotics, Mechatronics, the fields of Artificial Intelligence and so-called Autonomous Systems, among others, are liable to cut the understanding of man who is a social being free from the understanding of social relations; the conception of the human body free from the conception of individual actors; and conceiving of mechanics free from conceiving of the organism. Mechanical architectures and functioning that are constituents among others of human life structures and processes are analysed and hypostatised as sufficient for the comprehension of man ("technomorphism").

Or representatives of social sciences—not unlike those of other disciplines—might share a predilection to understand the whole world, including artifactual mechanics, by projecting characteristics of the social world onto the former: the machine is deemed man-like. Actor-Network Theory, Deep Ecology, New Materialism, Info-Computationalism and others are prone to blur the boundaries between the social sciences and the sciences of living things, between the latter and the sciences of physical things and, eventually, between the sciences of physical things and engineering sciences by attributing human features to any of those non-social disciplines. The conception of social forms is thought necessary for the comprehension of everything. That is blunt anthroposociomorphism.

Segregation might be made for the sake of either the identity of social science or that of engineering science: anthropocentric or, better, sociocentric positions traditionally distinguish the investigation of man as exclusive and belittle engineering undertakings, whereas trans- and post-humanistic positions argue for an imminent advent of a technological singularity that will make machines outperform man and thus the human race obsolescent. However, segregation might also promote the juxtaposition of social science and engineering science: they co-exist, any of them is worth as much as the other, and there are no grounds for giving supremacy to what counts as social forms or what counts as technological forms (relativism). Both are conceived as reciprocally exclusive ("heteromorphism").

No one of these options does establish true transdisciplinarity. In the case of technomorphism, social science gives up any autonomy and is invaded by engineering science. In the case of anthroposociomorphism, any autonomy of engineering science is forfeited, as social science projects its autonomy onto engineering science. In the case of heteromorphism, each discipline claims full autonomy over its own, self-contained area. A way out can be seen through an approach that assumes an interrelation of both disciplines in a systemic framework that grants (relative) autonomy to each of them according to their place in the overall framework. Both disciplines complement each other for the sake of a greater whole. That greater whole is achieved by framing both disciplines in a systems perspective, that is, by framing them as part of systems science. As such, social and engineering sciences combine for a common understanding of the systemic relationship of society and technology—of emerging techno-social systems. They make use of systems methodologies for empirically studying social systems and the artifactual in the context of technological applications implemented by social systems design. By doing so, they can form a never-ending cycle in which each of them has a determinate place: social systems science can inform engineering systems science by providing facts about social functions in the social system that might be supported with technological means; engineering systems science can provide technological options that fit the social functions in the envisaged techno-social system; social systems science can, in turn, investigate the social impact of the applied technological option in the techno-social system and provide facts about the working of technology. The social and the engineering parts of techno-social systems science are coupled so as to promote an integrated technology assessment and technology design cycle in a transdisciplinary sense.

4.2. A New, Systemic Picture of the World for Transdisciplinarity: Emergentism

Systems theories provide an ontology in which complex problems are pictured as complex because they take part in an overall interconnectedness of processes and structures that are constituted by self-organising real-world systems. Those systems bring about evolution and nestedness as emergent features of reality.

The world is pictured according to a multi-stage model of evolutionary systems [4] (see Figure 4).

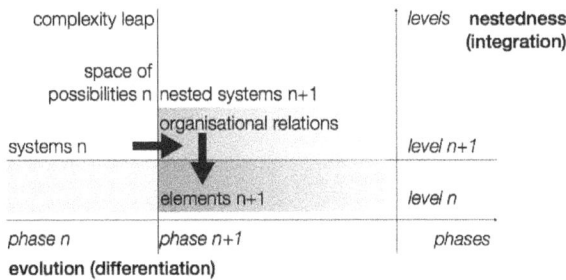

Figure 4. The multi-stage model of evolutionary systems.

Systems evolve during a phase n. Spontaneously, at a certain point in time, a leap in complexity emerges and one possibility out of the space of possibilities that are rooted in the reality of the systems during the phase n (which form the necessary condition for the transition to phase n + 1) is realised such that new organisational relations emerge. Those organisational relations realise a higher order in that they nest the old systems n as elements n + 1 of the new systems n + 1 during the phase n + 1. Thus they form another level n + 1 above the level n that is being reontologised, reworked, reshaped. Emergence of new systems in the course of evolution (differentiation) entails dominance of higher levels through the reconfiguration of what is taken over from the old (integration).

Emergentism (another term coined by Bunge) is an important ingredient of systems theories. It helps understand events and entities that function as less than strict deterministically. Transdisciplinarity must take into consideration less-than-strict determinism, which means that the mechanisms of the real world are not machine-like.

Emergentism provides an ontological superstructure for epistemological integrationism. Integrationism can integrate because evolution lets new features emerge.

Example. Let us consider the issue of anthroposociogenesis [5] as an example for the transcendence of natural and cultural science towards a true transdiscipline of anthropology by a systemic image of the evolutionary formation of society and humans.

In naturalistic accounts, the advent of human culture is owed to the development of nature. The former is the necessary result of the latter. Anatomic features such as cerebral growth, the descent of the larynx, bipedalism and the opposable thumb and the fingers, among others, are named as biological traits that seem to have caused psychic functions such as thought, the ability to speak and the ability to devise and manufacture artefacts, in particular tools.

One variety of culturalism would conflate the distinction between nature and culture as well, postulating that cultural features occur in pre-human life and overemphasising the continuity from natural to cultural evolution.

Another variety of culturalism would postulate a discontinuity between nature and culture and deny any comparability. Both spheres are considered to be ontologically independent of each other, and therefore, on that basis they might interact.

In contradistinction, emergentism sees nature as the necessary condition upon which culture as an evolutionary contingency could emerge. In turn, culture, as a supra-system, began to dominate via the higher-order structure from which it had taken its departure (see Figure 5). The key idea is that it is possibilities of co-operation that, when realised, made the difference in evolution. Evolutionary pressure unfolded a ratchet effect that yielded ever higher complex cooperation. Michael Tomasello characterises the life of our great ape ancestors as "mostly individualistic and competitive" [6] (p. 4). The ancestors were "individually intentional and instrumentally rational" [6] (p. 30) as great apes are still today. It was not before taking advantage of going beyond "individual intentionality" and adopting "more complex forms of cooperative sociality" [6] (p. 31) that early humans began to speciate. This brought about "shared intentionality" and enabled them to achieve shared goals. In particular, Tomasello hypothesises two steps towards a differentiated shared intentionality. A first step occurred "in the context of collaborative foraging" [6] (p. 33) around two million years ago with a culmination 400,000 years ago. This was a step from individual intentionality to joint intentionality, from the competitive sociality of great ape ancestors to ad hoc dyadic relationships with a "significant other" (G. H. Mead). Multiple and vanishing dyadic relationships formed in which early humans shared a joint goal. A second step led from early to modern humans some 200,000 years ago. They became "thoroughly group-minded individuals" [6] (p. 80) in larger groups, in which they "had to be prepared to coordinate with anyone from the group, with some kind of generic other" [6] (p. 81). This was a step from joint intentionality to collective intentionality, a step to the co-operation within a larger group organisation (culture) as the "generalised other" (Mead). Every step was a step of emergence of social systems in a new quality owed to the contingent realisation of possible social relations that represent a higher order of cooperation and reconfigure the old system. By accepting the interplay of emergence and dominance, the objects of natural science and cultural science reveal their systemic interrelationship and give way to a true transdisciplinary theory of the origins (and evolution) of human, social systems.

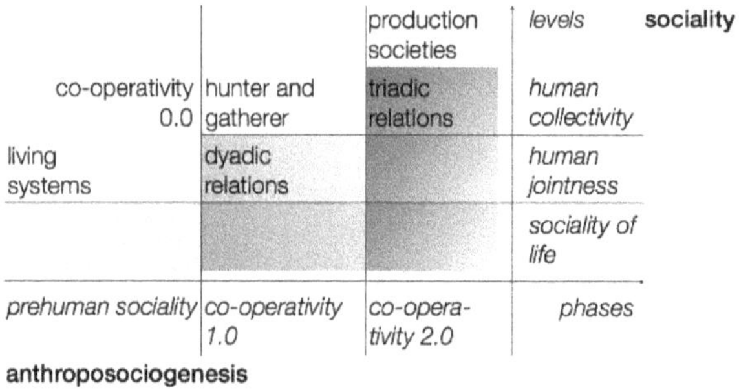

		production societies	*levels*	**sociality**
co-operativity 0.0	hunter and gatherer	triadic relations	*human collectivity*	
living systems	dyadic relations		*human jointness*	
			sociality of life	
prehuman sociality	co-operativity 1.0	co-opera-tivity 2.0	*phases*	

anthroposociogenesis

Figure 5. The emergence of cultural systems from natural systems through relations of joint cooperation and collective cooperation.

4.3. A New, Systemic World View for Transdisciplinarity: Synergism

Acting in the face of complex problems is based on praxiological assumptions about the interference with self-organising systems. Known mechanisms can be furthered or dampened according to what the goal shall be.

In the course of evolution, systems move on trajectories on which bifurcations occur. Bifurcations come up with a variety of possible future trajectories. Systems might not be in the position to avert devolution (a path that leads to the breakdown of the system) or they might be able to achieve a leap from the previous level of evolution on which they could enjoy a steady state onto a higher level which forms part of a successful mega-evolution (a breakthrough to a path that transforms the system) [7] (p. 314) [8] (pp. 103–104). Amplified fluctuations of parameters indicate possible and necessary punctuations (see Figure 6).

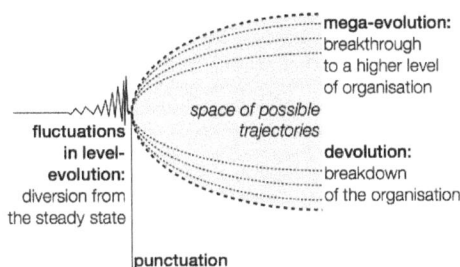

Figure 6. Bifurcations in systems evolution.

Self-organising systems have as raison d'etre the provision and production of synergetic effects [9]. If the organisational relations are not able any more to provide and help the elements produce synergy, the system will break down. Hindrances of letting synergy emerge are called frictions. Any social system is a social system by virtue of organisational relations of production and provision of the common good, that is, the commons are the social manifestation of synergy [10]. Hindrances of the commons supply are frictions that are systemic dysfunctions due to the suboptimal organisation of the synergetic effects. Any meaningful technology is oriented towards the alleviation of frictions and the advancement of synergy.

Thus systems technologies can help orient transdisciplinarity. Meaningful technology is technology endowed with meaning by

(1)   the participation of those affected in an integrated technology assessment and design process (that is, design builds upon assessment);
(2)   for the reflection of the expected and actual usage of technology: the assessment and design criterion is social usefulness, that is, the reflection of both;

    (a)   the adequacy to the purpose (utility; operational knowledge: know-how) and
    (b)   the purpose itself (the function technology serves; orientational knowledge: know why and what for).

The purpose is advancing the commons.

Synergism, the orientation towards synergy for every real-world system and towards the human value of the commons in the case of social systems—which is a world-view (weltanschauung) because it is value-laden—is the praxiological superstructure for emergentism. Synergy emerges, emergence brings about synergy.

Example. Let us turn to the crises that manifest themselves in the tension between social systems on the level of today's nation states as well as within every such social system regarding their subsystems such as, in particular, the cultural, the political, the economic, the ecological (eco-social) and the technological (techno-social) subsystems, which means that several branches of social sciences are challenged. The critical situation might be complicated or complex to analyse. Sometimes it is said that complicatedness—the presence of many factors that influence each other —is not an obstacle that we would not be able to overcome. However, the opposite might be true. The huge amount of options of interventions—proposed by different disciplines—might actually pose a practical difficulty. However, on the other hand, the situation we face nowadays is rather complex. In addition, if complexity enters the stage, problem-solving is also possible despite opposite views—in a true transdisciplinary effort.

Today, enclosures of the commons have been aggravated to such a degree that all of them morphed into global challenges. Global challenges drive an accumulation of crises that mark a decisive bifurcation. A Great Bifurcation lies ahead of humanity (see Figure 7) [11].

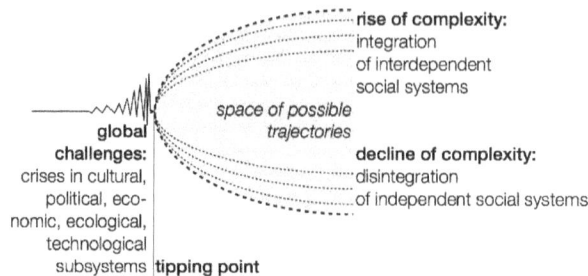

Figure 7. The Great Bifurcation in social systems evolution.

On the one hand, a possible change never seen before might be imminent—a transformation of the current state of civilisation into a new state that brings about a peaceful, environmentally sound and socially and economically just and inclusive world society by the integration of differentiated interdependent social systems. That rise in complexity is possible. If there is a mismatch between the complexity of a system and the complexity of the problems faced by the system, that system can catch up. It can solve the problem by activating the collective intelligence of the co-systems it is made up of and raise the complexity of its organisational relations or by activating the collective intelligence of co-systems of its own and raise the complexity of that supra-system in which they are nested in order to match or surpass the complexity faced. Intelligence is the capability of self-organising systems to generate that information which contributes in the best way to solving problems. The better their collective intelligence, that is, the better their problem-solving capacity and the better their capability to generate information, the better their handling of the crisis and the order they can reach. Higher complexity not only signifies a higher degree of differentiation. At least as importantly, it signifies a new quality of integration. Only a new level of integration can deal with an intensification of differentiation. The new system would disclose the commons. "It would be a Global Sustainable Information Society (GSIS).

GSIS requires,

(1)   for the first time in the history of our planet, on a higher social level—that is, globally—
(2)   a reorganisation of the social relations within and in between the interdependent social systems such that sociogenic dysfunctions with respect to the social, the social-ecological and the socio-technological realms can be contained—that is, a transformation into sustainable development—
(3)   through conscious and conscientious actors that are not only self- but also community-concerned—that is, under well-determined informational conditions." [11] (15)

On the other hand, a decline in complexity might be imminent, eventually leading to disintegration of allegedly independent social systems. Civilisation would fall apart.

Systemism is key to tackling complex problems such as the reciprocal influences of varying kinds of factors dealt with on the basis of varying disciplinary backgrounds within the social sciences in a transdisciplinary way. It is thus key to tackling global challenges and guiding the transformation of our social systems onto the next level of cooperation through a social systems design based upon social systems technologies.

5. Conclusions

In summary, systemic transdisciplinarity

(1) aims—by a systems world-view—at providing scientific knowledge for solving problems of frictions in the functioning of real-world systems, in particular in processes of the provision and production of the commons in social systems through meaningful systems technologies that support the re-organisation of social systems in order to safeguard sustainable development and rule out self-inflicted breakdowns;

(2) has as its scope the functioning of emergent real-world systems in the interconnectedness of their evolution and their nestedness, the scientific knowledge of which is a theoretical systems world picture needed for alleviating frictions;

(3) uses tools that generate scientific knowledge through a systems way of thinking by the method of equilibrating integration and differentiation for a proper understanding of how complexity grows.

Systemism has the power to transform the disciplines.

Conflicts of Interest: The author declares no conflict of interest.

References

1. Rosen, R. *Life Itself*; Columbia University Press: New York, NY, USA, 1991.
2. Bertalanffy, L.V. *General System Theory; With a Foreword by Wolfgang Hofkirchner & David Rousseau*; George Braziller: New York, NY, USA, 2015.
3. Hofkirchner, W. Four ways of thinking in information. *Triple-C: J. Glob. Sustain. Infor. Soc.* **2011**, *9*, 322–331.
4. Hofkirchner, W. *Emergent Information, a Unified Theory of Information Framework*; World Scientific: Singapore, 2013.
5. Hofkirchner, W. Ethics from Systems: Origin, Development and Current State of Normativity. In *Morphogenesis and the Crisis of Normativity*; Archer, M.S., Ed.; Springer: Dordrecht, The Netherlands, 2016; pp. 279–295.
6. Tomasello, M. *A Natural History of Human Thinking*; Harvard University Press: Cambridge, MA, USA, 2014.
7. Haefner, K. Information Processing at the Sociotechnical Level. In *Evolution of Information Processing Systems*; Haefner, K., Ed.; Springer: Berlin, Germany, 1992; pp. 307–319.
8. Oeser, E. Mega-Evolution of Information Processing Systems. In *Evolution of Information Processing Systems*; Haefner, K., Ed.; Springer: Berlin, Germany, 1992; pp. 103–111.
9. Corning, P. *The Synergism Hypothesis*; McGraw-Hill: New York, NY, USA, 1983.
10. Hofkirchner, W. Creating Common Good. The Global Sustainable Information Society as the Good Society. In *Morphogenesis and Human Flourishing*; Archer, M.S., Ed.; Springer: Dordrecht, The Netherlands, 2017; in print.
11. Hofkirchner, W. Information for a Global Sustainable Information Society. In *The Future Information Society, Social and Technological Problems*; Hofkirchner, W., Burgin, M., Eds.; Wolrd Scientific: Singapore, 2017; pp. 11–33.

systems

MDPI

Article

From Systems to Organisations

Maurício V. Kritz

National Laboratory for Scientific Computation, Av. Getúlio Vargas, 333, 25651-075 Petrópolis, RJ, Brazil;
kritz@lncc.br; Tel.: +55-24-2233-6007

Academic Editors: Gianfranco Minati, Eliano Pessa and Ignazio Licata
Received: 31 October 2016; Accepted: 10 February 2017; Published: 6 March 2017

Abstract: Warren Weaver, writing about the function that science should have in mankind's developing future, ideas and ideals, proposed to classify scientific problems into 'problems of simplicity', 'problems of disorganised complexity', and 'problems of organised complexity'—the huge complementary class to which all biological, human, and social problems belong. Problems of simplicity have few components and variables and have been extensively addressed in the last 400 years. Problems of disorganised complexity have a huge number of individually erratic components and variables, but possess collective regularities that can be analysed by resourcing to stochastic methods. Yet, 'problems of organised complexity' do not yield easily to classical or statistical treatment. Interrelations among phenomenon elements change during its evolution alongside commonly used state variables. This invalidates independence and additivity assumptions that support reductionism and affect behaviour and outcome. Moreover, organisation, the focal point in this complementary class, is still an elusive concept despite gigantic efforts undertaken since a century ago to tame it. This paper addresses the description, representation and study of phenomena in the 'problems of organised complexity' class, arguing that they should be treated as a collection of interacting organisations. Furthermore, grounded on relational mathematical constructs, a formal theoretical framework that provides operational definitions, schemes for representing organisations and their changes, as well as interactions of organisations is introduced. Organisations formally extend the general systems concept and suggest a novel perspective for addressing organised complexity phenomena as a collection of interacting organisations.

Keywords: system structure; organised complexity; organisation; models of organisation; whole-part graphs; synexions; organised sets; organisation interaction; information

---

1. Introduction

Natural phenomena stem from a collection of things that interact and, while interacting, change the aspects we observe in them. No change, no phenomenon. No interaction, no change. This observation can be used to make explicit who interacts and what is exchanged in the interactions that give rise to a phenomenon. This line of reasoning allows for classifying different phenomena by means of different concepts in an integrated manner [1–3]. Warren Weaver in 1948 [1] classified natural phenomena yielding to scientific investigation into three groups grounded on characteristics of 'things' and their interactions, but also on methods used to investigate them: 'problems of simplicity', 'problems of disorganised complexity' and 'problems of organised complexity'.

Phenomena in the first class have a small number of determinable things and variables. The motion of two or three balls on a billiard table, oscillations of two interacting pendulums, the motion of planets around the sun are typical examples of those. If the number of things and observable aspects increase, these problems become intractable not because of theoretical hindrances but due to computational or operational difficulties. For instance, when considering a large number of pendulums, billiard balls, or many planets simultaneously. Note that, even in these latter cases, understanding have been achieved by considering just two things at a time, in all sensible combinations.

Phenomena in the second class possess a huge number of erratic things and variables that interact in the same manner or whose interactions cannot be distinguished through changes in the observed aspects. Like in the first class, there is no restriction on who interacts with whom and the interaction possibilities remain unchanged throughout the course of the phenomena. These problems present collective regularities that can be investigated employing stochastic and statistical methods. Typical examples include: the motion of atoms in a volume of matter or gas, the motion of stars in the Universe, heredity, thermal behaviour, and traffic. In them, there are observable aspects, like temperature, pressure, vorticity, or birth-rates that only make sense for collectivities. While concerning motions and exchanges of energy and momenta, problems in these first two classes and methods for solving them have been the subject of physics.

In the 'organised complexity' class, not only the attributes of things change but also the number of interacting things and the nature and possibilities of their interactions change; what affects the dimension of the state-space and the description of the interactions [4–7]. Following Weaver, the singular characteristic of phenomena in the last class lies in the fact that they show the essential feature of organisation in both: components and interactions. To date, there are but a few definitions of organisation; none widely accepted nor used to explain life phenomena, the quintessence of 'organised complexity problems'. Organisation is something easily recognised but difficult to grasp. Perusing available examples in Weaver's 'organised complexity' class, some characteristics show up:

1. The number of variables is moderate but it is not possible to hold all but two or three variables with fixed values;
2. There are fundamental aspects that are non-quantitative or elude identification and measurement;
3. Aspects observed are entangled, invalidating assumptions about superposition and independence;
4. A collection of things interrelated in a stable and individuated manner may act as an aggregate thing, a whole, interacting with other things at the same or at different aggregation levels;
5. Interrelations and possible channels of interaction among elements change, affecting the phenomenon propensities or behaviour; and
6. Units of interaction 'adapt', 'learn', and 'fabricate' [8,9] other units of interaction adding new factors, aspects and variables, to the phenomenon description.

The 'problems of organised complexity' class encompass virtually all biological, health and social problems, extending to any phenomena that involve living-things as components [10]. Franklin Harold [11] distinguishes living from non-living through their capacity of maintaining, reproducing and multiplying "states of matter characterised by an *extreme degree of organisation*" (see also [12]). The characterisation of this class, though, does not primarily depend on the number of things or variable aspects involved. In his own words and emphasis [1]:

> They are problems which involve dealing simultaneously with a *seizable number of factors which are interrelated into an organic whole*.

This happens in a manner similar to composite systems and their collective dynamics, which leads to self-organisation, but involves also forbidden interactions, channels of interaction that change, and factors that are entangled and interdependent. These entangled factors encapsulate into units, adapt themselves and integrate several dynamic scales [6,8,13–17], leading to the picture described by Harold [11,14].

Quite a number of the problems enrolled as examples of phenomena in this class have been more or less successfully addressed nowadays. To this date, their mathematical description and treatment, that originated through associations with complex systems [18,19], employ methods akin to those used in the investigation of problems in Weaver's 'disorganised complexity' class, centring on formalisms of thermodynamics, state transitions and critical phenomena [20]. Note that, as a *state of matter*, living things are as organised as a crystal, as fluid as a liquid, but in no case aleatory as a gas or plasma. Properly re-phrasing several approaches employed to study living and 'organised complexity' phenomena, things in them may be rightly named *organised matter* and living phenomena considered as the interplay and dynamics of (material) organisations.

Despite their utter beauty and usefulness in explaining the appearance of organisation, the compatibility of organised matter dynamics with physical laws of out-of-equilibrium systems, the tendency of physical phenomena to self-organise, and what happens at the border of the (extended) "criticality zone" [20], these methods are of limited utility to describe entailments and what happens within 'organised complexity' phenomena—that is, to explain how organised-matter maintains, entails and evolves organisation.

In scientific enquiries, we primevally choose to describe interacting things in a phenomenon as members of some archetypical *thing-class*, most of which are associated with formalisms that support representing (modelling) the phenomenon and reasoning about it. Depending on which aspects are observed, on what is exchanged in interactions, and on what is being asked, typical *thing-classes* may be particles, fields, substances, bodies, fluids, molecules, organisms, individuals, populations, firms, organisations (e.g., social entities and human associations), ecosystems, or a mixture of them. Thing-classes highlight characteristics considered important in studying a phenomenon and for its explanation, based on correlates of it and previous experience. Thing-classes act as spectacles we use to observe, model and understand a phenomenon. Concomitantly, they constrain which aspects make sense, what can be observed, which questions can be posed, as well as what can be referred to and studied [3,21]. Nevertheless, it is well acknowledged that general systems theory and methodology [22,23] provide a way of formally representing and handling phenomena independently of the thing-classes chosen or the application domain.

Choosing a thing-class implies making hypotheses about the behaviour of things and their interactions and about which aspects are relevant to understand a phenomenon. Treating molecules as particles, we hypothesise that their geometric attributes like form, volume etc, plus chemical-affinity are not relevant to understand what is being observed. These choices constrain and freeze the way we look into Nature. Thus, one should use several perspectives wisely for the same phenomenon, adjusting them to our evolving questions and observation procedures in order to enlarge our perception. For instance, billiard balls over a table can be a collection of particles or of spherical bodies. The phenomenon is the same, the stand varies. Each perspective highlights and enlightens different facets of what is observed in their moving. Each stand moulds the set of questions that can be posed and answered under the point-of-view imposed by its choice. Different things in a phenomenon may be represented by elements of different thing-classes. A collection of interacting things can be considered to be of more than one thing-class bringing new enlightenment to a phenomenon, as in the case of the interplay between thermodynamics and statistical mechanics ([24] section 4.4). That these visions are related in certain situations, however restrictive, is a real wonder.

In this paper, I contend that to develop a more extensive, integrated, and encompassing attack to 'problems of organised complexity', earning a less *ad hoc* knowledge about the underlying phenomena, we need to enlarge our present collection of thing-classes with thing-classes that conform to Weaver's characterisation above and go beyond it; by interpreting phenomena as a collection of interacting organisations that alter the organisation of the things themselves and the connections between things (channels of allowed interactions) in a phenomenon while interacting. To accomplish this, organisation is approached from a novel and complementary stand that does not involve dynamics. Its definition and analysis employ relational mathematical tools grounded on 'sizeable numbers' and highlighting the role of interrelationships in the constitution of organisations, with no presuppositions about or reference to behaviour, context, and 'function'.

The purpose of this writing is to propose a generalisation for the concept of systems that could better aid taming the inherent complexity of 'organised complexity' problems, introducing a concept of organisation as a candidate for such thing-class. From the formal point of view, organisations are obtained from general systems apposing hierarchy and encapsulation to them. This provides a mathematical definition and model for the "systems of systems" idea, allowing to treat "systems of systems" as units of interaction [25]. Consequently, it also suggests an enhanced general systems perspective, the *organisation perspective*. This perspective hypothesises that things in a phenomenon are organisations, rather than particles, substances, individuals etc, that interact predominantly exchanging signals and a specific kind of in-formation (see Sections 3.2 and 5) that pictures in(side)formation. In this way, living things and phenomena may be seen as a fifth state of matter (organised-matter), that maintains, reproduces, multiplies, and enchains organisations [3], as proposed by F. Harold [11,14] and R. Rosen [8] (see also [3])

The investigation wherefrom this ideas emanate was initially inspired and driven by difficulties encountered while modelling ecological and biological systems with variable structure [4–7,13] whose extreme values, possible factors, or domain cannot be established in advance. Thus, examples and illustrations in this text are mostly centred on living subjects, despite being true that the idea of systems is subjacent to phenomena in all three Weaver classes, the 'organised complexity' class has an embracing character, and the concepts presented here have a wider application. Even with this restriction, the related literature is extremely vast. To keep bibliographic referring manageable, citations to work that focus on behaviour and analyse organisation and information from a dynamical stand were kept to a minimum. The historical account presented in Section 2 aims just to contextualise the present proposal and put bounds on what will be discussed rather than to picture past achievements. References supporting arguments are employed parsimoniously. Frequently, not all pertinent references were included in a citation. Finally, relations between dynamics and interaction graphs, a simple form of organisation as here addressed, were analysed in a previous writing of ours [26] and the references within this work should be accessed through it.

This work is structured as follows: (1) purpose, described in this introduction; (2) a short and far from exhaustive overview of occurrence of terms organisation and information in the literature of life systems and sciences with little reference to behaviour, in the next section; (3) the organisation concept and the accompanying framework, in the third section; (4) ontological considerations, justifications and exemplifications, in the fourth section; (5) seeing phenomena as interacting organisations, the organisation perspective, in the fifth section; (6) closing remarks, in the last.

2. Organisation and Information in Life Systems and Sciences

One of the most conspicuous characteristics of life phenomena is alternatively named architectural structure or organisation [11,12,27–29]. Both terms refer to the same idea—the relative position, connection, or interaction channel of things and thing components with respect to one another, that become hierarchically arranged as a consequence of encapsulation into wholes or units [30]. Organisation is a central characteristic of biological entities and biological phenomena [1,11,14,27,28,31,32]. Organisation appears everywhere. It may be a collective aspect, as in simple oscillatory chemical reactions, consensus bio-chemical setups of cells [33], and "self-organising" phenomena; or a structural, individual-centred aspect present in macro-molecules, cell compartments, and cellular functional modules [14,34,35].

Organisation is generally associated with material instances of biological entities: modules, organelles, vacuoles, tissues etc [32]. Nevertheless, biological processes manifest organisation as well, e.g., the cytoskeleton [36], network activation-deactivation assemblies [37], chromatin, chaperones complexes [38], dynamic self-assembly processes [14], etc. Organisations expand beyond organelles and cell inner structures into tissues and organs of multicellular organisms and further on into populations, societies and cultures. Despite this ubiquitousness, instantiation of analogous organisations at different scales present seemingly uncorrelated forms [2,39]. Organisation is also frequently associated with complexity in biological entities [1,27,40–42]. Despite its importance and the bridge it launches with the study of more general complex systems [43], this subject shall not be pursued here beyond the contents of Section 4.2. *Organising*, the process of arranging and evolving organisations, shall not be addressed either.

Organisation configurations in living systems are credited to affect interactions, outcomes, and properties such as stability, reactiveness, capture, infection, (organelle) multiplication, *stasis* and mitosis [11,14,44], as well as their own existence [29], even by those that approach the living from a molecular scale, without reference to scale integration. Cell components, like the cytoskeleton and its variants, encompass dynamical organisations that are continuously reassembled. That is, their elements and their stable relative positions and dependencies result from a well orchestrated collection of movements and rearrangements that continuously relocate components and substitute missing parts or create new organelles and structures, like vacuoles, lysosomes, filopodia, micro-tubules, or centroids [14,36,45,46].

Organisation, in its variegated though ill-defined to-be concepts, has been considered a distinctive characteristic of life entities and phenomena at least since the beginnings of last century [2,11,12,21,29,44,47]. Notwithstanding the impreciseness of the concept [48,49] organisation is central while considering quaternary structure of proteins, protein conformations and their effects on protein interactions [50], protein aggregates, motifs, and cellular functional units. It is helpful when considering spatial effects in biochemical networks and in many biological enquiries and explanations ([51], last sections), along with efforts to build a theoretical framework for biology [8,52–57] and chemistry [58].

An enormous amount of work has been produced in the last 100 years to refine the idea and identification of biological organisation, to understand its onset and to justify its possibility on physico-chemical grounds [2,21,22,37,39,40,44,59]. Explanatory efforts search to dissect organisation from several standpoints like being a consequence of self-organising dynamics [59,60], resulting from regulatory processes [52] or bursting out from information [40,61–63]. Only a handful of these efforts address organisation directly, trying to tame the concept by considering questions akin to "what is organisation?", by searching models for organisation, or by constructing models and arguments based directly on its properties [8,27–30,52,62–72]. Albeit, none of them provides a clear concept or definition of organisation [16,48,73]. In the present text, I take for granted that organisation is a fact of Nature, present a working definition and a mathematical model for organisation, and discuss some consequences of this approach. This is a variant of Niels Bohr suggestion of taking life, like quanta, as a given fact of Nature [74], unexplainable in terms of other natural facts.

Another idea tightly intertwined with life phenomena since its earliest developments is information. Information, in Shannon-Brillouin sense [51], has dominated biological discourse, being considered a central characteristic of living systems and associated phenomena [52]. Nevertheless, it is not as conspicuous, directly observable, and recognizable as organisation. Despite this, information has been considered the key observable attribute of the living albeit always recognised only during the analysis of a phenomenon and its interpretation.

Ontologically, information in biological phenomena has been associated with transmission of hereditary characters [75,76], regulatory (feedback) assemblages [52], immunology [77], and ecology [13,78]. Theoretically, it has been associated with biological structure and function as well as their adaptation to various stimuli [8,52,65], among other phenomena.

When associated to structure and function, it reflects the ability of organisms to perform tasks in response to environmental stimuli, which is allegedly characterised by the information contend of its organisations [61]. Recently, information is being credited as the main vehicle of biological interactions [79,80] and can be observed operating at the cellular scale [81].

Information is employed while searching for explanations concerning the appearance and resilience of biological organisation [40], for defining it formally [65], for defining life [61,82,83], and the consequent efforts to build biological theories (see [41] for a more detailed and critical survey). Moreover, information intervenes directly or indirectly in all essays to explain the living through computational metaphors [41,61,84–86].

In its commonly used sense, the employ of information in biological explanations is highly debatable [87,88] and information is erroneously taken as the key biological observable [51]. This is due to the persistence of using the term information with different, non-commutable meanings [30] and the misleading association between organisation, information, complexity, and computation that this overloading in meaning induces [51]. The definition of in-formation presented in Section 3.2 is non-numeric, grounded on the organisation concept introduced in Section 3.1, and inspired by the inner workings of the cell.

3. Theoretical Framework: The Organisation Perspective

Abiding to the view about natural phenomena delineated earlier, where phenomena stem from a collection of interacting things that alter what is observed, this section introduces *things* (organisations) and discusses what is exchanged in their *interactions* (in-formation). The framework to be presented in this section supports the representation of phenomena as a collection of interacting organisations that interact exchanging in-formation and changing their own organisation.

Aligning with the spirit of this special issue, this section shortens technical details in favour of examples and clarifications. The presentation of this framework will employ less formal and more illustrative arguments, in the hope of throwing light on basal concepts, letting the reader more at easy with a new form of seeing natural phenomena. Profiting from discrete and relational mathematical constructs, formal elements will be often introduced with the aid of figures, less formally but no less rigourously. Figures are symbols that may be more easily accessed by a multi-disciplinary audience.

3.1. Biological Organisation: A Minimalist Snapshot

As argued in Section 2, organisation is an important and widely used concept that has nevertheless no consensual definition. Organisation is recognised through patterns that appear among Weaver's *sizable number of factors* and are believed to form what he called *organic wholes*. Representations of this sort of phenomena are often called composite systems even when they behave linearly ([89] chapter 9) and do not present the aggregate behaviour and patterns that may be taken as "organic wholes".

At the beginnings of (general) system theory, organisation was mostly attached to the systems structure [10,23,30], conspicuous in its reaction term, slowly drifting to their dynamical behaviour with the maturation of concepts like attractors, basins of attraction, slow and center manifolds, homeostasis, homeomorphic indexes, perturbations, fluctuations etc [49,59,70,90]. That is, organisation came to be associated with self-organisation and emergent patterns in the dynamical behaviour of many-particle or many-component composite systems and, by extension, with complexity ([19] preface, [49] chapters 6 and 7). Collaterally and supported by the maturation of these concepts, a wealth of methods to analyse and illuminate these systems form various perspectives have been developed [19].

This approach, however, cannot easily handle hierarchical systems of variable structure so common in biology, ecology and other domains [4–7,66,91], that adapt to stimuli by changing the number of their state-variables, the quality and nature of their interactions, and where properties within a level cannot be observed or explained from information rooted in its lower-levels. These characteristics are the essence of Weaver's 'problems of organised complexity', underlying and justifying the approach presented in the sequel.

The organisation concept advanced below (Sections 3.1.1 and 3.1.2) is not a consequence of dynamics nor is it associated with the idea of function (biological or ecological) in any manner. It supports reasoning about organisations on their own, independently of any correlated phenomenon, accommodating organisation transformation and comparison (Sections 3.1.4 and 4.2). Nevertheless, organisations are tightly associated with dynamics, having the usual (dynamic) systems as special cases (see Section 3.1.3 below).

3.1.1. Organisations

Consider a large enough heap of bricks indistinguishable with respect to all relevant attributes, characteristics and factors. The bricks are thus identical although not the *same*. This heap constitutes a set of bricks. Choose 20 of these bricks at random. These 20 chosen bricks are still a set of bricks.

They form a subset of the heap, if you don't take them apart from the heap. If you do, there will be two heaps (sets) and their connection may have become imperceptible.

Ensuite, pick all bricks in this heap and build a house. The house has 6 pieces: a living room, a dining room, two bedrooms, a bathroom and a kitchen. In the middle, there is a hall giving access to the bedrooms and the bathroom. The kitchen is accessed from the dining room that is accessed from the living room. Walls, made with the bricks, divide and delimit each piece. Some walls have doors, some other have windows, a few may have both. After the house is built, there are no more bricks but walls and rooms. Is a house a set of bricks? Is a brick in a wall identical to a brick in another wall or to another brick somewhere else in the same wall?

Consider any of the walls of the house. Mark 20 bricks in it at random. Are these bricks as indistinguishable as before? They may be side by side and taking them out will make a hole in the wall; luckily the hole may become a window or a door, depending on its height, shape, and localisation. They may also be scattered throughout the wall; what may be a refreshing strategy. Or they may concentrate in a junction of two walls, weakening its structure and ruining the house. The bricks are in consequence not indistinguishable anymore. They are interchangeable (any two bricks may be swapped) but not indistinguishable; the effect of taking any (or a collection) of them out of the walls is not anymore the same. They became parts of a wall, which in turn became a part of the pieces of the house, and these latter parts of the house.

Organisation is what distinguishes a house from a set of bricks. It may be informally defined in the following way. An atom here refers to an epistemological or cognitive atom—something we cannot, or do not want to, inspect or subdivide.

Definition 1. *An Organisation is one of the following:*

- *an atom;*
- *a set of organisations;*
- *a group of organisations put somehow in relation to one another;*
- *nothing else.*

This definition purposefully lacks a better clarification of the expression *somehow in relation with one another*, since this may be instantiated in several ways. Notwithstanding, it includes as organisations sets of atoms, organisations, or both. In the case of the above example, for instance, bricks are atoms. Even if one is taken apart in the building process, we do not want to describe or analyse of what and how they are made of, independently of the relevance of their attributes. Nevertheless, we may consider them associated in many ways. A brick may be associated to another if they are in contact, sharing a face or a portion of a face, if they belong to the same wall, or yet to a specified region of a wall.

This idea of organisation may be applied even to the heap itself. A brick may be associated to another brick if they are close enough together and groups of bricks may be considered separately. The ones at the top of the heap or at its base, for instance. In this case, however, shaking the heap a little could provoke a complete upheaval in this organisation. In terms of organisation, the heap is not as stable as a house, although the heap is pretty stable as a set.

Note that this concept of organisation can distinguish facts which are difficult to be traced by current descriptions.

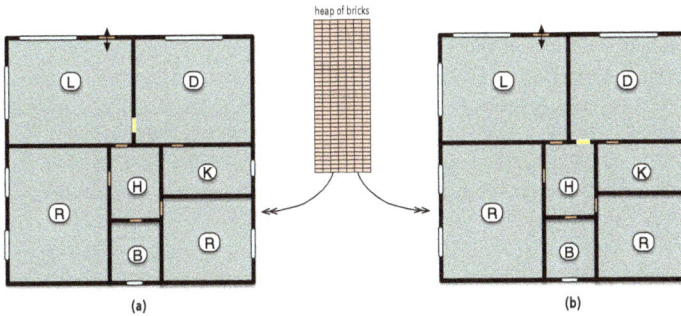

Figure 1. Two distinct houses obtained from exactly the same bricks. Due to a minimal position change, the yellow door connects the living and dining rooms in (**a**) and the hall and the dining room in (**b**).

The two houses depicted in Figure 1 differ only by the placement of a door in two distinct walls which are in contact. The number of bricks that occupy the volumes of the doors is the same, assuming that they have the same form and size. Moreover, their relative positions with respect to the walls of the house are very close to each other. Differences like these are difficult to be distinguished by either dynamical or statistical approaches. The energy and (statistical) information required to build either of the houses is the same. Moreover, the same set of bricks may have gone to one or another wall to erase the would-be door. The trajectories undergone by each brick from the heap to the walls can be traced by dynamical systems. Small perturbations in their dynamics led them from the same initial condition, the bricks heap, to different final conditions, the two houses.

Although similar, the organisation of both houses, given by the connectivity between rooms, is different; making one better suited for certain purposes than the other. For instance, the house in Figure 1a could be more easily used as a restaurant than the other one due to the accessibility of the kitchen from the dining room and of the dining room from the entrance via the living room, what clearly appears in the diagrams of Figure 2. Furthermore, the living-room may also be used to attend and direct costumers of the restaurant.

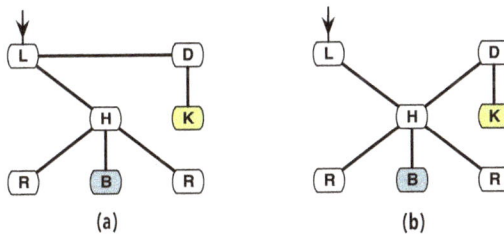

Figure 2. The ground access organisation of the two houses in Figure 1, respectively graphs (**a**,**b**). The new position of the door substantially changed their basic organisation.

3.1.2. A Mathematical Model for Organisations

Molecules, the ground stuff of living entities and processes [39], possess the characteristics of Weaver's 'organised complexity' class. Simple molecules may be understood through a small number of parameters and variables but not proteins nor any of the large biochemical molecules intervening in life phenomena. Protein molecules act as wholes when catalysing reactions, regulating biochemical processes, or chaperoning protein-folding although their parts are exchanged in protein construction and degradation processes. Spatial and chemical interrelations are crucial non-quantitative aspects of molecules. Protein dynamics (Vinson, 2009; Blanchoin et al., 2014) describe conformal changes in molecules that affect how a molecule interacts with other molecules or respond to stimuli from its environment.

Molecules are one of the simplest instances of organisation in Nature and fully conform to Definition 1, where chemical atoms are organisational atoms while chemical bonds instantiate the property 'in relation to one another', as hinted in Figure 3. Mathematically, molecules are depicted as graphs since long [92,93]; where nodes stand for chemical atoms and arcs for chemical bonds. Notwithstanding, we often describe molecules not by their atoms but in terms of other molecules, atom-groups or ions, to highlight chemical and structural properties, the way they interact chemically, or how they fit in a context. We refer to the hydroxyl group in alcohols, the carboxyl group in amino and other carboxylic acids, amino-acids in proteins and so on, treating them as units. This mode of describing molecules (Figure 3) is the essence of whole-part graphs (*wp*-graphs), the model for organisations described below.

To make these terms more precise, we need to restrain what may be considered as organisational atoms in modelling phenomena. Atoms in *wp*-graphs are required to be elements of a finite *admissible set* [94], **U**, established in advance. Admissible sets allow for treating certain sets as atomic organisations more conveniently [95]. Sets in **U** are not meant to be 'inspected' and differ from sets constructed by clause 2 of Definition 1. For instance, a brick as a set of clay particles is an element of **U** and as single element (atomic) organisation it is a member of the set 'heap of bricks', a trivial organisation under the definition that is being introduced here.

The **U** set need not to be the same in different representations, unless the organisations being represented interact or are related. **U** may also be typed; that is, its elements may be of different types, as when considering chemical substances like carbon, hydrogen, oxygen, nitrogen, etc. There is no restriction on the number of elements in **U**, or of one of its types. For instance, **U** may contain all (chemical) atoms in the Universe, letters of an alphabet, parts of a car, modules [96] and organelles of a cell etc. One may consider also an **U** consisting of both letters and chemical atoms if this helps understanding the phenomena in hand.

As hinted in Figures 3 and 4b, associations will be represented by hyper-graphs [97,98]. Arcs of graphs can connect at most two nodes. Edges of hyper-graphs may connect any number of nodes. The choice for hyper-graphs is justified for two reasons at least. First, it provides a simple and uniform description of delocalised bonds, or electron bonds embracing more than two atoms. For instance, the benzene ring is representable as a hyper-graph with 12 nodes $\{C_1, \cdots, C_6, H_1, \cdots, H_6\}$, 6 carbons and 6 hydrogens and 13 edges: six binding hydrogens to carbons $\{H_i, C_i\}$, six binding subsequent carbons $\{C_{i-1}, C_i\}, i = 1, \ldots, 6$, where $\{C_0 = C_6\}$, in the ring and one connecting all six carbon atoms $\{C_1, C_2, C_3, C_4, C_5, C_6\}$, as shown in Figure 3a.

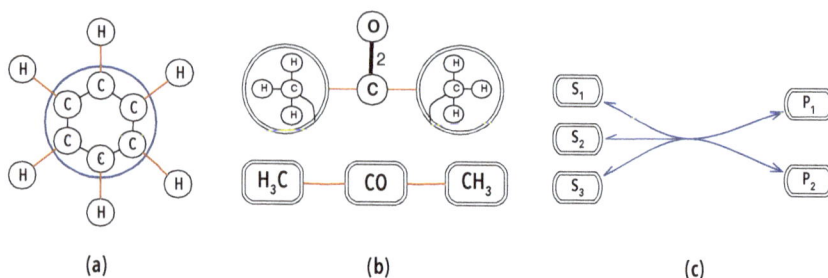

(a)   (b)   (c)

Figure 3. Molecular hyper-graphs: (**a**) the benzene ring (nodes, arcs, hyper-edge); (**b**) acetone (hyper-nodes, arcs); (**c**) a (bio)chemical reaction (hyper-nodes, 1 hyper-edge).

Second, chemical reactions are typical hyper-edge candidates, since they in general transform several substrates into several products. A minimal reaction takes 2 substrates into one product or vice-versa

(Figure 3c) due to mass conservation, unless conformal changes occur during chemical reactions. The enchainment of biochemical reactions results in networks that are indeed hyper-graphs [99].

Mathematically, a hyper-graph h is a pair of sets $\{N, A\}$ where N is the set of nodes and A the edge-set. The edge-set $A = \{a_i, i \in I, I \neq \emptyset\}$ is a collection of subsets of N, that is, $A \subset \wp(N)$ where $\wp(S)$ denotes the power set of the set S, satisfying:

$$\left.\begin{array}{rcl} a_i & \neq & a_j, (i \neq j), \\ a_i & \neq & \emptyset, (\forall i \in I), \\ \bigcup_{i \in I} a_i & = & N. \end{array}\right\} \tag{1}$$

Hypergraphs generalize graphs, in the sense that graphs are hypergraphs which edges have just 2 incident nodes, that is: $(\forall i \in I)[\#(a_i) = 2]$ [97]. To establish the framework this and all other definitions will be as generic and encompassing as possible. Restrictions, if any, shall be imposed in their instantiations at each phenomenon being modelled.

Hyper-graphs can be depicted with Venn diagrams, with hyper-edges, or as bi-partite graphs $\bar{h}$, as indicated in Figure 4. Bi-partite graphs have two groups of nodes and arcs connect nodes from one group to nodes of the other. When expressing hyper-graphs, one group of nodes is the hyper-graph node-set and the other the *names* of the hyper-edges.

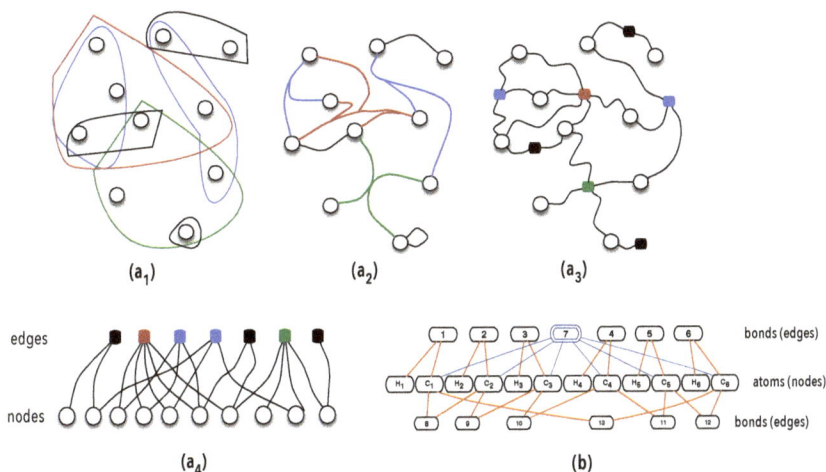

Figure 4. (a) The same hyper-graph drawn: as a Venn diagram (a_1); with type I Hyper-edges (a_2); with type II Hyper-edges (a_3); as a bipartite graph (a_4); (b) the benzene ring (Figure 3a) shown as a bipartite graph. Its strict sense hyper-edge is shown in blue.

To handle wholes and parts and the recursion of Definition 1, we will need some meta-elements besides **U** and (hyper)graphs: an enumerable set of meta-variables $V = \{v_0, v_1, v_2, \ldots\}$ and a collection of special *undistinguished* elements $\odot$. Meta-variables represent *voids*: places where organisations may grow, associate with other organisations, or detail their representation. For instance, docking sites in proteins, ion-binding sites, or polymerisation sites at the tip of filopodia [36] should be *voids*. Elements $\odot$ are unreachable constructive elements. They are hidden-names or 'hooks' of organisations that may become part of organisations with voids. Naming an organisation transforms it into a whole. Wholes being unreachable allows for swapping indistinguishable sub-organisations while preserving the overall organisation. Like when replacing bricks in a wall or when repairing DNA, proteins and organelles.

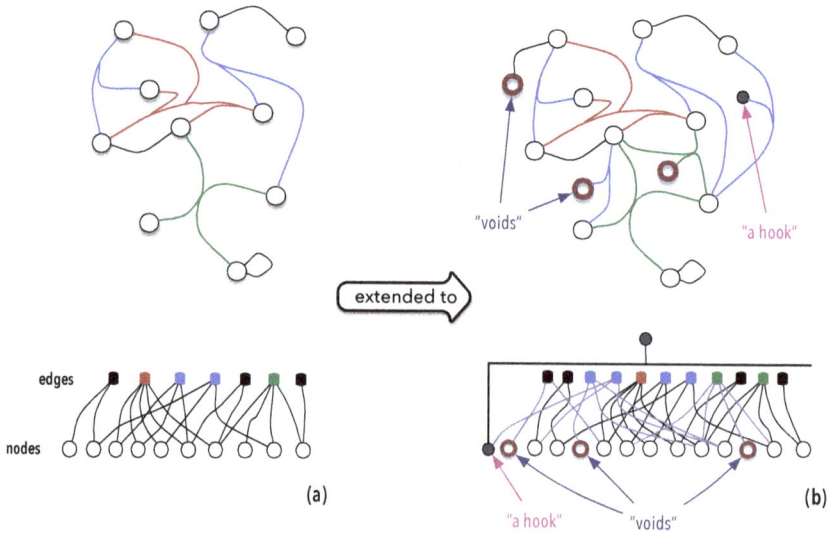

Figure 5. Extending hyper-graphs: the hyper-graph shown in (a) both with hyper-edges and as a bipartite graph is extended (b) by adding a 'hook' and some meta-variables (voids).

Hence, nodes will now be taken from the extended universal set $\mathbf{U}_\odot = \mathbf{U} \cup \mathbf{V} \cup \{\odot\}$. They may be organisational atoms, meta-variables (voids) or the $\odot$ element (hook). Hyper-graphs with nodes from $\mathbf{U}_\odot$ will be called *extended hyper-graphs*. Any hyper-graph may become *extended* by adding at least a meta-variable or a hook to it, as in Figure 5. The collection of all hyper-graphs will be denoted by $\mathcal{H}$, while the collection of extended hyper-graphs by $\boldsymbol{\mathcal{H}}$. That is,

$$\mathcal{H} = \{h = \{N, A\} \mid N \subset \mathbf{U} \wedge \#(N) < \infty \wedge A \subset \wp(N)\}, \tag{2}$$

$$\boldsymbol{\mathcal{H}} = \{h = \{N, A\} \mid N \subset \mathbf{U}_\odot \wedge \#(N) < \infty \wedge A \subset \wp(N)\}. \tag{3}$$

Clearly, $\mathcal{H} \subset \boldsymbol{\mathcal{H}}$.

Whole-part graphs are constructed by binding extended hyper-graphs with hooks or hidden-names to voids. This is achieved by "assigning" the generic $\odot$ element of a hyper-graph h^p to a meta-variable (void) in another hyper-graph h^w. This construction starts from the following operator prototype (see Figure 6):

$$\left. \begin{array}{l} \hookleftarrow : \boldsymbol{\mathcal{H}}^* \times \boldsymbol{\mathcal{H}}^\circ \longmapsto \{h' = h^w \hookleftarrow h^p\}, \\ h^w \text{ ni } v \quad \text{"="} \quad \odot \text{ in } h^p \end{array} \right\} \tag{4}$$

where $\boldsymbol{\mathcal{H}}^*$ denotes the class of all extended hyper-graphs that have meta-variables v as nodes, $\boldsymbol{\mathcal{H}}^\circ$ denotes the class of all extended hyper-graphs that have the special meta-element $\odot$ as a node, and **ni** is the mirror writing of u **in** h, a predicate identifying an element $u \in \mathbf{U}_\odot$ as a node of h.

An extended hyper-graph has just one $\odot$ element as node (one hidden-name) but may have many meta-variables v_i as nodes. Moreover, if several hyper-graphs are to be bound to h^w, there is in principle no special reason to privilege any binding order. Thus, the binding operator should be taken as a collateral binding of hyper-graphs h_i^p to h^w:

$$_m h^w \hookleftarrow < h_1^p, \dots, h_n^p > . \tag{5}$$

where m is the number of meta-variables in h^w, $n \leq m$ is the number of extended hyper-graphs to be bound to h^w. Without loss of generality, it is supposed that h_i^p binds to v_i.

Using the representation of hyper-graphs as bipartite graphs [98,100], binding hyper-graphs may be depicted pictorially, as in Figure 6.

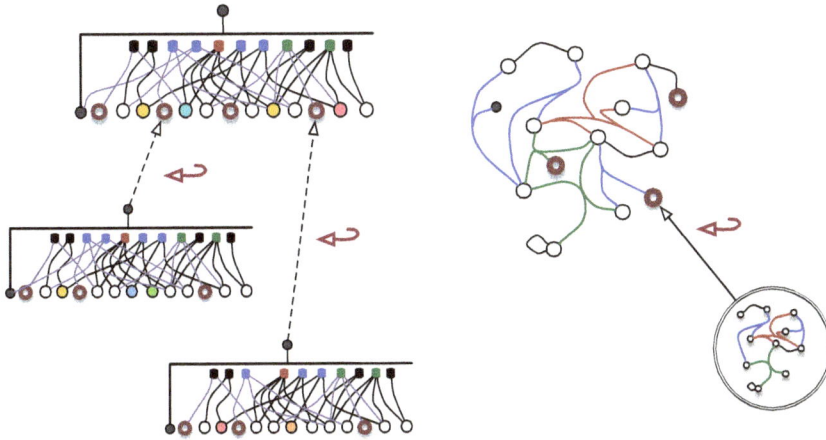

Figure 6. Binding extended hyper-graphs: extended hyper-graphs become "nodes" of extended hyper-graphs.

The binding operator has two possible interpretations suggested by common usage: as an 'encapsulator', enclosing organisations as elements of another organisation, or as 'hierarchy constructors' (Figure 7).

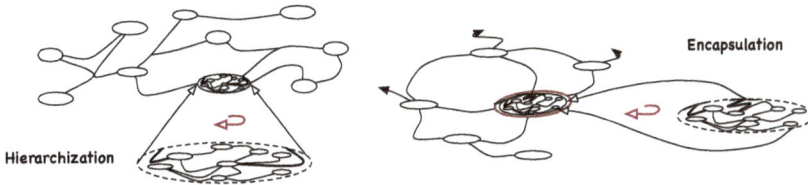

Figure 7. Binding Operator: as a hierarchy constructor and as an encapsulator.

To complete the construction of *wp*-graphs observe that, while there are still unbound meta-variables as nodes of any hyper-graph in the assemblage resulting from binding extended hyper-graphs to one another, it is possible to continue binding other hyper-graphs to them. These assemblages are the *wp*-graphs and are formalised by the following recursive definition:

Definition 2. *An object γ is a wp-graph, that is, $\gamma \in \Gamma$ if and only if:*

1. $\gamma \in \mathcal{H}$,
2. $\gamma \doteq {}_m h^\star \hookleftarrow <\gamma_1^\circ, \ldots, \gamma_n^\circ>$,
3. *nothing else.*

where γ° means that its "upmost" hyper-graph ${}_m h^\star$ of γ has a 'hook' as node; that is, $\mathrm{root}(\gamma^\circ) \in \mathcal{H}^\circ$. And the symbol $\doteq$ reads is build as or is given by and has a double interpretation: as a programming assignment during construction of wp-graphs and as a mathematical equality in Γ [95].

The fact that $\mathcal{H} \subset \mathcal{H} \subset \Gamma$ and a fixed point theorem for structures [101] warrant that Γ is non-void and well defined [95]. Note that Definition 2 makes no reference to sets nor atoms. The following lemma shows that Definition 2 indeed consider both, taking into account that atoms in Γ should be elements of $\mathbf{U}$. For the arguments it will be convenient to denote as $\wp_k(S)$ the collection of all subsets of S with exactly k elements, that is, $\wp_k(S) = \{s \subset S \mid \#(s) = k\}$.

Lemma 1. Γ *has the following properties:*

1. $\mathbf{U} \subset \Gamma$
2. $\wp_k(\Gamma^\circ) \subset \Gamma, (\forall k \in \mathbb{N}) [k < \infty]$

where $\Gamma^\circ = \{\gamma \in \Gamma^\circ \mid [\mathrm{root}(\gamma) \in \mathcal{H}^\circ]\}$.

Proof. (Ideas used in proofing case (1) intersperse the reasoning in (2).)

1. Note that since $\mathbf{U} \subset \mathbf{U}_\odot$ then $\wp(\mathbf{U}) \subset \wp(\mathbf{U}_\odot)$. Moreover, $\mathbf{U}$ can be identified with $\wp_1(\mathbf{U})$ and $\mathbf{U}_\odot$ with $\wp_1(\mathbf{U}_\odot)$ through the following injection

$$
\begin{aligned}
\mathbf{U}_\odot &\longrightarrow \wp_1(\mathbf{U}_\odot) \\
u &\longmapsto \{u\}
\end{aligned}
\tag{6}
$$

   and the immersion

$$
\begin{aligned}
\wp_1(\mathbf{U}) &\longrightarrow \mathcal{H} \\
\{u\} &\longmapsto \{\{u\}, \{\{u\}\}\}
\end{aligned}
\tag{7}
$$

   shows that $\mathbf{U} \subset \mathcal{H} \subset \mathcal{H} \subset \Gamma$.

2. If $g_{\mathsf{set}} \in \wp_k(\Gamma^\circ)$, then $g_{\mathsf{set}} = \{\gamma_1^\circ, \dots, \gamma_k^\circ\}$. Let $v_{\mathsf{set}} = \{v_1, \dots, v_k\} \subset V$ be a set of k meta-variables and let ${}_k h^\star = \{\{v_1, \dots, v_k\}, \{\{v_1, \dots, v_k\}\}\} \in \mathcal{H}$ be identified to v_{set} by the procedure in step 1. Then

$$
\gamma_{\mathsf{set}} \doteq {}_k h^\star \leftarrow\hookrightarrow < \gamma_1^\circ, \dots, \gamma_k^\circ >
\tag{8}
$$

   is an element of Γ.

   Moreover, γ_{set} and g_{set} are identified by arguments analogous to those of step 1. $\square$

The possibility of binding indefinitely new *wp*-graphs to meta-variables and the extensibility of any $h \in \mathcal{H}$ means that *wp*-graphs may grow forever and that adding details is unbounded in principle. Figure 8 displays typical elements (points) of Γ.

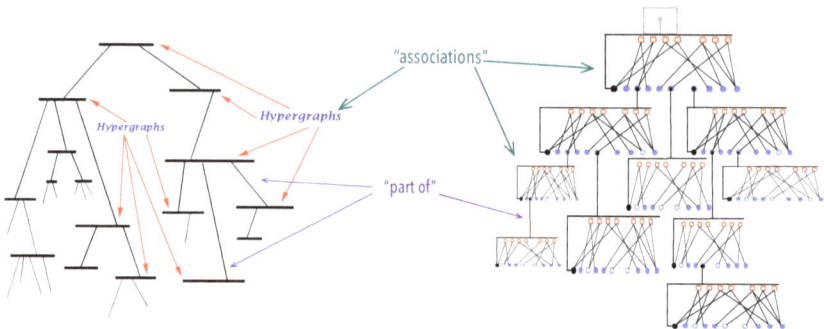

Figure 8. Typical *wp*-graph: localisation of associations and the part of relations in elements of Γ.

3.1.3. Synexions

The elements of $\mathbf{U}$ can be a variety of things: letters, chemical atoms, molecules, forms, objects, concepts, ideas etc. Whole-part graphs from Definition 2 may represent organisations whose atoms may be concrete, abstract or imaginary: words and paragraphs, molecules and macro-molecules, cultural entities, sketches, cognitive maps, pictures, mythological beings, ideas, etc. Otherwise, synexions are meant to represent organisations having existence reflected in terms of physical aspects. Straightforwardly, if a whole-part graph represents a molecule, a corresponding synexion would represent it with all volumes, angles, energy and vibrations possessed or defined by its parts and atoms. This intuition will guide the initial presentation of the concept. Nevertheless, it will be seen at the end of this section that usual dynamical systems are instances of synexions.

It is worth noting that the same object may be represented by way of different organisations, different elements of Γ (see Section 4.2), depending on what should be distinguished and on the questions addressed. For instance, the common organisation of a text is grounded on sequences of letters, words, phrases, sentences, paragraphs, sections etc. This organisation is fine for reading a text. However, if we are interested in mining key ideas in the text and associations among them, a better organisation could be as a cognitive map or a hypertext. The cell is another example. Its organisation may be based on its compartments and topological space-time relations among them or as a superposition of several bio-chemical networks grounded exclusively on chemical affinity. Molecules provide outstanding examples about organising an object in different ways, by grouping its atoms differently (Figure 3b).

The same subject of study may be seen as different organisations concurrently. Emotions may be concomitantly abstract and concrete—abstract while described in words or concrete if we consider the bio-chemical alterations and oscillations entailed with them. The organisations that may be associated with these two aspects of emotions are profoundly different.

Molecules, and all concrete organisations, occupy volumes in space and recognition of their organisation is moulded by our perception of spatial arrangements, dynamical stabilities-instabilities, and functional dependencies. Concrete organisations also vibrate and oscillate around a stable state, which means that they have extension in time as well as in space, being cylinders in space-time. Moreover, fluctuations in physico-chemical attributes may prevent, hamper, or facilitate the existence of certain organisations in favour of others (see Section 4.1.2). Therefore, we need means to represent organisations "embodied" in physical spaces. This will be achieved by associating volumes with *wp*-graphs in a way that preserves whole-part relations, as suggested in Figure 9.

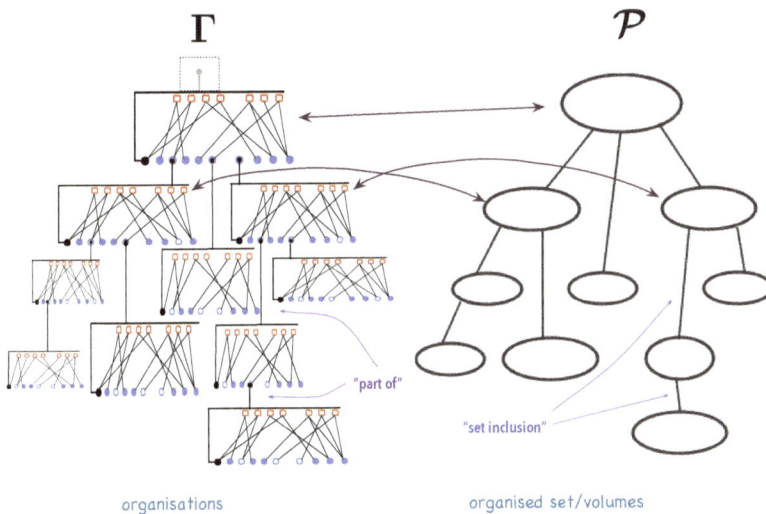

Figure 9. Synexions or Organised Sets (the relation partOf is defined in Section 3.1.4).

The results of these associations are the synexions (This term synthesises the key features of organised volumes. It is formed from the Greek verb συνεξω, which means *to hold or maintain as a whole*, and the particle ιoν, that means occurrence, instance. An alternative terminology could be *syntheions*, from συνθετοζ, that means composed by the union of parts, and ιoν, see [41,102].) or *organised volumes*. Synexions are associations between a $\gamma \in \Gamma$ and volumes or subsets of a "physical" space where set inclusion preserve whole-part relations. That is, if γ_1 is a sub-organisation of γ_2 then the volume associated with γ_1 is contained in the volume associated with γ_2.

Mathematically, let $\mathcal{P}$ be a collection of physical attributes including space and time that is endowed with the classical space-time structure [103]. That is, a space where molecules or any physical structure will not suffer conformal deformations when rotated or displaced. Synexions, denoted by $\mathcal{V}(\gamma)$, are recursive associations between volumes $\mathcal{V} \subset \mathcal{P}$ and sub-organisations γ° of $\gamma \doteq {}_m h^{\star} \hookleftarrow < \gamma_1^{\circ}, \ldots, \gamma_n^{\circ} >$ such that

$$\left. \begin{array}{ll} \mathcal{V}(\gamma) \subset \mathcal{P} & \text{if } \gamma \in \mathcal{H}, \\ \mathcal{V}(\gamma) = \bigcup_{i=1}^{k} \mathcal{V}(\gamma_i) & \text{if } \gamma = \{\gamma_1, \ldots, \gamma_k\}, \\ \mathcal{V}(\gamma) \supset \bigcup_{i=1}^{n} \mathcal{V}(\gamma_i) & \text{if } \gamma \cong {}_m h^{\star} \hookleftarrow s, \\ s = < \gamma_1^{\circ}, \ldots, \gamma_n^{\circ} >, \end{array} \right\} \tag{9}$$

where $n \leq m$.

Note that $\mathcal{V}(\gamma)$ is a one-to-many relation. To each $\gamma \in \Gamma$, there are many families of subsets of $\mathcal{P}$ that may be associated with it satisfying the constraint enforced by Equation (9). Any family of sets associated with γ may be uniformly expanded (fattened) or contracted (shrank) and still satisfy constraint (9), for instance, by changing temperature (vibration of parts), or by coherently displacing and deforming them in space-time, while preserving γ. The synexion space, $\mathcal{B}$, is the class of all associations between $\gamma \in \Gamma$ and subsets of $\mathcal{P}$, for any $\gamma \in \Gamma$. That is,

$$\begin{array}{ccc} \mathcal{B} : \Gamma & \longleftrightarrow & \wp(\mathcal{P}) \\ \gamma & \longleftrightarrow & \mathcal{V} \end{array} \tag{10}$$

Synexions are not sets in the usual sense. Subsets of $\mathcal{V}(\gamma)$ must also conform to conditions established by (9) and be formed from subsets of $\mathcal{V}(\beta)$, where β is a part of γ. Thus, we may have $\mathcal{V}(\gamma)_1 \cap \mathcal{V}(\gamma)_2 = \varnothing$ as organised volumes, even though $\mathcal{V}_1 \cap \mathcal{V}_2 \neq \varnothing$ as usual subsets of $\mathcal{P}$.

This property of synexions allows them to represent movements and deformations of organised things more effectively because it is possible to impose a kinematic behaviour to points in $\mathcal{V}(\gamma)$ that is constrained by Equation (9) to conform to the whole-part relation inherent to γ. This kinematic behaviour selectively changes and moves the volumes associated with parts of γ while preserving its organisational identity. That is, if β_1, β_2 are two parts of γ, $(\forall t)[\mathcal{V}_t(\gamma) \supset \mathcal{V}_t(\beta_i)], i = 1, 2$. Moreover, if β_1 partOf β_2 then $(\forall t)[\mathcal{V}_t(\beta_1) \subset \mathcal{V}_t(\beta_2)]$ and, conversely, if β_2 partOf β_1 then $(\forall t)[\mathcal{V}_t(\beta_2) \subset \mathcal{V}_t(\beta_1)]$. Otherwise, $(\forall t)[\mathcal{V}_t(\beta_2) \cap \mathcal{V}_t(\beta_1) = \varnothing]$. Cell-motion [45,104] is a good example of this feature, since organelles and cell-parts deform and move with the cell and within the cell without intercepting themselves, nor destroying the inner organisation of the cell. Synexions support the disentanglement of physical from organisational changes, which has far reaching consequences. They provide a bridge between organisation and (usual) physico-chemical dynamics. They enforce dynamical coherence: characteristic times and distances of parts are smaller than those of wholes. A cell cannot undergo mitosis before all its parts are duplicated, including the nucleus and the outer membrane [14].

Remark 1 (Elements of $\mathcal{B}$). *This rather informal and rigourless note presents some of the simplest elements of $\mathcal{B}$. The association between dynamical systems and interaction graphs is discussed in detail and with the due rigour in [26]. As discussed in this work, any dynamical system*

$$\left. \begin{array}{l} \frac{d\vec{x}}{dt}(t) = \vec{F}(\vec{x}(t)), \\ \vec{F} : \mathbb{R}^n \longrightarrow \mathbb{R}^n, \end{array} \right\} \tag{11}$$

can be associated with a graph $g_{\vec{F}} \in \mathbf{G}_n$, the set of all graphs with n nodes. Since $\mathbf{G}_n \subset \mathcal{H} \subset \mathcal{H} \subset \Gamma$ for any atom-set $\mathbf{U}$ containing the names of variables in the dynamical system given by (11), $g_{\vec{F}} \in \Gamma$. Furthermore, the collection of orbits of a dynamical system of the class (11) is a subset $\mathcal{O}_{\vec{F}}$ of $\mathcal{P}$ whenever the dynamical system represents a natural phenomenon. The specific nature of $\mathcal{O}_{\vec{F}}$ is intimately dependent on $\vec{F}$ but is often a variety or a set of chronicles [8], and is tightly related to properties of $g_{\vec{F}}$ [26].

From another stand, the mapping that associates a dynamical system $\vec{F}$ with its interaction graph $g_{\vec{F}}$ is not injective and there are many $\vec{F}_a, \vec{F}_b$ such that $g_{\vec{F}_a} = g_{\vec{F}_b}$. Then

$$\mathcal{O}(g_{\vec{F}}) = \bigcup_{\vec{F}_*} \mathcal{O}_{\vec{F}_*}, \forall \vec{F}_* \text{ such that } g_{\vec{F}_*} = g_{\vec{F}} \tag{12}$$

is an element of $\mathcal{B}$, since $g_{\vec{F}} \in \mathcal{H}$ and $\mathcal{O}(g_{\vec{F}}) \subset \mathcal{P}$, satisfying the first case of Equation (9). Clearly, $\mathcal{O}_{\vec{F}}(g_{\vec{F}})$ also belongs to $\mathcal{B}$, for any dynamical system given by (11). That is, interaction graphs are (simple) organisations and the orbits of dynamical systems associated with it are $\mathcal{P}$-volumes in the sense considered above.

If $\vec{F}$ depends on a parameter τ, $\mathcal{O}(g_{\vec{F}_\tau})$ is an "orbit" in $\mathcal{B}$ and exemplifies transformations of synexions. However, parameter dependency in dynamical systems do not in general affect its dimension n. When it does, it does in general by reducing the value of n while trapping orbits into sub-varieties of $\mathcal{O}_{\vec{F}_\tau}$. Generic transformations $\mathcal{T}:\mathcal{B} \longrightarrow \mathcal{B}$ are not subject to this constrain and systems of variable structure [4–7,66,91] can be straightforwardly represented with the concourse of transformations in $\mathcal{B}$.

3.1.4. Further Basic Constructions

Spaces Γ and $\mathcal{B}$ have interesting properties. Synexions are material organisations but not the only "concrete" organisations. Special organisations like sets, lists, trees, S-expressions and other data-structures can be identified as sub-classes of Γ [105]. Transformations, operations and relations can be defined in Γ and $\mathcal{B}$. Albeit a proper discussion about mathematical operations, predicates and properties of Γ and $\mathcal{B}$ being outside the scope of this writing, developing a few of them here will better illustrate the *wp*-graph and synexion constructs. Those below are restricted to Γ and support arguments in some of the following sections. Analogous transformations can be defined in $\mathcal{B}$. Two of them are illustrated in Figure 10.

$$\mathsf{T}(\cdot) : \Gamma \longrightarrow \Gamma$$

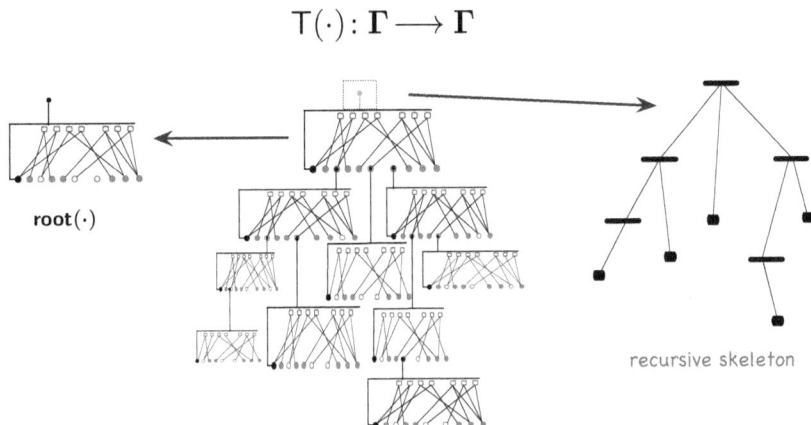

root$(\cdot)$

recursive skeleton

Figure 10. Transformations in Γ: the root$(\cdot)$ and the recursive skeleton Skt$(\cdot)$ of γ.

Definition 2 implicitly defines a mapping in Γ, $root(\cdot)$ (see Figure 10). The operator $root: \Gamma \longrightarrow \mathcal{H}$ is properly defined as:

$$\left.\begin{array}{rcl} root(\gamma) & = & \gamma, \text{ if } \gamma \in \mathcal{H}. \\ root(\gamma) & = & {}_m h^\star, \text{ if } \gamma \doteq {}_m h^\star \hookleftarrow < \gamma_1^\circ, \ldots, \gamma_n^\circ >, \; n \leq m. \end{array}\right\} \qquad (13)$$

To define equality, observe first that if $h_1, h_2 \in \mathcal{H}$, or $h_1, h_2 \in \mathcal{H}$, $h_1 = h_2$ in the set-theoretical sense, that is, $N(h_1) = N(h_2)$ and $A(h_1) = A(h_2)$ as sets.

Definition 3 ((*Equality*)). *Two wp-graphs* $\gamma_1, \gamma_2 \in \Gamma$ *are equal* $(\gamma_1 = \gamma_2)$ *iff:*

1. $\gamma_1, \gamma_2 \in \mathcal{H}$ *and* $\gamma_1 = \gamma_2$, *as elements of* $\mathcal{H}$;
2. $\gamma_1, \gamma_2 \in \Gamma \backslash \mathcal{H}$ *and*

   - $root(\gamma_1) = root(\gamma_2)$,
   - $(\forall 1 \leq i \leq n)[\gamma_{1i}^\circ = \gamma_{2i}^\circ]$,

where, as in Logic, [s] *means that* s *is* **true**.

Note that, following Definition 2, $\gamma_1, \gamma_2 \in \Gamma \backslash \mathcal{H}$ means that $\gamma_1 \doteq {}_{m_1} h_1^\star \hookleftarrow < \gamma_{11}^\circ \ldots \gamma_{1n_1}^\circ >$ and $\gamma_2 \doteq {}_{m_2} h_2^\star \hookleftarrow < \gamma_{21}^\circ \ldots \gamma_{2n_2}^\circ >$. Furthermore, $root(\gamma_1) = root(\gamma_2)$ implies that $n_1 = n_2$, $m_1 = m_2$, and ${}_{m_1} h_1^\star = {}_{m_2} h_2^\star$ as elements of $\mathcal{H}$.

An important thing to remark about Definition 2 is that it induces a whole-part relationship in Γ, which is highlighted by the predicate $partOf: \Gamma \times \Gamma \longrightarrow \{T, F\}$, where the token partOf should be taken as a single stropped (mathematical) symbol denoting this predicate. This naming method, common in the theory of programming languages, shall be used for creating symbols suggestive of their semantics for uncommon mathematical entities defined in Γ and $\mathcal{B}$. The relation partOf is recursively defined as:

Definition 4. *The values of* γ *partOf* $\beta = partOf(\gamma, \beta)$ *for* $(\gamma, \beta) \in \Gamma \times \Gamma$ *are given by:*

1. *If* $\gamma, \beta \in \mathcal{H}$, *then* $partOf(\gamma, \beta) = [\gamma = \beta]$, *as elements of* $\mathcal{H}$.
2. *If* $\gamma, \beta \in \Gamma \backslash \mathcal{H}$ *then* $partOf(\gamma, \beta) = T$ *if either condition holds:*

   - $(\exists 1 \leq i \leq n_2) [\gamma = \beta_i^\circ]$ *or*
   - $(\exists 1 \leq i \leq n_2)[partOf(\gamma, \beta_i^\circ) = T]$.

3. *Else,* $partOf(\gamma, \beta) = F$.

Hierarchy is one of the most conspicuous characteristics of organisation. It was intentionally left out while developing Definition 1 and its mathematical model (Definition 2). Nevertheless, as suggested in Figures 8–10, hierarchy is an intrinsic characteristic of any *wp*-graph induced by its recursive construction and attached to the whole-part relationship inherent in Γ-elements. The operator $Skt: \Gamma \longrightarrow rT$, where rT is the class of recursive trees in Γ [105], points it out. To define it, let $M(\gamma) \subset U_\odot \backslash U = V \cup \{\odot\}$ be the set of all meta-variables (meta-nodes) used to construct γ (Definition 2). It is important to note that meta-variables in $root(\gamma)$ and in any γ', such that $[\gamma'$ partOf $\gamma]$, are *different* even thought the notation may be duplicated to make the reading more intuitive.

Definition 5. *The recursive skeleton function of a wp-graph (Skt) is given by:*

1. *If* $\gamma = h \in \mathcal{H}$, $Skt(\gamma) = \{N, A\}$, *where* $N = N(h) \cap M$ *and* $A = \{N\}$;
2. *if* $\gamma \in \Gamma \backslash \mathcal{H}$, *then*

$$Skt(\gamma) = Skt({}_m h^\star) \hookleftarrow < Skt(\gamma_1^\circ), \ldots, Skt(\gamma_n^\circ) >,$$

3. *nothing else.*

That is, $\mathrm{Skt}(\gamma)$ is a replica of γ where all nodes from $\mathbf{U}$ and related associations have been erased.

The last Γ-transformation to be introduced here is the *connection network*. It is in a certain sense a counterpart for Skt, since it erases the meta-elements $M(\gamma)$ from γ leaving only nodes from $\mathbf{U}$ and the associations relative to them. This is achieved by substituting the binding $v_i" = "\odot$ **in** h_i°, $v_i \in N(_m h^\star)$, defined in Equation (4), by a normal node $hc_i \in \mathbf{U}$ mimicking this hierarchical connection. For the sake of simplicity, the definition presented here will be restricted to organisations which hierarchies have at most two levels, that is organisations such that $\mathrm{Skt}(\gamma)$ is a tree of just one or two levels, by omitting the final recursion step. If $n < m$ in Definition 2, there are in $\mathrm{root}(\gamma)$ meta-variables $v_j, n < j \le m$ that are not bound to any γ_j°. Likewise for any $\mathrm{root}(\gamma')$, such that $[\gamma' \text{ partOf } \gamma]$. Let $M(\gamma), M_{fr}(\gamma)$ and $M_{bd}(\gamma)$ denote respectively the set of all meta-variables, unbounded meta-variables and bounded meta-variables in γ. For $\gamma \in \mathcal{H}$, $M(\gamma) = N(\gamma) \backslash \mathbf{U}$ and, in general, $M(\gamma) = M_{fr}(\gamma) \cup M_{bd}(\gamma)$. Furthermore, let $A_j(h)$ denote the set $\{a \in A(h) \mid v_j \in a$, where $v_j \in M\}$ and $A_\odot(h)$ the set of all $\{a \in A(h) \mid [\odot \in a]\}$.

Definition 6. *The values of the* connection network *function,* $\mathrm{Cnt} \colon \Gamma \longrightarrow \mathcal{H}$*, under the restrictions above, are given by:*

1. *If* $\gamma \in \mathcal{H}$*, then* $\mathrm{Cnt}(\gamma) = h = \{N, A\}$*, where*

$$N = N(\gamma) \backslash M_{fr}(\gamma) \text{ and } A = \{a \backslash M_{fr}(\gamma) \mid a \in A(\gamma)\}.$$

2. *If* $\gamma \doteq {}_m h^\star \longleftarrow <h_1^\circ, \ldots, h_n^\circ>$*,* $h_i^\circ \in \mathcal{H}$*,* $n \le m$*, then* $\mathrm{Cnt}(\gamma) = \{N, A\}$*, where*

$$N = ((\cup_{j=1}^n N(h_j^\circ)) \cup N(_m h^\star)) \backslash M(\gamma)) \cup \{hc_1, \ldots, hc_n\}, \tag{14}$$

$$A = A^{up} \cup (\cup_{i=1}^n A_i^{lo}), \tag{15}$$

and the latter are modifications of $A(_m h^\star)$ *and* $A(h_i^\circ)$ *given, respectively, by:*

$$A^{up} = \{a' \mid a' = (a \backslash M(\gamma)) \cup (\cup_{j \in \{k \mid [v_k \in a]\}} \{hc_j\})\}, \text{ and} \tag{16}$$

$$A_i^{lo} = \{a' \mid a' = \{hc_i\} \cup (a \backslash M(h_i^\circ)), \forall a \in A_\odot(h_i^\circ)\}. \tag{17}$$

Cnt is non-injective. A rough idea of the set Cnt^{-1} can be obtained for organisations with a two-level hierarchy through the following reasoning. A given hyper-graph represent the channels of possible interactions in a phenomenon. Any hyper-graph (network) can be organised in several ways by partitioning the network into sub-networks and encapsulating these as aggregate units of interaction. Network-partitions interact with other nodes as composite nodes by means of collective aspects (like temperature or pressure) and channels of interaction encompassing all possible interactions of encapsulated nodes with nodes "external" to it, i.e., nodes in other partitions. Network partitions are obtained by partitioning their node-sets and rearranging its arcs accordingly. This stand will be used in Section 4.2 to estimate the size of Cnt^{-1}.

Lastly, the following observation will be supportive of several examples and arguments. A *process* is a collection of 'states' (enchainments or entailments of natural events) that cohere the enchainments with ordered temporal moments when instantiated into a physical space [106]. When life phenomena is described in $\mathcal{B}$, life processes are a collection of enchainments in $\mathcal{B}$ with a condition about their immersion in the time axis, e.g., biochemical pathways. Hence, life processes naturally include actual organisations as part of their states. That is, spaces Γ and $\mathcal{B}$ allow for considering and handling organisations of processes which states can contain organisations as components [3]. This is of great relevance for biological phenomena since chemical processes, as entailments of chemical reactions and substrates, are processes in this sense. Organisations of (bio-)chemical processes arise by considering two chemical processes to be associated whenever they exchange substrates or influence each other.

Processes can be considered as wholes whenever they present homeostasis, or any other form of permanence or recurrence, and have a distinctive functional character.

3.2. In-formation

Organisations convey information. This is clear when organisations are texts or pictures and their information-content is conveyed to humans. But even microorganisms have memory, process information, anticipate, coordinate tasks and make decisions [33,107–109]. Thus, information-driven interactions are a distinctive feature of biological systems at all scales, as discussed in [79,80], from distinct standpoints. They provoke changes in internal organisations and behaviour of a thing, that is, they provoke in(side)formations.

Existing information concepts focus on transmission of information, having a statistical nature and requiring a large enough set of messages, known in advance. Therefore, their application to scales or contexts where the number of intervening "things" and factors is at most moderate and far from homogeneous or isotropic is delicate, *ad hoc*, and can only provide hints about the phenomenon propensities. The widespread employ of the term *information* has introduced overload and bias in its meaning since long (see [30], footnote 1 on p. 194) and its utilisation is often inadequate [51].

From a basal point of view, information need not be quantifiable [80,110,111], despite the usefulness of measures in comparing things and in describing natural phenomena. A novel, non-quantitative, concept of in-formation is introduced in the sequel. It can be employed at all (biological) scales [2] in an integrated manner and is related to how organisations change. Being grounded on synexions, it is ontologically bounded to biological interpreters and observers. In this sense, in-formation is closer to molecular processes and changes at the molecular scale that *instantiate* signal processing, memory, reactiveness and anticipation in cells and tissues; and covers most relevant aspects of information in biological phenomena. It also accommodates information exchange at the sub- and supra-cellular levels, being useable in other domains as well. Exchanged in-formation in (biological) interactions is retrospectively observed and identified thanks to changes in organisations and behaviour.

Information in the sense to be presented is not a measure, measurable, or quantifiable. It targets the etymological roots of the word information: *in-formare*, or "to form inside." From an organisational stand, it addresses information at Level B (meaning) and C (effectiveness) of Shannon's Communication Theory more directly than at the commonly addressed Level A (transmission) ([112] chapter 1). Moreover, transmission of in-formation does not require identifiable *senders* nor a fixed number of messages. Nor is it constrained by the pre-definition of a set of signals and messages. Notwithstanding, usual measures of information-transmission can be recast from the in-formation concept below once there is a sender besides a receiver and the set of messages can be determined *a priori*. In the sequel it will be *assumed* that any biological entity is represented as a 'synexion' or organised set (volume), $\mathcal{V}(\gamma)$, for a properly chosen atom set **U** and $\gamma \in \Gamma$.

3.2.1. Perceptions

The concept of in-formation to be presented is grounded on an ontological *hypothesis*. Namely, that all biological entities 'perceive' and, by extension, so do organisations that represent them. The purpose of this section is to clarify the use of this term since it has here a rather specific meaning.

Perceptions are strongly intertwined with signals. Biological entities often have a living boundary that filters and transduces incoming signals. Let us call this kind of boundary *skin*. In individual organisms, 'skins' occupy a connected physical region and are part of its organisation, dividing the world in two regions: *inside* and *outside* the entity [8,113]. They also help the maintenance of particular homeostatic internal conditions. The "outside" region immediately around, together with anything it contains that may interact with the entity itself, is the entity's *environment*. Cell membranes are straightforward examples of 'skins'. Notwithstanding, there are organisations inside cells, in cellular matrices, in the mucosa-epidermis complex of multi-cellular organisms and in collective entities that are 'skins' in the above sense and not easily recognised as such.

Definition 7. *A signal is any perturbation (sudden variation) in the environment conditions, concentrated in time and space, that propagates eventually encountering a biological entity or another appropriate receiver.*

Encounters have the usual meaning of two or more things being at the same neighbourhood in space at the same moment. This definition includes as signals: travelling molecules or bodies, local variations in pressure, temperature and concentrations, waves etc. Due to space-time continuity, whenever a signal encounters an entity it reaches its outer boundary (skin) first.

In biological and 'organised complexity' phenomena, signals are expected to provoke drastic and disruptive changes in the structure of the systems involved, since this is how their organisations are altered; changes that eventually rend them them unrecognisable. Indeed, biological systems are signal-amplifiers par excellence [74]. Perceptions are effects that a signal has upon the biological organisation or organised receiver it encounters.

Definition 8. *Perception is a two stage process: it has an imprint moment and a recall moment:*

▶ *Imprint*
   Any signal σ encountering sensory apparatuses (in the skin) of a biological entity $\mathcal{V}(\gamma)$ at time t_σ and transmitted into it provokes (localised) changes in its organisation.
▶ *Recall*
   Moreover, if another signal σ' encounters the same biological entity at time $t \geq t_\sigma$ and tends to provoke the same change in the organisation of $\mathcal{V}(\gamma)$ as signal σ, σ' is recognised as being the same signal as σ.

Hence, *perception* is an action rather than an entity, organisation or fact and results in imprints (see Figure 11). Imprints, that are organisational alterations, may decay over a short time, remain for longer periods, or become part of the organisation. Signals σ and σ' need not be exactly the same perturbations of the environment, taking all attributes into account. With respect to the perception process of a synexion $\mathcal{V}(\gamma)$, however, σ and σ' will be considered to be the "same" signal, whenever they provoke the same imprint. This is dependent on the complexity of $\mathcal{V}(\gamma)$ and of the signals in the imprint class associated with the same alteration in $\mathcal{V}(\gamma)$.

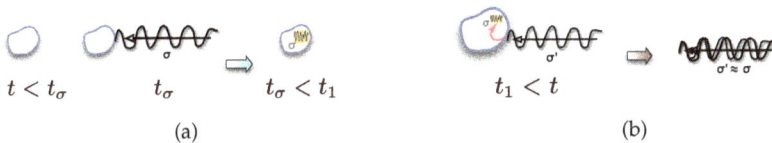

$$t < t_\sigma \qquad t_\sigma \qquad t_\sigma < t_1 \qquad\qquad t_1 < t$$

(a)   (b)

Figure 11. Perception: imprint (**a**), recall (**b**).

This sharpens the idea of similarity of signals (see Remark 2). Mistaking strictly different signals related to a sole imprint as the same signal is part of the perception process. Therefore, we say that a long lasting imprint is a model for σ and its similarity class.

3.2.2. In(side)formation and Interpretation

Imprints are changes in the organisation of a synexion $\mathcal{V}(\gamma)$ but do not enforce alterations in behaviour. Modifications in $\mathcal{V}(\gamma)$ may affect only the associated volumes $\mathcal{V}(\gamma) \mapsto \mathcal{V}'(\gamma)$, only its organisation $\gamma \mapsto \gamma'$, or both $\mathcal{V}(\gamma) \mapsto \mathcal{V}'(\gamma')$. Generally, signals provoke initial changes in volumes $\mathcal{V}(\cdot)$ (physico-chemical processes) that eventually migrate to changes in its organisation γ. In brains, the first relate to the electro-magnetic activity and are likely to decay; while the second and third involve synaptic and organisational changes and are long lasting. Depending on the level of detail, changes in organisation imply changes in the organised volumes as well. Good examples of this are the cellular signalling system and the sensory-nervous-brain systems in multi-cellular organisms (see Sections 4.1.1 and 4.1.2). Imprints that do not provoke changes cannot be detected.

Imprints that change the behaviour of biological entities will be called in-formation. The definition below employs observers, that are biological entities themselves. The role of observers, notwithstanding, is primarily to acknowledge that some change has happened. They are needed to detect and compare changes. Thus, any artefact that make special observations and compare them is an observer. Their role will be greatly clarified while refining our understanding of what is in-formation.

The definition of in-formation below relies on the following ontological hypothesis that is suggested by living phenomena and entities:

Hypothesis 1.

A: *Any biological entity or process may be represented in B.*
B: *All biological and life-related entities or processes perceive.*
C: *Perceptions are unique for a given biological entity or process—same signal, same imprint;*
D: *Biological organisations emit signals that uniquely characterise them, that is, they may be recognised by means of the signals they emit.*

Hypothesis 1C and 1D are not strictly necessary to define in-formation. They are relevant though while considering information-exchange in biological interactions, for rendering in-formation as a usable concept, and for clearly understanding its biological consequences.

Hypothesis 1A is the kernel of the organisation perspective. Notwithstanding, the representation of biological elements and processes as synexions is not unique, nor coerced in any manner. Any biological entity A may be represented by synexions $\mathcal{B}(\gamma)$ and $\mathcal{B}'(\gamma')$ reflecting different organisations, levels of detail or perspectives of study. Changing the synexions that retract A enriches our perspective in the same way as seeing matter distinctly as a cloud of particles, a body, or a substance does. Proteins may be seen as two organisations at least: a sequence of amino-acids while studying folding or as an assemblage of secondary domains and docks while studying function. Even so, one representation as a synexion is enough to discuss about exchange and interpretation of signals.

The imprint, $\text{iprt}_{\mathcal{B}(\gamma)}(\sigma)$ of a signal σ in $\mathcal{B}(\gamma)$ may be different from its imprint in synexions $\mathcal{B}(\gamma')$ or $\mathcal{B}(\gamma')'$, even when all these synexions retract the same entity A. In consequence, the similarity classes of σ may be distinct in each representation of a biological entity as a synexion. This multiplicity accommodates the representation of different levels of detail and different dynamical states. Our ability to understand information-driven interactions will depend on how coherently organisational changes ascribed to perceptions are represented and this can only be solved by reference to the signal itself or its source. The organisation framework makes this subtle point explicit but there will always exist a compromise between the complexity of synexions and their reliability as representations of things in natural phenomena.

Furthermore, it is well accepted nowadays that cells perceive and remember [107,108]. There is a clear association between conditions in the environment and activation-deactivation patterns of biochemical switches in the cell nucleus, maintained by the signalling cell system [15,114]. Hypothesis 1B, however, goes beyond that extending the perception concept down towards the sub-cellular scale and above towards the non-organismic entities and collectivities scale. The synexions framework accommodates in-formation other than that processed by neural systems or DNA transcription and intergeneration transmission, that are the only forms of biological information generally considered until recently [80].

Before advancing further, it is worth making the following observations:

- A signal σ reaching two different biological entities *a* and *b*, represented by $\mathcal{B}^1(\gamma_1)_a$ and $\mathcal{B}^2(\gamma_2)_b$ may provoke different imprints in their organisations, even if $\mathcal{B}^1$, $\mathcal{B}^2$ and γ_1, γ_2 are similar. However, if the collection of signals associated with imprints $\text{iprt}_{\mathcal{B}^1(\gamma_1)}(\sigma)$ and $\text{iprt}_{\mathcal{B}^2(\gamma_2)}(\sigma)$ is the same, that is, if any signal leading to the first imprint will also lead to the second imprint, the perception should be considered same for $\mathcal{B}^1(\gamma_1)_a$ and $\mathcal{B}^2(\gamma_2)_b$;

- Stabilised imprints are *models* for signals or collections of signals;
- A travelling molecule is a signal, because it is a localised variation in density, mass and other aspects concentrated in time and space;
- Pressure and concentration variations, being more diffuse and collective perturbations of environment attributes, may not be always taken as signals. This suggests that environmental variations depend on scale sensitivities as well as the complexity of the perturbation and the perceiver to be considered as signals. Signal and perception are thus relative concepts;
- Encountering is always due to relative motion. Either the signal propagates or the organism is moving and reaches a resting obstacle that acts as a perturbation in the perceived environment. What is important is that signal and organism approach each other in space and time for an encounter to occur.

Besides signals and perception, two or more special biological entities will be required to specify in-formation. One shall be termed interpreter, the others will be observers.

Definition 9. *Given a signal σ and at least two biological entities $\mathcal{I}$ and $\mathcal{O}$ in $\mathcal{B}$, σ will be termed an in-formation if the following occurs* conjointly:

$$\text{Signal } \sigma \text{ reaches } \mathcal{I} \text{ at time } t_\sigma \text{ and } \mathcal{I} \text{ perceives it;} \tag{18}$$

$$\mathcal{O} \text{ perceives both } \mathcal{I} \text{ and } \sigma, \text{ before, at, and after time } t_\sigma; \tag{19}$$

$$\text{At a latter time, } t' \geq t_\sigma, \tag{20}$$

$$\text{iprt}_{\mathcal{O}}(\mathcal{I}_{t'}) \text{ differs from iprt}_{\mathcal{O}}(\mathcal{I}_t) \text{ for } t \leq t_\sigma.$$

That is, if $\mathcal{O}$ perceives changes in $\mathcal{I}$, after its encounter and interaction with σ.

Thence, $\mathcal{O}$ says that σ is an in-formation for $\mathcal{I}$ and that $\mathcal{I}$ has interpreted σ. The observer $\mathcal{O}$ acknowledges changes and the interpretation of σ by $\mathcal{I}$. The observer $\mathcal{O}$ is not needed otherwise and $\mathcal{I}$ may be the observer itself if it is complex enough to perceive its own perceptions, that is, create a model for them, and maintain a model of itself. Since $\mathcal{O}$ is in $\mathcal{B}$, so are its perceptions (models) of $\mathcal{I}$ and σ. Both perceptions are sub-synexions of $\mathcal{O}$ and have extensions in time as much as it does. The perceptions of $\mathcal{I}$ and σ, though, need to extend beyond $t' \geq t_\sigma$ for $\mathcal{O}$ be able to detect differences between its anticipation of $\mathcal{I}$ at t' and its new perception of $\mathcal{I}$ at t'.

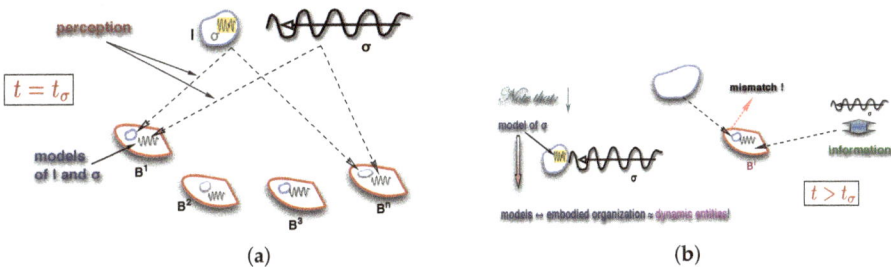

Figure 12. in-formation: Observing a signal perception (**a**); recognizing σ as in-formation (**b**).

Figure 12 portrays the information-interpretation arrangement, highlighting observers (named $\mathcal{B}^i$ in the figure), models of $\mathcal{I}$ and σ, and its dependence in time. The fact that in-formation causes changes in organisations is crucial for the organisation perspective (see Section 5), that also relies on the following premiss:

Hypothesis 2. *Every whole is an interpreter of its parts.*

It should be noted though that Definition 9 does not, strictly speaking, describe information. Instead, it describes how to *recognise* that a signal is in-formation—through changes provoked by the signal in an organisation whose behaviour changes with respect to what was anticipated [8,115] with respect to it—suggesting a procedure to observe it.

This is somewhat analogous to energy which is a property attached to configurations in fields and can only be observed indirectly, through its effects in the components of a $\mathcal{P}$-phenomenon. In-formation, likewise, is essentially immaterial and intangible and can only be acknowledged through its effects in 4-dimensional (i.e., extending in $(\vec{x}, t)$) $\mathcal{P}$-organisations.

4. Ontological Considerations

Subsections of this section contain arguments and discussions relative to employing the organisation-information theoretical framework to understand natural phenomena, as well as justifications for modelling decisions inspired by their observation as a collection of interacting organisations.

4.1. Space, Time and In(side)formation

The definition of in-formation provided above requires the immersion of organisations in space-time, and makes explicit use of space-time events and models immersed in space-time. In the sequel, arguments supporting the dependence of in-formation on space, time and $\mathcal{I}$ are presented at several scales and domains. The examples below are simple and far from extensive. Their aim is to justify modelling decisions and clarify the constructs. Recently, though, a wealth of biological and biomedical scientific investigations provided many examples of the spatial and temporal nature and dependence of living components and phenomena (see, for instance, the last two sections of [51]).

4.1.1. Cognitive Domain

Talking about issues of cognition and models starts with humans; bringing the analysis closer to our usual sense of information. Cognition, seen as the acquisition and incorporation of in-formation and perceptions in the sense of previous definitions, is not restricted to humans. In multi-cellular organisms, a part of their organisation is specialised as a signal processing system. It also handles imprints resulting from perception process and any response or reaction to them. Complex signals, coming out of complex organisations, are processed by the sensorial-signal systems. Imprints are mostly associated with the brain and nervous system in multi-cellular organisms, although not being restricted to them. In cells, an elaborated and complex network of reactions centred around DNA, the chromatin, and nucleotides adapt and respond to variegated signals, recording them and changing gene-expression as well as behaviour [11,15,108,116]. The examples in this section nevertheless refer to human cognition.

Signals are apparently affected by the organisation of their sources. The organisation of sources seems to be reflected in imprints associated with signals emanating from them, at least partially. In the following discussion, association of parts is mainly given by topological proximity (neighbourhood), while the whole-part hierarchy by encapsulation of groups into unities.

In texts, that are sequential objects, the vicinity is given by collaterality, grouping letters into words. Words side by side form phrases. Phrases side by side, with punctuation marks interspersed, form sentences, and so on. However, parenthesis, notes and footnotes relief a bit texts from dependence on sequential vicinity, by enclosing their contents into units. Technical texts have other forms of encapsulating and naming text unities that may be referred later or earlier in the text, thus indirectly re-appearing at several points of the text.

The same coalescent mechanism appears in figures and pictures, although their inherent two or three dimensions (2D or 3D) make the idea of neighbourhood much richer from a practical point of view. To exemplify how neighbouring associate things, let's consider two simple geometrical objects, here taken as wholes and as logical atoms in the universe **U**: a circle (◯) and a line segment (|). The circle can be rotated at will without appearing to be modified but the line segment will present different inclinations if rotated. Bringing the circle and line segment together (see Figure 13) in different manners will reveal important characteristics of information.

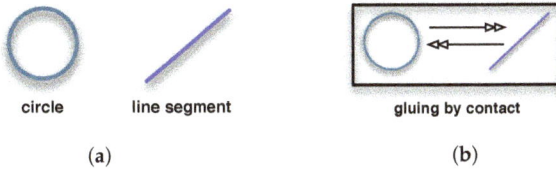

circle   line segment   gluing by contact

(a)   (b)

Figure 13. Geometrical objects (**a**); joining them by contact into a whole (**b**).

The line segment may remain tangent to the circle after joining or may transect it and that the circumference of the circle may touch or cross the line segment at different distances form its centre. These different possibilities of contact will result in different tokens, symbols and signs that may convey different meanings, if they convey information at all. Each circle-glued-with-segment forms now a visual unit (Figure 14), a whole in the sense of both Γ and $\mathcal{B}$.

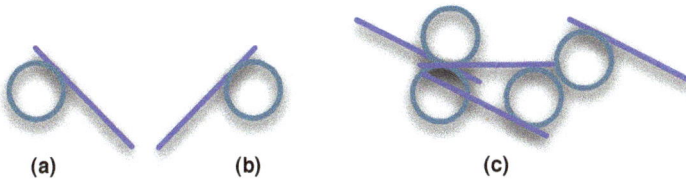

(a)   (b)   (c)

Figure 14. Circle-line-segment visual unit (**a**); a mirrored circle-line-segment visual unit (**b**); and a collection of them (**c**).

Not all wholes convey information, but there are cases where a whole may not convey information, or convey a different information, due to the manner it relates to its surroundings. For instance, it may be difficult to recognise anything or associate a meaning to the circle-segment wholes as they appear in Figure 14. There is no easy clue in the visual units of images in Figure 14a,b or in those in the the the heap of the image in Figure 14c that help us recognizing them. What, by Definition 8, means associating these units with previous imprints (known signals).

Considering Figure 14a,b together as unique image, it becomes possible to vaguely identify this new whole as a pair of (angry) eyes… if one has seen *lots* of animations. But what if we straight these units and line them up like in Figure 15?

Figure 15. Letters.

Don't they become immediately recognizable as letters {p,q,d,b}? Implicit subliminal visual references to the borders of the paper, that provide a sense of verticality, and from one whole to the other enforce their identification as letters. Anyhow, a group of letters like a syllable can be rotated to any degree and be severely distorted remaining recognizable, similarly to a group of chemical atoms forming a molecule. This observation and Hypothesis 2 suggest why molecules often have different functions in cells.

Relative distances between letters are strong topological clues enforcing the recognition of letter-groups as units or wholes, reducing the relevance of clues related to the environment. Similar phenomena occur in time when considering sounds and music. Groups of sounds or musical notes are more stable signals to our perception than scattered individual sounds or notes. It is more difficult to make known music or meaningful words unrecognizable than individual sounds or uncorrelated sound sequences.

Other visual units can be formed out of a circle and a line segment, as shown in Figure 16.

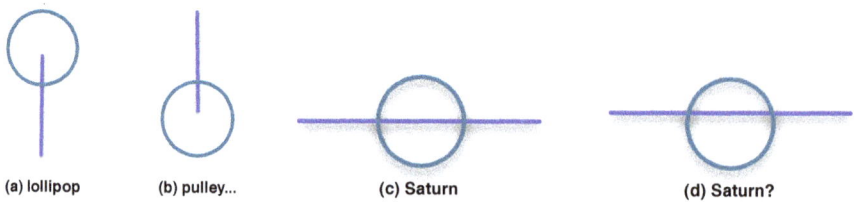

| (a) lollipop | (b) pulley... | (c) Saturn | (d) Saturn? |

Figure 16. Non-letter visual units out of a circle and line segment. See text for details.

Some will be easily recognised like those in Figure 16a–c; while others like Figure 16d will not, even if the sketches in (c) and (d) differ due to very small relative displacements of one part in relation to the other. Possible meanings for the resulting symbols are indicated by labels within the picture.

From another stand, there are visual units that defy interpretation no matter what is done with them, like the image of Figure 17a in two dimensions.

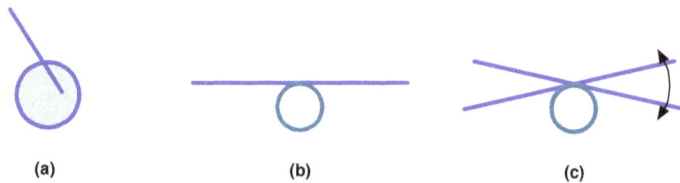

| (a) | (b) | (c) |

Figure 17. Visual circle-segment wholes difficult to recognise. (a) non-sensical image; (b) purposeless thing; (c) swinging **b**. See text for details.

The image of Figure 17b may initially defy recognition. However, shaking or moving it a little bit, as indicated in Figure 17c will make them recognizable as the sketch of a see-saw. This recalls the importance of movement and the time-component in models and imprints, while recognizing many things and phenomena, particularly life-phenomena. Synexions are space-time objects and tubes resulting from the displacement of any spatially organised volume along time is a synexion. Hence, Definitions 8 and 9 contemplate cases where the recognition and interpretation of objects can only be made along time, while moving. The fact that certain changes are anticipated help the recognition of changes by observers $\mathcal{O}$ in Definition 2.

Visual assemblages that form wholes and have been recognised may be associated in distinct forms resulting in new units or wholes, as suggested in Figure 18.

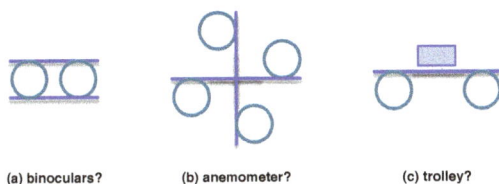

(a) binoculars? (b) anemometer? (c) trolley?

Figure 18. More complex (made from previous units) visual wholes difficult to recognise. Images (a) and (b) are composed exclusively from units in Figure 14a–b, while (c) adds a rectangle-like visual unit.

This process may be carried out indefinitely resulting in visual units composed of other visual units which recognition helps the recognition of the whole unit. Abstracting from top-down or bottom-up stands, the explicit recursion in Definition 2 and Equation (9), inherited by Definitions 8 and 9, allows for the appearance of however complex organisations, signals, imprints and information. Cells present several space-time organisations that conform to these definitions, but their *biological function*, including molecules, motifs, or modules, may change depending on their localisation in the cell, their conformation and their motion.

4.1.2. Molecular Scale

Simple molecules are straightforwardly represented in Γ (Section 3.1.2) and $\mathcal{B}$ (Section 3.1.3). Following the same procedure, proteins and other large molecules can also be represented in Γ and $\mathcal{B}$ as huge, difficult to comprehend, plain hyper-graphs. Notwithstanding, using the now established hierarchical structure of proteins (see, for instance, [104] and the articles at http://www.proteinspotlight.org), proteins may be easily represented in Γ. Given that the basic constituent parts of a protein are amino-acids rather than atoms, a protein can be depicted as a sequence of the hyper-graphs representing amino-acids. Sequences of hyper-graphs belong to Γ [105]. Thus, proteins have at least two representations in Γ.

Yet, proteins are often depicted in terms of secondary structures and other familiar molecular components. Considering secondary and ancillary protein structures as nodes and the chemical bonds holding together these molecular structures as arcs, proteins may be represented Γ in several ways, where connectivity always reflects chemical bonds and hierarchy isolates identifiable protein domains and sub-units. The nodes of this hyper-graph are hyper-graphs themselves retracting (sub)molecules and by Definition 2 this assemblage belongs to Γ. Therefore, any assemblage of secondary protein structures is also straightforwardly modelled in Γ.

Their representation in $\mathcal{B}$ is also immediate, since volumes occupied by atoms in space do not interpenetrate and their nearby positions unite the atom-volumes into molecule-volumes respecting constraint (9). Apart from that, protein components vibrate coherently, extending these volumes into space-time. The space-time volumes of the atoms of a protein and the partial union of them by amino-acids and subdomains provide the required volumes of $\mathcal{V}(\gamma)$ letting synexions represent proteins. It is also clear from the previous discussion that any macro-molecule has many different representations $\gamma \in \Gamma$, depending on which sub-units are being considered as wholes. Consequently, they also have multiple representations in $\mathcal{B}$. Anyway, protein folding is a transformation from Γ, or $\mathcal{B}$, into itself.

At any point in time, a protein may be *activated* or *deactivated* by a signal reaching it. Activation and deactivation are due to changes in protein organisation either in Γ, if new chemical bonds are formed or small molecules become attached to the protein, or just in $\mathcal{B}$, due to a re-arrangement in the tertiary structure caused by changes in the relative positions of its constituent parts. These organisational changes modify their "function", that is, the manner a protein chemically behaves and reacts to external stimuli. Therefore, a perturbation in the environment, be it a travelling molecule or variations in distribution of energy or mass concentrations, may cause an alteration in the organisation of a protein and change its behaviour. Under Definitions 8 and 9, the (chemical) perturbation is a signal that provokes a re-organisation of the protein changing its chemical behaviour. Therefore, the perturbation is an in-formation for the protein.

An example of organisation alteration due sole to changes in space and not in Γ, is the pigmentation protein chameleonine. Chameleonine is the name given to the protein that change the colour of certain spots down the spine of *Chamaeleo differensis* individuals when temperature changes and is apparently present only on them [117]. Changes in environmental conditions, in this case temperature variations, provoke a change in the colour of the tissue where it resides. Temperature variations really change only the energy of its components that reside in the vibrations of the protein's atoms. Chameleonine responds to this variation with a deformation in its spatial configuration due to structural stresses. As a result of this new (spatial) organisation, it changes behaviour reflecting, or re-emitting, a different light frequency. The signal in this case is temperature variation that, at the molecular scale, means a change in the mean number of molecules hitting a chameleonine molecule per time slice, and the consequent amplification of the vibration of its parts. This example conspicuously shows the need to include space and time in the conceptualisation of in-formation.

Chiral isomers further highlight spatial dependence of perception and information. This rich subject will not be discussed here, but a few observations illustrate important aspects of perception and information. Chiral isomers commonly respond equally to simple stimuli like temperature or to simple substrates in chemical reactions. However, depending on the complexity of molecules involved, on the environment where they are immersed or on the complexity of the signals reaching them, the effect in the organisation of wholes of which they are parts may be dramatically distinct, provoking deforming developmental diseases in humans, like the infamous thalidomide, particularly when chiral isomers are the signals [118,119]. Effective life phenomena is strongly dependent on essentially one of the chiral isomers.

From another stand, many biochemical organisations and structures in cells are not static. They are grounded on interacting homeostatic chemical processes, particularly when cycling. Hence, biological phenomena are really build on organisations of processes. A now canonic example of this character of theirs are the bacteria molecular motors. There are pictures of these ingenious engines available. However, these "structures" do not appear in engine snapshots at fixed moments in time. They arise from the superimposition of several snapshots at distinct times ([45] Figure 2), meaning that a flux of chemical substrates cooperate to give existence to these "motors". In terms of organisations and synexions, notwithstanding, fluxes are naturally represented as process-organisations and may be considered either as observers or interpreters of signals that revert turning and "swimming" directions, for instance.

4.1.3. Cellular Scale

At the cellular scale, organisations are more conspicuous in eukaryotes than prokaryotes, despite the complex organisation of biochemical processes existing in the latter. The preferred reference to organisations in eukaryotes in the sequel, though, is just a matter of explanatory simplification.

Looking into the internal organisation of cells, there are organelles, special "tissues" like the endoplasmic reticulum and the variegated membranes, recurrent and stable molecular agglomerates, as well as a plethora of sustained biochemical processes that serve a variety of purposes in the cell. Among them we find transportation systems that using vacuoles protect substrates from reacting with chemicals existing in the cell before they reach a certain place or organelle in the cell. Signalling systems, extending from plasmatic to nucleus membranes, are responsible to transmit changes in the state of chemical switches in the cell membrane, that act as sensors, to the nucleus. There, the nucleotide-DNA-chromatin-nucleosome system retains the conditions of membrane-switches in the form of genetic inhibition-activation patterns [116] and more elaborated settings, like the CRIPSP-Cas systems, even alter DNA loci [108,120]. These nuclear reorganisations are imprints following Definition 8.

This travelling-wave system changes the switching status of a collection of molecules that surrounds the DNA and maintains some portions of the genes active while inhibiting others. Besides messenger RNA, that transports in-formation, there are also signalling pathways from the nucleus to the cytoplasm that activate or deactivate protein synthesis, that are determined by the inhibition-enhancing patterns of DNA sites. Therefore, it is not DNA that really controls cell activities but a complex formed by DNA and nucleotides that "switch" genes on and off. The messenger and transcriptions systems transmit these patterns of free DNA to appropriate sites in the cell. Also in this case, environmental perturbations that change the status of membrane sensors are signals as in Definition 7, while the DNA-nucleotide complex acts, in turn or concomitantly, as a memory (the organisation that concentrate imprints), a decider, or a transducer and may affect cell behaviour, that is, DNA-nucleotide-chromatin complex interprets (Definition 9) changes in the state of membrane-switches.

To see biotic-interactions as exchange of in-formation, we need to understand the effects of differences in time propagation between two wholes. In prokaryotes, transportation is due mainly to diffusion in heterogeneous media. Diffusion takes little time to bring molecules from one extreme of the cell to another, due to the small volume of the latter. Eukaryotes are about 10 times as lager as prokaryotes. To get equivalent transfer and reaction rates and characteristic times in larger cells, there must be some facility to accelerate and "direct" diffusion to the proper places. This "facility" is organisation. More precisely, dynamical organisations.

4.1.4. Physiology and Behaviour

Comments in this subsection refer to multi-cellular organisms. Toward larger scales, the organisation of multi-cellular organisms pretty resembles that of eukaryotes and prokaryotes from the right perspective. Following J. G. Miller [2], living systems are composed of nineteen systems, independently of scale. For instance, there is the nervous (electrical signaling) system, the immune (repair) system, the digestive system, the motor system, the boundary (environmental interface) system, the memory and learning systems. In spite of being analogous, they may however appear in completely distinct forms from one scale to another.

Anyhow, there are systems in larger organisms that are not as easily paired among scales as those enrolled above, like the endocrine system (chemical signaling?) or boundary systems in ecosystems and societies. Moreover, the categorisation in nineteen systems is somewhat arbitrary, since some of them may be considered as one single thing while others further subdivided. Also, focusing on the nervous and endocrine systems, we see that two signaling systems with remarkably different characteristic propagation times co-exist, one based on electro-chemical wave propagation and the other on diffusion and transport. Nonetheless, phenomena occurring within the limits of these systems are much richer and complex in multi-cellular organisms than in cells and, by extrapolation, in phenomena at larger scales.

In contrast to systems, multi-cellular and more complex organisms are better represented as a collection of superimposed organisations symbiotically cooperating by exchanging matter, energy and in-formation, particularly the latter. The characterisation of which organisations should be used in a representation will always be arbitrary to a certain extent, as in the case of systems. Finding (biologically) sound heuristics to support this task will greatly enhance our knowledge about phenomena involving living things. Clearly the macroscopic size of multi-cellular and more complex organisms impose a strong dependence on space and time for any exchange within and among the organisations representing living systems. Therefore, a mandatory guideline is the rendering of categorised organisations compatible in terms of characteristic times, volumes and frequencies while discovering organisations. Equation (9) proposes a possible guideline with respect to these concerns.

When we acknowledge exchange of information as a distinguishing characteristic of living matter, the necessity of this compatibility greatly promotes the use of space and time in the definition of in-formation.

4.1.5. Cultural Domain

Contrary to knowledge that may be individual, culture is a collective phenomenon. Notwithstanding, human purposeful collectivities (like enterprises, firms, industries, households, schools, communities etc), collective phenomena (like person-to-person contacts that propagate diseases, gossips, information and matter; or like cultural centers and cultural networks attracting and educating people) and human creations (like science, music, art and literature) are all founded on organisations and information as pictured in Sections 3.1 and 3.2.

Purposeful human collectivities are formed by a collection of people that associate with each other with a (pre-)determined objective. Some organisations, like firms, industries, enterprises and (big) households, spontaneously or not organise themselves by having a hierarchy where smaller groups report to and/or are commanded by others. Such organisations are the basic action units in a cultural domain and have been extensively studied since long [121]. Most of them comply at least approximatively if not strictly to Definition 1.

Knowledge may be represented by classes of signals and their imprints organised as a consequence of associations and abstractions which are formed out of multiple interpretations and experiences, in the lines suggested by the discussion of Section 4.1.1. Culture also may be viewed as an organisation (in the sense of Definition 1) composed of concepts, ideas, taboos, individual knowledge, the imprints of a man or population etc. For instance, in a piece of music, art or literature there are often references, usually implicit, to another work in the same class or even in another class. There is music inspired by literature and folkloric wisdom, literature that refers to lyrics and so on. Literature is itself a web of veiled references to pieces of other literature.

In science, referencing is made explicitly; and this differentiates it from the other cultural expressions. Books, although sequentially arranged, may be as tortuous as any folding molecule due to backward and forward cross-referencing [8, Note to the reader]. In science, literature or music, one can only appreciate the beauty and depth of a piece of work if one has good acquaintance with a large number of other pieces of work, at least in the same cultural class or domain of discourse. Even within the explicit referential system of science, one has to know and understand the referred material to properly perceive and understand the work that referred to them.

Culture, however, differs from individual knowledge in two important ways. Instead of residing in the memory of a person, it is registered in books, writings, unspoken cultural premises, "learning by seeing" or by "experiencing" and many other extra-organic media and conveyors. Furthermore, the interpretation of this collection of signals and imprints is not made by an individual in particular but, instead, by the whole community that produces and retains the culture. This means that many of the writings and extra-organic, non-individual, media contain ideas and discussion about elements of the culture itself, being self-reflexive and self-referential to a much higher extent than knowledge.

The nature of science is not much different from culture. Its findings (atoms in **U**) are also organised in the manner of Definition 1 and its interpretation is a collective enterprise. The distinctions reside primarily on rules, methods and underlying paradigms about how to collectively develop scientific knowledge and the relying on observation as a conflict resolver. The language of science, even when it doesn't use mathematics, for instance, is developed in such a manner that differences in interpretation are minimised, and arguing procedures are standardised and their rules accessible to everyone. In this sense, the collective doing of science is more self-conscious than that of culture. Culture is produced more instinctively and intuitively than science. In Science, there are a more definite and explicit purpose to be followed and methodological rules to abide to.

4.2. Organisation and Complexity

The intertwining among living phenomena, organisation, information, language and complexity has been acknowledged since long [52,61,122–126], as well as the importance of innumerous concepts introduced by system thinking to deal with them [52,54,56,124]. Nevertheless, a consensual concept of complexity, whilst important, is still elusive [43]. Until now, there is no widely accepted definition of complexity, which meaning depends on the domain of inquiry. Those more widely used resort to concepts in computation, Shannon-Brillouin information or code-interpretation for their definition [41,51,127]; not to systems or organisation concepts. A few largely neglected efforts to connect information and organisation do, however, exist: the work of H. Atlan [40,62,65,77] and I. Walker [63]. This section discusses how complexity inserts itself in the organisation framework introduced above (Section 3).

The first thing to note is that there is no single way to represent things that compose phenomena in Γ or $\mathcal{B}$. Different representations may reflect different emphasis, different details, or distinct perspectives, standpoints, purpose, and questioning. To see this, consider again the example of the house (Section 3.1). A possible organisation for the house in Figure 1a, o_c, is given by the graph in Figure 2a and belongs to Γ, since undirected graphs are special cases of hyper-graphs [97] and hyper-graphs belong to Γ (Definition 2, case 1). In $o_c \in \Gamma$, the house-rooms and entrance are atomic elements and are the only nodes, i.e., $\mathbf{U} = \{L, D, K, H, B, R_1, R_2, Ent\}$. The graph o_c is such that:

$$\left. \begin{array}{rcl} N(h) & = & \mathbf{U} \\ A(h) & = & \{\{Ent, L\}, \{L, D\}, \{D, K\}, \{L, H\}, \{H, B\}, \{H, R_1\}, \{H, R_2\}\} \end{array} \right\} \tag{21}$$

The arcs of $A(o_c)$ represent the access (doors) between the rooms and the external access.

Other organisations for this same house can be constructed by resourcing to meta-variables and different levels of detail, as indicated in Figure 19.

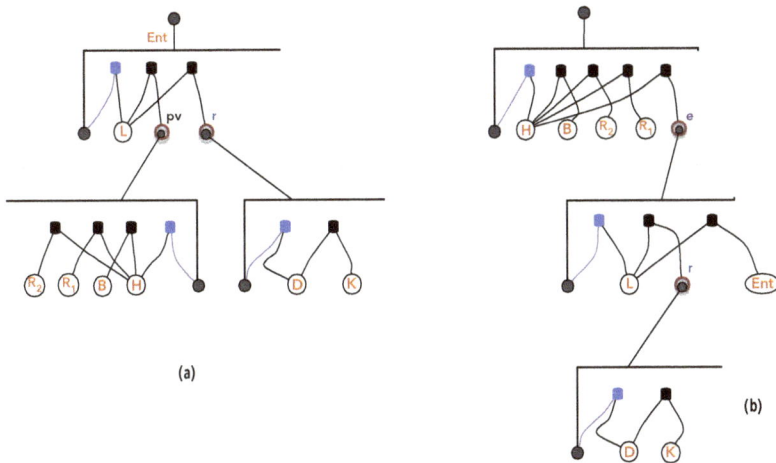

Figure 19. Two alternative organisations of house in Figure 1a with different levels of detail: (**a**) has two levels and (**b**) has three. Objects represented are the same, but the perception of what is primary or secondary is different.

The first is a two-level organisation and the second a three-level organisation. Their expressions in Γ refer to the universal set $\mathbf{U}$ of Equation (21) and are given in the sequel. The formal specification for organisation o_a, pictured in Figure 19a, is:

$$
\begin{aligned}
\mathbf{U}_\odot &= \mathbf{U}\backslash\{Ent\} \cup \{\odot, v_r, v_{pv}\}, \\
h_a^\star &= \{\{\odot, L, v_r, v_{pv}\}, \{\{\odot, L\}, \{L, v_r\}, \{L, v_{pv}\}\}, \\
h_r^\circ &= \{\{\odot, D, K\}, \{\{\odot, D\}, \{D, K\}\}, \\
h_{pv}^\circ &= \{\{\odot, H, B, R_1, R_2\}, \{\{\odot, H\}, \{H, B\}, \{H, R_1\}, \{H, R_2\}\}, \\
v_r &\ "=" \ \odot \text{ in } h_r^\circ, v_r \text{ in } h^\star, \\
v_{pv} &\ "=" \ \odot \text{ in } h_{pv}^\circ, v_{pv} \text{ in } h^\star;
\end{aligned}
\tag{22}
$$

while for organisation o_b, of Figure 19b, it is:

$$
\begin{aligned}
\mathbf{U}_\odot &= \mathbf{U} \cup \{\odot, v_r, v_e\}, \\
h_b^\star &= \{\{\odot, H, B, R_1, R_2, v_e\}, \{\{\odot, H\}, \{H, B\}, \{H, R_1\}, \{H, R_2\}, \{H, v_e\}\}, \\
h_e^\circ &= \{\{\odot, Ent, L, v_r\}, \{\{\odot, L\}, \{L, v_r\}, \{L, Ent\}\}, \\
h_r^\circ &= \{\{\odot, D, K\}, \{\{\odot, D\}, \{D, K\}\}, \\
v_e &\ "=" \ \odot \text{ in } h_e^\circ, v_e \text{ in } h^\star, \\
v_r &\ "=" \ \odot \text{ in } h_r^\circ, v_r \text{ in } h_e^\circ.
\end{aligned}
\tag{23}
$$

Any of the three organisations o_a, o_b and o_c can be further detailed by adding more levels. For instance, any room ρ in $\mathbf{U}\backslash\{Ent\}$ can be turned into a meta-variable v_ρ and bound to copies h_ρ of the hyper-graph h_{4w} below that describes four connected walls, one of which has a door. Its nodes and arcs are given by:

$$
\begin{aligned}
N(h_{4w}) &= \{\odot, W_1, W_2, W_3, W_4\}, \\
A(h_{4w}) &= \{\{\odot, W_p\}, \{W_1, W_2\}, \{W_2, W_3\}, \{W_3, W_4\}, \{W_4, W_1\}\}
\end{aligned}
\tag{24}
$$

where W_i represent the walls, and the arcs represent their bindings at the corners. The wall $W_p, 1 \leq p \leq 4$, with the door is associated with $\odot$.

It is important to note that while creating this new level of detail the walls are duplicated in each room as part of the organisation. The *concrete* instances of organisations may collapse distinct parts of them in one object. This highlights the fact that there is a lot of freedom in the immersion of *wp*-graphs in the physical reality, or else, in $\mathcal{B}$. Otherwise, walls in Figure 1 may be double-walls, with an air cushion in-between, or even be separated by larger spaces, the organisation of the house remaining the same, although not its instantiation in $\mathcal{B}$.

Another important thing to note is that, except for the intermediate nodes hc_{pv}, hc_r and hc_3, which can be erased from $Cnt(\cdot)$ without loss of connectivity,

$$
Cnt(o_a) = Cnt(o_b) = o_c.
$$

This illustrates that the mapping Cnt (Definition 6) is non-injective and there is always more than one multi-level (hierarchical) organisation which can be associated with a given network under no matter which heuristics. Hence, network topology is largely insufficient to determine the organisation of bio-chemical networks or of connection-diagrams of other complex phenomena.

This observation leads to considering the question: "How many ways are there to hierarchically organise a network?". Or else, what is the size of $Cnt^{-1}(h)$, for $h \in \mathcal{H}$? A more in-depth discussion of this problem is outside the scope of this text. However, a hint about its magnitude can be obtained by inspecting Definition 6, even if it is constrictive and considers only a two-level hierarchy.

From Equations (14)–(17), it is clear that $N(Cnt(\gamma))$ and $A(Cnt(\gamma))$ are constructed respectively as unions of the node-sets and arc-sets of $root(\gamma)$ and parts of γ and that neither nodes nor arcs can appear concomitantly at more than one level or more than one part of γ. If they do appear, like the walls in the example above, they are different instances of the same organisation and must be considered distinct in Γ; that is, a γ_1 can never be equal to, or the same as, a copy of it that is part of another *wp*-graph $\gamma \in \Gamma$. Furthermore, from the definition of hyper-graph [41,97],

$$N(h) = \cup_{a \in A(h)}\, a$$

and it is always possible to recover $N(h)$ from $A(h)$, for all $h \in \mathcal{H}$.

On that account and from an inspection of case 2 of Definition 6, it is clear that, to construct any $\gamma \in \mathrm{Cnt}^{-1}(h)$: $A(h)$ needs to be partitioned into v_p subsets, $2 \le v_p \le \alpha_h$, $\alpha_h = \#(A(h))$; $v_p - 1$ meta-variables assigned to the partition that will be $\mathrm{root}(\gamma)$; and the assignments of Equation (4) established. Since v_p is a parameter and, for each partition, there is a circular choice about which partition will be $\mathrm{root}(\gamma)$ and which partition will be bound to which meta-variable,

$$\#(\mathrm{Cnt}^{-1}(h)) \ge \sum_{v_p=2}^{\alpha_h} v_p!\, P(v_p, \alpha_h), \tag{25}$$

where $P(q, \kappa)$ is the number of ways the integer κ can be partitioned into q distinct integers [128]. Considering that, for any reasonably representative network $h \in \mathcal{H}$, α_h is anywhere in the range from v_h^2 to 2^{v_h}, $v_h = \#(N(h))$, $\#(\mathrm{Cnt}^{-1}(h))$ is astronomically large for any useful set of nodes and arcs.

By weighting atoms and counting interrelations against maximum relevance and total number of possible interconnections under variegated guidelines, a wealth of complexity measures can be defined for organisations modelled in $\gamma \in \Gamma$, that help comparing the many $\gamma \in \mathrm{Cnt}^{-1}(h)$. The family of complexity measures presented in the sequel illustrates this. It has well defined, finite values for each $\gamma \in \Gamma$, including self-referential ones. Self-referential *wp*-graphs can be obtained by, e.g., considering the nodes of h_{4w} as variables $w_1, \ldots, w_4$ and binding copies of h_{4w} to them indefinitely but finitely [100].

The family of measures below is defined for organisations that may be constructed or observed. It is thus computed over the observed organisations rather than metered directly in phenomena. Despite being inherently non-determinant, it can be used to help choosing the best representation of constituents of phenomena in terms of organisations, as indicated in the end of this section. It is based on the following heuristics: (1) conformity to the whole-part relation (Definition 4), in the sense that the complexity of a part should be smaller than that of the whole, (2) possibility of gauging relative contributions of atoms and parts to the overall organisation without biases, and (3) sets of organisations (Definition 1) should also have a complexity assigned to them.

The relative contributions may be dependent on external factors like the 'purpose' or 'function' of the organisation as well as in the domain of application of the model and, thus, are subject to ontological guidelines. For instance, the rooms of the house in Section 3.1.1 may have different degrees of importance (weighting) depending if it is intended to become a residence, an office, or a restaurant. On top of that, organisations representing visual elements in Figure 17 may be used to discuss them from a cognitive perspective (Section 4.1.1) or a cultural perspective (Section 4.1.5). It is likely that the parts and atomic elements composing the visual units will have completely different degrees of importance in each analysis.

So, let $\omega : \mathbf{U}_\odot \longrightarrow [0, a] \subset \mathbb{R}$ be a *given* weight-function defined for the elements of $\mathbf{U}_\odot$, such that ($\forall\, \gamma \in \Gamma$):

$$\left.\begin{array}{rcl} \omega(\odot) & = & 0, \\ \omega(v) & = & 1, \text{ if } v \in M_{fr}(\gamma), \\ \omega(v) & = & \lambda, \text{ if } v \in M_{bd}(\gamma), \end{array}\right\} \tag{26}$$

where λ relates to the inter-level significance and is such that $\sum_{l=1}^{\infty} \lambda^l \le \infty$. Contributions are generally evaluated with respect to a single γ. Because of this, it is enough to consider $a = 1$ since there exits several procedures to normalise ω in a way that it reflects only relative contributions of atoms *within* a γ, due to the finiteness of node-sets in γ.

The formulas for complexity measures $\Xi : \Gamma \longrightarrow \mathbb{R}_+$ in this family will be presented case by case for the three classes of organisations induced by Definitions 1 and 2:

$$\Xi(h), \quad \text{if} \quad h \in \mathcal{H}, \tag{27}$$

$$\Xi(\gamma_1,\ldots,\gamma_l), \quad \text{if} \quad \gamma = \{\gamma_1,\ldots,\gamma_l\}, \text{ and} \tag{28}$$

$$\Xi(_m h^\star \hookleftarrow <\gamma_1^\circ,\ldots,\gamma_n^\circ>) \quad \text{if} \quad \gamma \in \Gamma \backslash \mathcal{H}. \tag{29}$$

It will help to use a few shorthands. For $h \in \mathcal{H}$, let v_h and α_h be as above (Equation (25)) and

$$C(h) \quad = \quad \frac{\alpha_h}{2^{v_h}}, \tag{30}$$

$$\text{nbh}(n) \quad = \quad \#(\cup_{a\in A(h)|n\in a}\, a) - 1, n \in N(h), \tag{31}$$

$$\mathcal{I}_{h,\omega}(n) \quad = \quad \frac{\text{nbh}(n)\omega(n)}{v_h \sum_{i=2}^{v_h} C_{v_h-1}^{i-1}}, n \in N(h). \tag{32}$$

where $\#(S)$ is the cardinality of S and C_k^p is the number of possible combinations of k elements in groups with p elements. The value $C(h)$ is an estimate of the connectivity of h and $\mathcal{I}_{h,\omega}(n)$ of the influence of node n on its neighbours. Moreover, $v \hookleftarrow \gamma$ means that $\text{root}(\gamma) \in \mathcal{H}^\circ$ and γ is bound to meta-variable v (see Equation (4)).

That stated,

$$\Xi(h) = \frac{1}{v_h} \sum_{n\in N(h)} \omega(n) + C(h) + \sum_{n\in N(h)} \mathcal{I}_{h,\omega}(n), h \in \mathcal{H}. \tag{33}$$

If $\gamma \in \Gamma \backslash \mathcal{H}$, $\gamma \doteq {}_m h^\star \hookleftarrow <\gamma_1^\circ,\ldots,\gamma_n^\circ>$, $\text{root}(\gamma) = {}_m h^\star$ and its complexity is:

$$\Xi(\gamma) = \frac{1}{v_h} \sum_{n\in(N(h)\cap\mathbf{U})} \omega(n) + \sum_{v\in M_{bd}(\text{root}(\gamma))} \omega(v)\Xi(\gamma_v^\circ) + C(h) + \sum_{n\in N(h)} \mathcal{I}_{h,\omega}(n), \tag{34}$$

where $v \hookleftarrow \gamma_v^\circ$. For $\gamma = \{\gamma_1,\ldots,\gamma_l\}$, $\{\Xi(\gamma_1),\ldots,\Xi(\gamma_l)\}$ is a set of positive real values and any monotone statistics (*sum, mean, max* etc), for instance, provides a complexity measure that abide to the properties below. That is, if $E : \mathbb{R}^l \longrightarrow \mathbb{R}$ is monotone for each argument,

$$\Xi(\gamma) = E(\Xi(\gamma_1),\ldots,\Xi(\gamma_l)). \tag{35}$$

Each measure Ξ in this family is such that:

- The more associations there are in a *wp*-graph, higher its complexity is;
- $\Xi(\gamma) > \Xi(\gamma_1), (\forall \gamma_1, \gamma \in \Gamma) \mid [\gamma_1 \text{ partOf } \gamma]$;
- The more detailed an organisation is, that is, deeper the hierarchical levels go or bigger the number of its parts is, higher its complexity is;
- The finiteness of $\Xi(\gamma)$ for all $\gamma \in \Gamma$ and the existence of self-similar $\gamma \in \Gamma$, require that the contribution of deeper levels in the hierarchy decays rapidly, e.g., $\sum_{l=1}^{\infty} \lambda^l \leq \infty$.

Therefore, since the same natural object or entity may be associated with distinct organisations, it is *a fortiori* possible that natural objects have several (organisational) complexities. Thence, complexity in Γ is a concept associated with the description and representation of a phenomenon in terms of organisations rather than with the phenomenon itself [41,129]. Moreover, $\Xi(\gamma'), \gamma' \in \text{Cnt}^{-1}(h)$ can be used as an indicator to find the "most convenient" representation (organisation) of some-thing in $\text{Cnt}^{-1}(h)$, where $h \in \mathcal{H}$ retract its *raw* connections.

Remark 2 (On the complexity of signals and imprints:).
The discussion about imprints just after Hypothesis 1 can be illuminated by the following observations

1. By definition, $\Xi(\mathcal{B}(\gamma)) = \Xi(\gamma), \forall \mathcal{B}(\gamma) \in \mathcal{B}$;
2. If $\mathcal{B}_{prt}(\beta) = iprt_{\mathcal{B}(\gamma)}(\sigma)$ then $\Xi(\beta) \leq \Xi(\gamma)$, since $[\beta\ partOf\ \gamma]$;
3. $iprt_{\mathcal{B}(\gamma)}(\sigma) = iprt_{\mathcal{B}(\gamma)}(\sigma')$ does not imply that $\Xi(\sigma) = \Xi(\sigma')$;
4. It is possible that $\mathcal{B}_{prt}(\beta) = iprt_{\mathcal{B}(\gamma)}(\sigma)$ and $\Xi(\sigma) > \Xi(\beta)$, or even that $\Xi(\sigma) > \Xi(\gamma)$.

4.3. Organisation Perspective in Action: Re-Thinking Flows

Thinking in terms of organisations and organisational changes can bring new approaches to traditional scientific modelling. The investigation of cellular cytoskeleton, in particular flows and mechanical effects in plasma-membrane protrusions, is an active field of experimentation [36,46,130] and modelling [131]. These investigations focus on actin-polymerisation versus disassembly determining the retrograde flow of actin-filaments, its stability and force through actin-binding myosin motors. Mechanical and dynamical effects on the plasma-membrane result primarily from these phenomena. Notably, the cytosol is considered as a backcloth substrate, although actin diffusion, translation and sequestration in the cytosol may have important regulatory contributions to the actin network dynamics. Filopodia are long membrane protrusions containing parallel bundles of F-actin whose dynamics is governed by the regulation of polymerisation/disassembly processes. How the required amounts of G-actin are delivered within the challenging filopodial structure is a fascinating question. Existing models build on diffusion as the key actin-deliver mechanism [132,133].

In collaboration with A. Prokop and C.A. de Moura [134], we started to include a further hypothesis, where a mixture of cytosol and G-actin circulates changing organisation at the tip and bottom of the filopodium (Figure 20), from an incoming diluted solution into an outgoing tube formed by cytosol trapped into the actin filament bundle. The tip polymerisation drives a steady back-flow of the F-actin filament bundle at the core of the filopodium, with cytoplasm caught between the filaments (Phase B). This volume "outflow" is replaced by a compensatory influx of a diluted solution of G-actin molecules in the cytoplasm (Phase A) towards the tip of the filopodium, that occurs between the F-actin filament bundle and the cell membrane. At the tip of the filament the flow of the diluted solution blends conveniently towards the polymerisation points, guided by molecular organisations around it.

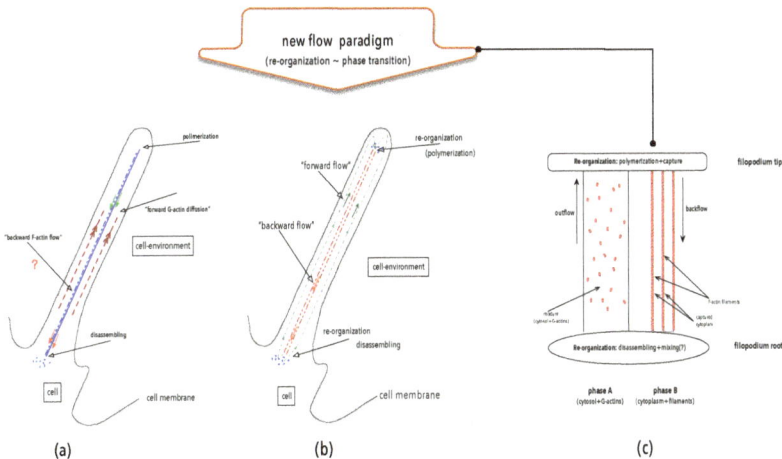

Figure 20. Filopodia: **(a)** Diffusion Flow, **(b)** Mixture Flow, **(c)** Re-organisation Diagram.

The latter hypothesis is suggested by the organisation perspective. However, the models and observations needed to verify it are quite different from those under the diffusion view. Despite its hydrodynamical elegance, this alternative cannot arise under the diffusion perspective because cytosol displacements are disregarded, and only diffusive movements represented. A new starting point, the organisation perspective, is needed to suggest this alternative and the models and observations required. This perspective also suggests that, at a second stage, the actin flow together with the polymerisation and disassembling processes are to be considered as a biological organisation unit responding to cellular regulatory processes and signals coming from both the cell's interior and exterior.

5. Interacting Organisations: Biological and Complex Phenomena

What makes living phenomena so singular? Things in physical phenomena move and exchange mass, momenta, and energy. When a cell phagocytes another cell, does it absorb just energy and mass, or does it intake in-formation as well, as suggested in several recent publications about cell immunology? When a firm "phagocytes" another (mergers & acquisitions), does it earns just material and monetary assets, or does it income know-how and knowledge as well? Presently, self-organisation and information help our understanding of systems out of equilibrium and sustain the perception that organisation is a natural phenomenon. They do not, however, help our understanding about the absorption of in-formation and knowledge suggested above, nor about the different effects of signals and imprints.

Also, neither organisation nor information is thought to affect/effect exchanges or cause specific behaviour [20,24,127]. Nor are they considered as possible and valid components of system states, promoting propensities and patterns of behaviour. In chemical phenomena, new substances are created when portions of molecules, organisations as well, are permuted among molecules, re-organising them. Chemical reactions re-organise substances into other substances, that appear and disappear. Molecular organisation determines which reactions are possible and which substances become which. Nevertheless, substrates and products remain the same, as well as their possible interactions, and organisation is but a parametric descriptive factor fixed once-for-all, as a natural law, in present day theoretical efforts towards living and complex phenomena.

Notwithstanding, outcomes in living phenomena require specific types of matter-energy in adequate amounts [32] that are properly interrelated and positioned [14]. Conformation states of molecules change, changing the reactions its substance can perform. Another distinctive aspect of living phenomena is that interactions exchange in-formation besides mass and energy [79,80]. Hence, the network of possible interactions is altered by the in-formation exchanged and by dynamics itself, as a consequence of conformal changes in molecules, changes protein activation states, the intensity of regulatory reactions, and so on. The present framework naturally allows for considering organisations as components of a system's state by modelling a phenomenon in $\mathcal{B}$ instead of $\mathcal{P}$.

Molecules are the simplest non-trivial organisations (see Section 3.1.2). At the molecular and intra-cellular scales, modules [96] may be interchanged with organisations, as here understood. At higher, more aggregate scales, it is not clear that observable aggregate modules [135] equate with organisations. Nor is it sharply clear that that near decomposability [121,136] will be applicable at smaller scales. Yet, organisations are aggregates of interrelated things that act as an unit at a higher, more aggregate scale and whose aggregation affects and effects behaviour at the higher scale.

The organisation perspective hypothesises that living and general complex phenomena result from a collection of interacting organisations, which interact exchanging in-formation, besides organised mass-energy assemblages. Moreover, interactions change the interacting organisations what, in turn, changes how organisations react as wholes. This is expressed by Definition 9. Hypothesis 2 realises emergence. The following arguments further illustrates the soundness of these conjectures.

The organisation framework straightforwardly allows for considering a dynamics of organisations, where changes in organisations (architectures) affect dynamical possibilities and propensities while dynamical stresses induce changes in organisations. Let's consider the smallest living unit—a cell. An eukaryotic cell for explanatory reasons.

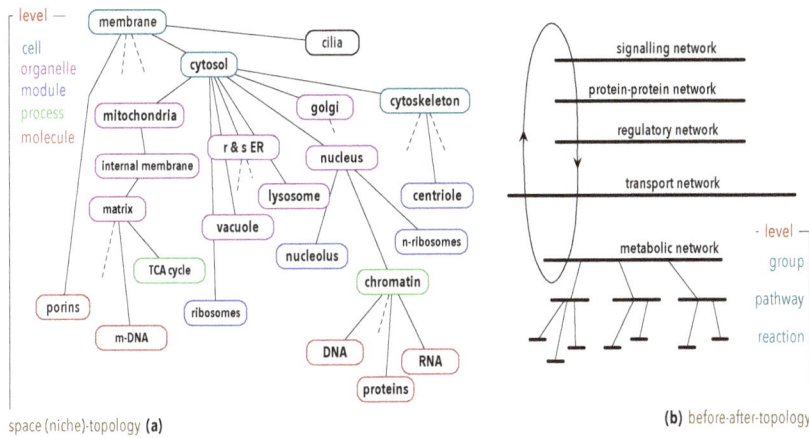

Figure 21. Cellular Organisation: topology bound (**a**) and re-action bound (**b**). Both are synexions (see Section 3.1.3).

Figure 21 displays sketches of two cell-views as organisations. One is grounded on space-topology and *A* is hierarchically "below" *B* if *A* is contained or occurs inside *B* [104,137]. The other is based on (chemical) affinity. The conspicuous incompleteness of each is due to very different reasons. It is easy to grasp that the *wp*-graph in Figure 21a is indeed a synexion, since our perception of organelles include their volumes in space and the partOf relation represented is induced by set inclusion. The number of molecules in each unit depicted grows from bottom to top. This representation clearly conforms to Equation (9) and is a synexion. Although there is no time, and things like vibrations, relative displacements etc, depicted in the figure, they do exist and may become part of the model. As suggested by this figure, synexions accommodate different levels of detail and some branches may reflect more detail than others.

Figure 21b displays biochemical networks of the cell schematically, the metabolic one with a little more detail. Cellular networks are hyper-graphs [99]. Although almost never made explicit, they are also synexions since molecules occupy volumes in space and chemical reactions require time. Or, at least, the sub-network that is "active" under any given conditions is a synexion, since they spread in space and time. First, because chemical reactions have characteristic times that may be regulated by interactions among cellular bio-chemical networks. Hence, each pathway or cycle has a tunable characteristic time. Second, because reactions and substrates are localised in specific regions of the cell. Unfortunately, information about frequencies, rates and localisation are still largely missing in biochemical data-bases nowadays, even though this has been changing in the last years [114,138].

Many essays in literature investigate the organisation of cellular networks [100]. Explicitly or not, they represent networks as bi-partite graphs [139] and their results are often in the form *wp*-graphs. But their results are not synexions since in general they do not use information about space-time localisation or characteristic times, distances and frequencies, even when available. Discarding this sort of information may hamper our understanding of life's liveness due to several reasons. To cite one, molecules of substrates and residues of proteins need to *fit* the space-time niche where they are, e.g., by adjusting vibrations. Besides that, connections between chemical reactions occurring in distinct regions of a cell are often fake, since different molecules of the same substrate take part in each of them and there is no real connection between the reactions [100,140].

This indicates how important it is to map the many networks in Figure 21b onto one another and onto the organisation sketched in Figure 21a. The constraint given by Equation (9) is an important guide to obtain data that support more realistic biochemical networks. It is also an important designing clue towards a coherent representation of living entities in Γ and $\mathcal{B}$ from the structural and dynamical stands.

From the organisation perspective, biological entities intake organisations, decompose them into simpler organisations (catabolism), storing the energy released, and rearrange these simpler organisations into parts of their own (anabolism) with little expenditure of energy, outputting matter and energy that is not usable. Instead of being completely catabolised, assimilated organisations may resonate with definite portions of the assimilating entity, being perceived and provoking imprints that may eventually induce behavioural alterations [141]. A good example is the assimilation of viruses by prokaryotic cells as immunological memories [108]. However, absorbed nucleic-acids may distort the cell's behaviour [142–144], often making it destroy itself. In both cases, which may occur alongside, biological interactions entail organisational changes and in-formation exchange. Hypotheses 1, particularly C and D, are necessary to make this description sound. This kind of reasoning allows for addressing questions about the input of in-formation and knowledge during phagocytosis or the merging of firms and other human organisations, as long as the complexity of catabolic results is large enough to retain meaningful signals.

Two concepts in physics and chemistry are important to understand dynamics, stasis and stable dynamical regimes—fluctuations and perturbation by virtual changes. The first is ontological and effectively occur in phenomena inducing swiftness and ability to change. The second is methodological and enables ideal inspection of alternative virtual behaviours, not observed but possible, and the subsequent questioning about why Nature has chosen the path we observe. Fluctuations are ubiquitous in living phenomena. Lively proteins flop between stable conformal states, altering interaction possibilities [50]. Changes in population density are identified by quorum-sensing set-ups that oscillate around chemical equilibrium points [33]. Complex oscillations between cellular modules and processes that occur during cell division determine form and localisation [14], and so on. That is, they manifest *organisation-fluctuations*.

Spaces Γ and $\mathcal{B}$ provide tools that allow for defining organisational difference and distance between $\gamma_1, \gamma_2 \in \Gamma$ as well as neighbourhoods around organisations in Γ, that may be processes like the TCA cycle. Two points in Γ are close together if they differ only near the leaves of their recursive skeletons; that is, near the bottom of the diagrams in Figures 9 or 10. Elements $\mathcal{V}(\gamma)$ of $\mathcal{B}$ present two forms of fluctuation. They can oscillate around a state in $\mathcal{P}$, which means that their instantiations $\mathcal{V} \subset \mathcal{P}$ (the "phase-space" retract) oscillate around a stable "set-value" in $\wp(\mathcal{P})$ or γ cyclicly exchanges some of its lower-complexity parts with elements available in its Γ-neighbourhood, like when a protein docking-site is constantly being flip-flopped. One of the main features of the formalism underlying this perspective is the ability to enforce distinguishing organisational fluctuations in a complex $\mathcal{P}$ from dynamics [134]. Nevertheless, a would-be methodology of *virtual behaviours* (counterpart of virtual displacements) needs to encompass evolutionary perspectives to be effective in explaining how the actual organisation of 'problems of complexity' came to be.

6. Conclusions

This work addresses the description of organisation and suggests means to use organisation as a tool to understand and explain living as well as other complex phenomena. Concomitantly, it enlarges the general systems approach, sharpening systemic terms like: interacting parts, communication, wholeness, modular organisation, whole-part relationship, hierarchical systems and so on. The elements in **U** may be time-functions or space-time-functions that chronicle observable aspects of phenomena [145] and compose system states ([8] chapter 4), offering other means to obtain immersions in $\mathcal{B}$, that do not necessarily comply with Equation (9). Thus, it also supports considering organisations that are permanently being rebuilt by steady substitution of their hierarchically lower parts, as if the walls W_i of Equation (24) were a collection of dynamical systems permanently substituting their bricks.

It introduces mathematical spaces for modelling organisations and defines in-formation grounded on the organisation concept and the tools provided by these spaces. These spaces accommodate organisational fluctuations and the in-formation concept extends the usual Shannon-Brillouin information; addressing the Shannon-Weaver levels of meaning and effectiveness [112] with no *a fortiori* reference to senders, the pre-established set of messages or "intention". Organisations and in(side)formation augments our tools to address questions about the acquisition of imprints, know-how and knowledge during interaction of organisations, phagocytosis and mergers included.

The complexity of organisation is central in these enquiries since an imprint, being part of an organisation, is bound to have a complexity smaller than that of the organisation. Being dynamical (like all elements in $\mathcal{B}$), it is prone to oscillation and destabilisation. On the other side, any imprint is a candidate to become a model and a model, knowledge; depending on relative complexities, *stasis*, and dynamical relations.

Definition 9 looks anticipatory. Although only observers are required to anticipate the interpreter's behaviour to acknowledge change, the basic property enabling anticipation is the immersion of Γ in $\mathcal{P}$, i.e., the space $\mathcal{B}$, and thus any synexion with internal models may anticipate. Imprints in an interpreter may be rock-stable, hampering change. Organisation fluctuations distort perceptions, destabilises and fluidises imprints and models, perturbing interpretations. The organisation perspective thus support investigations about the role of imprecision, fuzziness, adaptation, anticipation, emotion and moods in adaptation, learning and cognition [115,146–148] during transfer of information at any scale and encompasses all three Shannon-Weaver levels [112].

The introduced framework is able to represent (biological) organisations from a relational stand. To define information in this context, organisations were immersed in a physical fabric endowed with space-time, resulting in "concrete" organisations that associate structural organisation with the dynamics of usual changes (Section 3.1.3). It presents examples and arguments supporting and justifying this immersion and a statement about using the organisational perspective for studying living phenomena. One advantage is that a mathematics can be developed in Γ (and $\mathcal{B}$) that contemplates transformations of organisations, relations among organisations and other mathematical tools, where theorems can be proven and relational reasoning sharpened. It enables reasoning about organisations independently of any ontological references or immersions in $\mathcal{P}$, unveiling properties intrinsic to the relational aspects of organisations (Sections 3.1.4 and 4.2). Furthermore, it supports the distinction and identification of organisation changes amid dynamics in $\mathcal{P}$. This text extends and improves previous presentations of the above definitions [106,149,150].

Although inspired on life phenomena, the present concept of biological organisation extends beyond life and may be used at the molecular and supra-organismic scales [10,14,28,66,121]. General systems researchers think organisations as a collection of interacting components, alluding only informally to hierarchy and treating organisation components as deciders [30]. On the other hand, the definition of systems is meant to capture organisation [23,30,124]. General systems are special cases of organisations as above, since a set of interacting components can be formalised as mathematical relations [23], that are associated with hyper-graphs without isolated nodes [98] and, thus, are exactly the organisations delineated in case 1 of Definition 2. Hence, the term organisation may be used instead of the term system wherever it appears, providing a solid ground to handle systems of varying structure [4], even with unknown bounds and domains for the structural variations.

Definition 2 formalises the hierarchy inherent in organisations, extending the former definitions in many ways. It gives a formal meaning to terms like "system within a system" and "system composed of systems", so common in the literature. Mappings and relations can be defined in Γ, allowing for comparing organisations, transforming them into one another, and considering organisations as *the same* while portions of theirs vary (Sections 3.1.4 and 4.2). The concepts and tools here introduced, therefore, add to the system thinking framework [124].

There are many ways of building bio-mathematical worlds, life-like universes and theoretical explanations to study and understand life and richer phenomena, that develop along variegated reasoning directives [10,41,53,55,57,60,65,82,151–154]. Some consider singular characteristics of life, like organisation or reproduction, trying to explain them in terms of Shannon-Brillouin information, computational metaphors, bio-semiotics, or self-organising dynamical systems [8,41,60,84,155]. They draw on partial analogies, focus on the onset of life, or target the definition of life and organisation. Many go beyond and aim towards a theoretical biology [8,56,152,156,157]. Of the latter, two stand out: Rosen's (M,R)-systems and Gánti's Chemoton Theory [55].

It is not yet known whether these approaches relate to or can be rephrased in Γ and $\mathcal{B}$. Probing into possibilities, the following can be said about the last two. The Chemoton represents a would-be pre-biotic living entity, being formed by three interconnected cyclic processes, represented by symbols standing for unspecified molecules, enclosed by a boundary molecular circle plus a hypothesis about diffusion of food and waste through the enclosing circle. Cyclic processes and the molecular circle can be straightforwardly represented in Γ (see Section 3.1). The diffusion of waste and nutrients through the boundary circle can be approximatively represented in $\mathcal{B}$, even without the instantiation of the Chemoton in $\mathcal{B}$. Representing the Chemoton in $\mathcal{B}$, however, requires the specification of physical and chemical properties for each (molecular) node. Chemical properties will affect the relative position of would-be molecules in the cycles, as well as the length and stabililty of the cycles. Physical properties should determine which molecules compose the cycles. Thence, expressing the Chemoton in $\mathcal{B}$ amounts to solving Gánti's riddle.

Rosen addressed different questions. His formalism [8,68] is directed to express features that could unquestionably identify and distinguish life, independently of how living entities are instantiated. Despite both approaches being relational, it is unclear how to map his categorical diagrams into Γ or $\mathcal{B}$. Without resource to mathematical underpinnings, the following provides a possible interpretation of Rosen's ideas under the organisation perspective.

Block diagrams unite components into wholes. Their homologies in the organisation perspective are thus the hyper-graphs at the root and forks of a γ's recursive skeleton, $\text{Skt}(\gamma)$, the root acting also as a closure. Rosen's basic components $f : A \to B$, $A \xrightarrow{f} B$, or $H(A, B)$ [8, Section 5K]. may be atoms in Γ or $\mathcal{B}$, depending on how the sets A and B are constructed. The functors Φ and β are not in Γ but could be transformations from Γ into Γ.

From a formalistic standpoint, Γ and $\mathcal{B}$ are closer to Rashevsky's (1954) initial ideas, grounded on graphs and discrete topology. Notwithstanding, the present approach departs from the majority of existing approaches, perhaps radically, in what it abstracts from "biological function" or "origins of life" phenomena and questions, instead of attempting to represent and answering them. Arcs and bindings in whole-part graphs represent more directly observable and identifiable relations: spacial contiguity, chemical bonds or affinity, preferences, channels of interaction, and so on; instead of function which is a difficult observable [158], often only recognised *a posteriori* and in relation to external interactions with elements of the organisation's environment. Biological function is a many-to-many relationship. This means that a biological entity may play several functional roles, while the same biological function may be fulfilled by different entities. Having hyper-graphs as a fundamental building block, the present perspective does support the organisation of pathways [100] and the identification functional modules in a many-to-many fashion [34,35].

Complexity science, see [11, Chap 10] and [49], adaptive systems [153,159] and systems biology [160] are revivals of the "general systems theory", initiated by von Bertalanffy around the 1930s [52] to address problems in biology [22,52], that were boosted by cybernetics concepts [136]. Organisations may be employed in these fields in substitution of the systems concept, considering interactions among them and adding the benefit of reasoning about systems while organisations globally, instead of in a case by case, *ad hoc*, and behaviour centred manner.

For instance, with the tools offered by general systems theory, it is not possible to describe a collection of interacting (eco)systems, maintaining their identity and internal dynamics while considering interactions and dynamics among them, as it is not possible to consider a collection of interacting organelles (biochemical systems) in a cell, maintaining their identities in the same way. It is necessary to smash these phenomena into a huge and all-encompassing (dynamical) system and be compelled to analyse the behaviour of the latter, loosing sight of the inherent hierarchy in the phenomena and the fact that organelles are often standalone entities encapsulated by permeable membranes that modulate their interactions with rest of the cell. Organisations, synexions in particular, help to keep behavioural levels separate and subsystems as units, studying each separately or in conjunction in several ways.

It allows also for discussing organisations as such, independently of any dynamics, as when discussing the relation of organisation and complexity in Section 4.2. Furthermore, techniques already developed for composite systems can support the disentanglement of dynamics between the various hierarchy levels of organisations and creation of aggregated states for sub-organisations. Moreover, mathematical investigations about properties of Γ and $\mathcal{B}$, may develop a coherent and encompassing platform to address Weaver's 'problems of organised complexity'. It allows for extending formalisations to chemistry [56], aligning with Gánti's proposal of addressing questions of how can raw materials and simple organisations assemble into larger organisations [161], maintain themselves and evolve, addressing also more complex entities like communities, societies and beyond [2], by using the concept of modules [135].

It is worth emphasising that the framework presented above adds but does not supersedes or substitute any of the previous essays. Instead, it offers a formalism where they can possibly be rephrased and brought together providing richer pictures of complex features and traits. Even communication channels and Shannon-Brillouin information can be regained in Γ or $\mathcal{B}$, if the sender is made explicit and the set of exchangeable messages is known and fixed in advance. Yuri Lazebnik in 2002 [162] ingeniously claimed for an unambiguous biological language. Such language is vital for biotechnology [109] and for constructing (more fundamental) biological theories [17]. In physical and chemical phenomena, only the attributes of 'things' change. Interactions and the possibilities of interaction do not change, or change mildly and are considered fixed. In living and complex phenomena, on the contrary, 'interactions' (or relations) change wildly, even more often than 'things'. It is my hope that the organisation framework and the perspective it introduces, bringing interactions and things to the same level of attention, will serve as a basis for developing such language and help boosting the description of inter-level relations [163] and the concomitant addressing of proximal and ultimate explanations [164] by allowing the consideration of a dynamics of relations associated with the usual dynamics, as discussed in [26]. If it comes to assist the establishment of a philosophy and basic principles for the living and complex sciences [16,165] will be an added bonus. The importance of having a consensual philosophy and well-established principles is addressed in [22,23,124] in an unsurpassable manner.

Acknowledgments: These ideas and framework have been being developed and polished during almost three decades. During this time, I had the pleasure to be assisted by a great number of colleagues and people that supported the long standing process of consolidating and testing them. They helped through discussions, advice, key questions, and aids of variegated kind. The list presented in the sequel is certainly incomplete, due to misfirings in synapses of my memory-brain. My heartfelt apologies to those not in the list but who should be, many of whom read previous versions and drafts of this paper, providing criticisms that greatly improved the text. I would like to express my deepest gratitude to the whole group of Mathematical Modelling of Knowledge Diffusion, headquartered at LNCC and the Federal University of Bahia, to the Faculty of Life Sciences of the University of Manchester, as well as, to W. Aber, D.S. Alves, P. Antonelli, J.-Y. Béziau, L.H. Coutinho, A.C. Gadelha Vieira, M. Grinfeld, D.C. Krakauer, R. Lins de Carvalho (In Memoriam), F. Lopes, C. Menezes, S. McKee, A.C. Olinto, N. Papavero, M.M. Peixoto, A.A. Pinto, J.C. Portinari, A. Prokop, D.L. Robertson, M. Trindade dos Santos, J.-M. Schwartz, Iain W. Stewart, C.M.M. Vianna, S. Webb, my present and former students, and persons who attended the many talks, short-courses, and normal courses I presented on this topic or on "modelling techniques" posing wise questions. Last but not least, M.C.F. Bittencourt, M. Butturini and I.U. Cavalcanti invaluable suggestions improved the style, expressiveness, organisation, and quality of this writing. Any remaining flaws are, nevertheless, my sole responsibility.

I would also like to acknowledge the financial support of the following agencies and programs, through several grants and financial support along the development of this work: CAPES (specially fellowship No. 10313/88-2), PCI/LNCC, Wellcome Trust (specially grant 097820/Z/11/B) and FAPERJ (specially grant No. 101.261/2014).

Conflicts of Interest: The authors declare no conflict of interest. Financial sponsors had no word whatsoever in any phase of this study.

References

1. Weaver, W. Science and Complexity. *Am. Sci.* **1948**, *36*, 536–544.
2. Miller, J.G. *Living Systems*; McGraw-Hill Book Co., Inc.: New York, NY, USA, 1978.
3. Kritz, M.V. Boundaries, Interactions and Environmental Systems. *Mec. Comput.* **2010**, *29*, 2673–2687.
4. Mohler, R.R.; Ruberti, A. (Eds.) *Recent Developments in Variable Structure Systems, Economics and Biology*; Lecture Notes in Economics and Mathematical Systems; Springer: Berlin/Heidelberg, Germany, 1978; Volume 162.
5. Junk, W.J. (Ed.) *The Central Amazon Floodplain: Ecology of a Pulsating System*, Springer: New York, NY, USA, 1997.
6. De Ruiter, P.C.; Wolters, V.; Moore, J.C. (Eds.) *Dynamic Food Webs : Multi-species Assemblages, Ecosystem Development, and Environmental Change*; Theoretical Ecology Series; Academic Press: Boston, MA, USA; Elsevier: Amsterdam, The Netherlands, 2005.
7. Pascual, M.; Dunne, J.A. *Ecological Networks: Linking Structure to Dynamics in Food Webs*; Studies in the Sciences of Complexity, Santa Fe Institute, Oxford University Press: New York, NY, USA, 2006.
8. Rosen, R. *Life Itself: A Comprehesive Inquiry into the Nature, Origin, and Fabrication of Life*; Complexity in Ecological Systems Series; Columbia University Press: New York, NY, USA, 1991.
9. Hofmeyr, J.H.S. The biochemical factory that autonomously fabricates itself: A systems biological view of the living cell. In *Systems Biology: Philosophical Foundations*; Elsevier B.V.: Amsterdam, The Netherlands, 2007; pp. 217–242.
10. Miller, J.G. Living Systems: The Organization. *Behav. Sci.* **1972**, *17*, 1–182.
11. Harold, F.M. *The Way of the Cell: Molecules, Organisms and the Order of Life*; Oxford University Press: Oxford, UK, 2001.
12. Lillie, R.S. Living Systems and Non-living Systems. *Philos. Sci.* **1942**, *9*, 307–323.
13. Ulanowicz, R.E. On the nature of ecodynamics. *Ecol. Complex.* **2004**, *1*, 341–354.
14. Harold, F.M. Molecules into Cells: Specifying Spatial Architecture. *Microbiol. Mole. Biol. Rev.* **2005**, *69*, 544–564.
15. Alon, U. *An Introduction to Systems Biology: Design Principles of Biological Circuits*; Mathematical and Computational Biology, Chapman & Hall/CRC: London, UK, 2007.
16. Boogerd, F.C.; Bruggeman, F.J.; Hofmeyr, J.H.S.; Westerhoff, H.V. *Systems Biology: Philosophical Foundations*; Elsevier: Amsterdam, The Netherlands, 2007.
17. Krakauer, D.C.; Collins, J.P.; Erwin, D.; Flack, J.C.; Fontana, W.; Laubichler, M.D.; Prohaska, S.J.; West, G.B.; Stadler, P.F. The challenges and scope of theoretical biology. *J. Theor. Biol.* **2011**, *276*, 269–276.
18. Anderson, P.; Jeldtoft-Jensen, H.; Oliveira, L.P.; Sibani, P. Evolution in complex systems. *Complexity* **2004**, *10*, 49.
19. Roberts, A.J. *Model Emergent Dynamics in Complex Systems*; Mathematical Modeling and Computation, SIAM, Society for Industrial and Applied Mathematics: Philadelphia, PA, USA, 2015.
20. Bailly, F.; Longo, G. Extended Critical Situations: The Physical Singularity of Life Phenomena. *J. Biol. Syst.* **2008**, *16*, 309–336.
21. Watson, D.L. Biological Organization. *Q. Rev. Biol.* **1931**, *6*, 143–166.
22. Mesarović, M.D. (Ed.) *System Theory and Biology*; Springer: New York, NY, USA, 1968.
23. Klir, G.J. *Facets of Systems Science*, 2nd ed.; Plenum Press: New York, NY, USA, 2001.
24. Balescu, R. *Equilibrium and Non-Equilibrium Statistical Mechanics*; A Wiley-Interscience Publication, John Wiley & Sons: New York, NY, USA, 1975.
25. Rosen, R. On interactions between dynamical systems. *Math. Biosci.* **1975**, *27*, 299–307.
26. Kritz, M.V.; dos Santos, M.T. Dynamics, Systems, Dynamical Systems and Interaction Graphs. In *Dynamics, Games and Science II*; Peixoto, M.M., Rand, D., Pinto, A.A., Eds.; Springer: Berlin, Germany, 2011; pp. 507–541.
27. Ashby, W.R. Principles of the Self-Organizing System. In *Principles of Self-Organization: Transactions of the University of Illinois Symposium*; von Foster, H., Zopf, G.W., Jr., Eds.; University of Illinois, Pergamon Press: London, UK, 1962; pp. 255–278.
28. Rosen, R. Biological Systems as Organizational Paradigms. *Int. J. Gen. Syst.* **1974**, *1*, 165–174.
29. Varela, F.G.; Maturana, H.R.; Uribe, R. Autopoiesis: The Organization of Living Systems, Its Characterization and a Model. *Biosystems* **1974**, *5*, 187.
30. Miller, J.G. Living Systems: Basic Concepts. *Behav. Sci.* **1965**, *10*, 193–237.

31. Auger, P. The Methods and Limits of Scientific Knowledge. In *On Modern Physics*; Clarkson N. Potter, Inc. Publisher: New York, NY, USA, 1961; pp. 79–108.

32. Miller, J.G. Living systems. *Curr. Mod. Biol.* **1971**, *4*, 55–256.

33. Miller, M.B.; Bassler, B.L. Quorum Sensing In Bacteria. *Ann. Rev. Microbiol.* **2001**, *55*, 165–199.

34. Stoney, R.A.; Ames, R.M.; Nenadic, G.; Robertson, D.L.; Schwartz, J.M. Disentangling the multigenic and pleiotropic nature of molecular function. *BMC Syst. Biol.* **2015**, *9*, S3.

35. Oyeyemi, O.J. Modelling HIV-1 Interaction with the Host System. Ph.D. Thesis, University of Manchester, Manchester, UK, 2016.

36. Prokop, A.; Beaven, R.; Qu, Y.; Sánchez-Soriano, N. Using fly genetics to dissect the cytoskeletal machinery of neurons during axonal growth and maintenance. *J. Cell Sci.* **2013**, *126*, 2331–2341.

37. Wang, H.; Zheng, H. Organized modularity in the interactome: Evidence from the analysis of dynamic organization in the cell cycle. *IEEE/ACM Trans. Comput. Biol. Bioinform.* **2014**, *11*, 1264–1270.

38. Brownridge, P.; Lawless, C.; Payapilly, A.B.; Lanthaler, K.; Holman, S.W.; Harman, V.M.; Grant, C.M.; Beynon, R.J.; Hubbard, S.J. Quantitative analysis of chaperone network throughput in budding yeast. *Proteomics* **2013**, *13*, 1276–1291.

39. Peacocke, A.R. *An Introduction to the Physical Chemistry of Biological Organization*; Clarendon Press: Oxford, UK, 1983.

40. Atlan, H. *Entre le Cristal et la Fumée. Essay sur l'Organization du Vivant*; Éditions du Seuil: Paris, France, 1986.

41. Kritz, M.V. Creating Bio-Mathematical Worlds. P&D Report 29/95, LNCC/MCT, Petrópolis, 1995. In Proceedings of the 13th European Meeting on Cybernetics and Systems Research, Vienna, Austria, 9–12 April 1996.

42. Bizzarri, M.; Palombo, A.; Cucina, A. Theoretical aspects of Systems Biology. *Prog. Biophys. Mol. Biol.* **2013**, *112*, 33–43.

43. Kitto, K. High end complexity. *Int. J. Gen. Syst.* **2008**, *37*, 689–714.

44. Waddington, C.H. (Ed.) *Biological Organization, Cellular and Sub-cellular*; Pergamon Press: London, UK, 1959.

45. Berg, H. Motile behavior of bacteria. *Phys. Today* **2000**, *53*, 24–29.

46. Blanchoin, L.; Boujemaa-Paterski, R.; Sykes, C.; Plastino, J. Actin Dynamics, Architecture, and Mechanics in Cell Motility. *Physiol. Rev.* **2014**, *94*, 235–263.

47. Needham, J. On the dissociability of the fundamental processes in ontogenesis. *Biol. Rev.* **1933**, *8*, 180–233.

48. Rosen, R. *Structural And Functional Considerations in the Modelling of Biological Organization*; Technical Report 77 25; The Center for Theoretical Biology, SUNY: Buffalo, NY, USA, 1977.

49. Schweitzer, F. (Ed.) *Self-Organization of Complex Structures: From Individual to Collective Dynamics*; CRC Press, Taylor and Francis Group, LLC.: Boca Raton, FL, USA, 1997.

50. Vinson, V.J. Proteins in Motion. *Science* **2009**, *324*, 197.

51. Longo, G.; Miquel, P.A.; Sonnenschein, C.; Soto, A.M. Is information a proper observable for biological organization? *Prog. Biophys. Mol. Biol.* **2012**, *109*, 108–114.

52. von Bertalanffy, L. *General Systems Theory*; Allen Lane The Penguin Press: London, UK, 1968.

53. Fontana, W.; Buss, L. The Barrier of Objects: From Dynamical Systems to Bounded Organizations. In *Boundaries and Barriers*; Casti, J.L., Karlqvist, A., Eds.; Addison-Wesley Publishing Company, Inc.: Reading, MA, USA, 1996; pp. 55–115.

54. Letelier, J.C.; Soto-Andrade, J.; Guíñez Abarzúa, F.; Cornish-Bowden, A.; Luz Cárdenas, M. Organizational invariance and metabolic closure: Analysis in terms of systems. *J. Theor. Biol.* **2006**, *238*, 949–961.

55. Cornish-Bowden, A. Tibor Gánti and Robert Rosen: Contrasting approaches to the same problem. *J. Theor. Biol.* **2015**, *381*, 6–10.

56. de la Escosura, A.; Briones, C.; Ruiz-Mirazo, K. The systems perspective at the crossroads between chemistry and biology. *J. Theor. Biol.* **2015**, *381*, 11–22.

57. Montévil, M.; Mossio, M. Biological Organization as Closure of Constraints. *J. Theor. Biol.* **2015**, *372*, 179–191.

58. Dittrich, P.; di Fenizio, P.S. Chemical Organisation Theory. *Bull. Math. Biol.* **2007**, *69*, 1199–1231.

59. Kauffman, S.A. *At Home in the Universe: The Search for Laws of Self-Organization and Complexity*; Oxford University Press: New York, NY, USA; Oxford, UK, 1995.

60. Kauffman, S.A. *The Origins of Order: Self-Organization and Selection in Evolution*; Oxford University Press: New York, NY, USA; Oxford, UK, 1993.

61. Rashevsky, N. Life, information theory, and topology. *Bull. Math. Biol.* **1955**, *17*, 229–235.

62. Atlan, H. Application of information theory to the study of the stimulating effects of ionizing radiation, thermal energy, and other environmental factors. Preliminary ideas for a theory of organization. *J. Theor. Biol.* **1968**, *21*, 45–70.

63. Walker, I. *The Evolution of Biological Organization as a Function of Information*; Editora INPA: Manaus, Brazil, 2005.

64. Rashevsky, N. Topology and life: In search of general mathematical principles in biology and sociology. *Bull. Math. Biol.* **1954**, *16*, 317–348.

65. Atlan, H. On a formal definition of organization. *J. Theor. Biol.* **1974**, *45*, 295–304.

66. Pahl-Wostl, C. *The Dynamic Nature of Ecosystems, Chaos and Order Entwined*; John Wiley & Sons: Chichester, UK, 1995.

67. Maturana, H. The Organization of the Living: A Theory of the Living Organization. *Int. J. Hum. Comput. Stud.* **1999**, *51*, 149–168.

68. Louie, A.H. (M,R)-Systems and their Realizations. *Axiomathes* **2006**, *16*, 35–64.

69. Kineman, J.J. Relational Science: A Synthesis. *Axiomathes* **2011**, *21*, 393–437.

70. Baas, N.A. On structure and organization: An organizing principle. *Int. J. Gen. Syst.* **2013**, *42*, 170–196.

71. Baas, N.A. On higher structures. *Int. J. Gen. Syst.* **2016**, *45*, 747–762.

72. Hellerman, L. The Animate—Inanimate Relationship. *Int. J. Gen. Syst.* **2016**, *45*, 734–746.

73. Baas, N.A. Self-organisation and Higher Order Structures. In *Self-organisation of Complex Structures: From Individual to Collective Dynamics*; Schweitzer, F., Ed.; CRC Press: Boca Raton, FL, USA, 1997; pp. 71–81.

74. Bohr, N.H.D. Light and Life. *Nature* **1933**, *131*, 457–459.

75. Maynard Smith, J. The Concept of Information in Biology. *Philos. Sci.* **2000**, *67*, 177–194.

76. Adami, C. Information theory in molecular biology. *Phys. Life Rev.* **2004**, *1*, 3–22.

77. Atlan, H.; Cohen, I.R. Self-organization and meaning in immunology. In *Self-Organization and Emergence in Life Sciences*; Feltz, B.; Crommelinck, M.; Goujon, P., Eds.; Springer: Dordrecht, The Netherlands, 2006; pp. 121–139.

78. Hauhs, M.; Lange, H. Ecosystem dynamics viewed from an endoperspective. *Sci. Total Environ.* **1996**, *183*, 125–136.

79. Jablonka, E.; Lamb, M.J. Evolution in Four Dimensions: Genetic, Epigenetic, Behavioral, and Symbolic Variaton in the History of Life. In *Life and Mind: Philosophical Issues in Biology and Psycology, A Bradford Book*; The MIT Press: Cambridge, MA, USA, 2005.

80. Roederer, J.G. *Information and its Role in Nature*; The Frontiers Collection; Springer: Berlin, Germany, 2005.

81. Adamatzky, A.; Armstrong, R.; Jones, J.; Gunji, Y.P. On creativity of slime mould. *Int. J. Gen. Syst.* **2013**, *42*, 441–457.

82. Chaitin, G. To a Mathematical Definition of 'LIFE'. *SIGACT News* **1970**, *4*, 12–18.

83. Chaitin, G.J. Toward a Mathematical Definition of "Life". In *Maximum Entropy Formalism*; Levine, R.D., Tribus, M., Eds.; M.I.T. Press: Cambridge, MA, USA, 1979; pp. 477–498.

84. Pattee, H.H. Simulations, Realizations, and Theories of Life. In *Artificial Life*; Number VI in SFI Series in the Sciences of Complexity; Langton, C., Ed.; Addison-Wesley Publishing Company, Inc.: Redwood City, CA, USA, 1989; pp. 63–77.

85. Atlan, H.; Koppel, M. The cellular computer DNA: Program or data. *Bull. Math. Biol.* **1990**, *52*, 335–348.

86. Emmeche, C. The Computational Notion of Life. *Theoria* **1994**, *9*, 1–30.

87. Griffiths, P.E. Genetic Information: A Metaphor in Search of a Theory. *Philos. Sci.* **2001**, *68*, 394–412.

88. Jablonka, E. Information: Its interpretation, its inheritance, and its sharing. *Philos. Sci.* **2002**, *69*, 578–605.

89. Thaller, B. *Advanced Visual Quantum Mechanics*; Springer Science+Business Media Inc.: New York, NY, USA, 2005.

90. Scott Kelso, J.A. *Dynamic Patterns: The Self-Organization of Brain and Behaviour*; The MIT Press, A Bradford Book: Cambridge, MA, USA, 1999.

91. Bruggeman, F.J.; Westerhoff, H.V.; Boogerd, F.C. BioComplexity: A pluralist research strategy is necessary for a mechanistic explanation of the "live" state. *Philos. Psychol.* **2002**, *15*, 411–440.

92. Lloyd, E.K. Counting Isomers and Isomerizations. In *Graph Theory and Its Applications: East and West, Proceedings of the First China_USA International Graph Theory Conference, Jinan, China, June 9–20, 1986*; Annals of the New York Academy of Sciences; Capobianco, M.F., Guan, M., Hsu, D.F., Tian, F., Eds.; The New York Academy of Sciences: New York, NY, USA, 1989; Volume 576, pp. 377–384.

93. Dobrowolski, J.C. The chiral graph theory. *MATCH Commun. Math. Comput. Chem.* **2015**, *73*, 347–374.

94. Barwise, J. *Admissible Sets and Structures*; Springer: Berlin, Germany, 1975.

95. Kritz, M.V. *On Biology and Information*; P&D Report 25/91; LNCC/MCTI: Petrópolis, Brazil, 1991.

96. Hartwell, L.H.; Hopfield, J.J.; Leibler, S.; Murray, A.W. From molecular to modular cell biology. *Nature* **1999**, *402*, C47–C52.

97. Berge, C. *Graphs and Hypergraphs*; North-Holland: Amsterdam, The Netherlands, 1973.

98. Schmidt, G.; Ströhlein, T. *Relations and Graphs: Discrete Mathematics for Computer Scientists*; EACTS Monagraphs on Theoretical Computer Science; Springer: Berlin, Germany, 1993.

99. Klamt, S.; Haus, U.U.; Theis, F. Hypergraphs and Cellular Networks. *PLoS Comput. Biol.* **2009**, *5*, e1000385.

100. dos Santos, M.T.; Kritz, M.V. On the Hierarchical Organization of Metabolic Networks: An Underlying Mathematical Model. In *Proceedings of the International Symposium on Mathematical and Computational Biology (BIOMAT 2005)*; Mondaini, R.; Dilão, R., Eds.; E-papers Serviços Editoriais Ltda.: Rio de Janeiro, Brazil, 2006; pp. 221–241.

101. Manna, Z. *The Mathematical Theory of Computation*; McGraw-Hill Co.: New York, NY, USA, 1974.

102. Bailly, A. *Dictionnaire Grec Français*, 26th ed.; Librairie Hachette: Paris, France, 1963. (1e Édition, 1894).

103. Arnol'd, V.I. *Mathematical Methods of Classical Mechanics*; Vol. 60, *Graduate Texts in Mathematics*; Springer: Berlin, Germany, 1978.

104. Alberts, B.; Johnson, A.; Lewis, J.; Raff, M.; Roberts, K.; Walter, P. *Molecular Biology of the Cell*, 4th ed.; Garland Science: New York, NY, USA, 2002.

105. Kritz, M.V. On Relations between i-graphs and Data Structures. *Logique Analyse* **1996**, *39*, 153–164.

106. Kritz, M.V. Biological Organizations. In *Proceedings of the IV Brazilian Symposium on Mathematical and Computational Biology (BIOMAT IV)*; Mondaini, R., Ed.; BIOMAT Consortium, E-papers Serviços Editoriais Ltda.: Rio de Janeiro, Brazil, 2005; Volume 2, pp. 89–103.

107. Mitchell, A.; Romano, G.H.; Groisman, B.; Yona, A.; Dekel, E.; Kupiec, M.; Dahan, O.; Pilpel, Y. Adaptive prediction of environmental changes by microorganisms. *Nature* **2009**, *460*, 220–224.

108. Nuñez, J.K.; Lee, A.S.Y.; Engelman, A.; Doudna, J.A. Integrase-mediated spacer acquisition during CRISPR-Cas adaptive immunity. *Nature* **2015**, *519*, 193–198.

109. Nielsen, A.A.K.; Der, B.S.; Shin, J.; Vaidyanathan, P.; Paralanov, V.; Strychalski, E.A.; Ross, D.; Densmore, D.; Voigt, C.A. Genetic circuit design automation. *Science* **2016**, *352*, aac7341.

110. Stonier, T. *Information and the Internal Structure of the Universe*; Springer: London, UK, 1990.

111. Fenzl, N.; Hofkirchner, W. Information Processing in Evolutionary Systems. In *Self-organisation of Complex Structures: From Individual to Collective Dynamics*; Schweitzer, F., Ed.; CRC Press: Boca Raton, FL, USA, 1997; pp. 59–70.

112. Shannon, C.E.; Weaver, W. *The Mathematical Theory of Communication*; University of Illinois Press: Urbana, IL, USA, 1949.

113. Hoath, S.B. Considerations on the Role of the Skin as the Boundary of an Autopoietic System. In Proceedings of the Biology, Language, Cognition and Society: International Symposium on Autopoiesis, Belo Horizonte, Brazil, 18–21 November 1997.

114. Santra, T.; Kolch, W.; Kholodenko, B.N. Navigating the multilayered organization of eukaryotic signaling: A new trend in data integration. *PLoS Comput. Biol.* **2014**, *10*, e1003385.

115. Nadin, M. Anticipation and dynamics: Rosen's anticipation in the perspective of time. *Int. J. Gen. Syst.* **2010**, *39*, 3–33.

116. Iyer, K.V.; Pulford, S.; Mogilner, A.; Shivashankar, G.V. Mechanical activation of cells induces chromatin remodeling preceding MKL nuclear transport. *Biophys. J.* **2012**, *103*, 1416–1428.

117. Gerritsen, V.B. Moody wallpaper. *Protein Spotlight* **2003**, *33*. Available online: http://citeseerx.ist.psu.edu/viewdoc/download?doi=10.1.1.616.8546&rep=rep1&type=pdf (accessed on 1 March 2017).

118. Ai, C.; Li, Y.; Wang, Y.; Chen, Y.; Yang, L. Insight into the effects of chiral isomers quinidine and quinine on CYP2D6 inhibition. *Bioorg. Med. Chem. Lett.* **2009**, *19*, 803–806.

119. Levinson, H.; Mori, K. The Pheromone Activity of Chiral Isomers of Trogodermal for Male Khapra Beetles. *Naturwissenschaften* **1980**, *67*, 148–149.

120. Barrangou, R.; Marraffini, L.A. CRISPR-Cas Systems: Prokaryotes Upgrade to Adaptive Immunity. *Mol. Cell* **2014**, *54*, 234–244.

121. Simon, H.A. *The Sciences of the Artificial*, 3rd ed.; The MIT Press: Cambridge, MA, USA, 1996.

122. Lenneberg, E.H. *Biological Foundations of Language*, reprint ed.; Robert E. Krieger Publ. Co.: Malabar, FL, USA, 1984. (Original edition and copyright: John Wiley & Sons, Inc., 1967.)

123. Casti, J.L.; Karlqvist, A., Eds. *Complexity, Language, and Life: Mathematical Approaches*; Vol. 16, *Lecture Notes in Biomathematics*; Springer: Berlin, Germany, 1986.

124. Weinberg, G.M. *An Introduction to General Systems Thinking*; Dorset Hause Publishing: New York, NY, USA, 2001.

125. Bosch, O.; Maani, K.; Smith, C. *Systems Thinking—Language of Complexity for Scientists and Managers*; The University of Queensland; Queensland, Australia, 2007; pp. 57–66.

126. Corning, P.A.; Szathmáry, E. "Synergistic selection": A Darwinian frame for the evolution of complexity. *J. Theor. Biol.* **2015**, *371*, 45–58.

127. Zurek, W.H. Complexity, Entropy, and the Physics of Information; In *Santa Fe Institute Studies in the Sciences of Complexity*; Perseus Publishing: Cambridge, MA, USA, 1990.

128. Watkins, J.J. *Number Theory: A Historical Approach*; Princeton University Press: Princeton, NJ, USA; Oxford, UK, 2014.

129. Rosen, R. Complexity and System Descriptions. In *Systems, Approaches, Theories, Applications*; Hartnett, W., Ed.; D. Reidel Publishing Co.: Dordrecht-Holland, The Netherlands, 1977; pp. 169–178.

130. Shao, X.; Li, Q.; Mogilner, A.; Bershadsky, A.D.; Shivashankar, G.V. Mechanical stimulation induces formin-dependent assembly of a perinuclear actin rim. *Proc. Natl. Acad. Sci. USA* **2015**, *112*, E2595–E2601.

131. Ditlev, J.A.; Mayer, B.J.; Loew, L.M. There is More Than One Way to Model an Elephant. Experiment-DrivenModeling of the Actin Cytoskeleton. *Biophys. J.* **2013**, *104*, 520–532.

132. Erban, R.; Flegg, M.B.; Papoian, G.A. Multiscale Stochastic Reaction–Diffusion Modeling: Application to Actin Dynamics in Filopodia. *Bull. Math. Biol.* **2013**, *76*, 799–818.

133. Mogilner, A.; Rubinstein, B. The physics of filopodial protrusion. *Biophys. J.* **2005**, *89*, 782–795.

134. De Moura, C.A.; Kritz, M.V.; Leal, T.F.; Prokop, A. Mathematical-computational Simulation of Cytoskeletal Dynamics. In *Modeling and Computational Intelligence in Engineering Applications*; Santiago, O.L., da Silva-Neto, A.J., Silva, G., Eds.; Springer: Berlin, Germany, 2016.

135. Callebaut, W.; Rasskin-Gutman, D. *Modularity: Understanding the Development and Evolution of Natural Complex Systems*; The Viena Series in Theoretical Biology; The MIT Press: Cambridge, MA, USA, 2005.

136. Simon, H.A. The Architecture of Complexity. *Proc. Am. Philos. Soc.* **1962**, *106*, 467–485.

137. Ehret, C.F. Organelle Systems and Biological Organization. *Science* **1960**, *132*, 115–123.

138. Herrgård, M.J.; Swainston, N.; Dobson, P.; Dunn, W.B.; Arga, K.Y.; Arvas, M.; Büthgen, N.; Borger, S.; Costenoble, R.; Heinemann, M.; et al. A consensus yeast metabolic network reconstruction obtained from a community approach to systems biology. *Nat. Biotechnol.* **2008**, *26*, 1155–1160.

139. Lambert, A.; Dubois, J.; Bourqui, R. Pathway Preserving Representation of Metabolic Networks. *Comput. Graph. Forum* **2011**, *30*, 1021–1030.

140. Kritz, M.V.; dos Santos, M.T.; Urrutia, S.; Schwartz, J.M. Organizing Metabolic Networks: Cycles in Flux Distribution. *J. Theor. Biol.* **2010**, *265*, 250–260.

141. Keung, A.J.; Khalil, A.S. A unifying model of epigenetic regulation. *Science* **2016**, *351*, 661–662.

142. Breinig, F.; Sendzik, T.; Eisfeld, K.; Schmitt, M.J. Dissecting toxin immunity in virus-infected killer yeast uncovers an intrinsic strategy of self-protection. *Proc. Natl. Acad. Sci. USA* **2006**, *103*, 3810–3815.

143. Schmitt, M.J.; Breinig, F. Yeast viral killer toxins: Lethality and self-protection. *Nat. Rev. Microbiol.* **2006**, *4*, 212–221.

144. Rodríguez-Cousiño, N.; Maqueda, M.; Ambrona, J.; Zamora, E.; Esteban, R.; Ramírez, M. A new wine Saccharomyces cerevisiae killer toxin (Klus), encoded by a double-stranded rna virus, with broad antifungal activity is evolutionarily related to a chromosomal host gene. *Appl. Environ. Microbiol.* **2011**, *77*, 1822–1832.

145. Zadeh, L.A.; Polak, E. *System Theory, Vol. 8, Inter-University Electronics Series*; TATA McGraw-Hill Publishing Co. Ltd.: Bombay/New Delhi, India, 1969.

146. Louie, A.H. Anticipation in (M,R)-systems. *Int. J. Gen. Syst.* **2012**, *41*, 5–22.

147. Burstein, G.; Negoita, C.; Kranz, M. Postmodern Fuzzy System Theory: A Deconstruction Approach Based on Kabbalah. *Systems* **2014**, *2*, 590–605.

148. Igamberdiev, A.U. Anticipatory dynamics of biological systems: From molecular quantum states to evolution. *Int. J. Gen. Syst.* **2015**, *44*, 631–641.

149. Kritz, M.V. *Biological Information and Knowledge*; Relatório de P&D 23/2009; LNCC/MCT: Petrópolis, Brazil, 2009.

150. Kritz, M.V. Biological Organization, Biological Information, and Knowledge. *bioRxiv* **2014**, *2014*, 012617.

151. Johnson, J. A Theory of Stars in Complex Systems. In *Complexity, Language, and Life: Mathematical Approaches*; Lecture Notes in Biomathematics; Casti, J.L., Karlqvist, A., Eds.; Springer: New York, NY, USA 1986; Volume 16, pp. 21–61.

152. Bergareche, A.M.; Ruiz-Mirazo, K. Metabolism and the problem of its universalization. *BioSystems* **1999**, *49*, 45–61.

153. Holland, J.H. Studying complex adaptive systems. *J. Syst. Sci. Complex.* **2006**, *19*, 1–8.

154. Letelier, J.C.; Cárdenas, M.L.; Cornish-Bowden, A. From L'Homme Machine to metabolic closure: Steps towards understanding life. *J. Theor. Biol.* **2011**, *286*, 100–113.

155. Emmeche, C. A Bio-semiotic Note on Organisms, animals, machines, Cyborgs, and the Quasi-autonomy of Robots. *Pragmat. Cogn.* **2007**, *15*, 455–483.

156. Luz Cárdenas, M.; Letelier, J.C.; Gutiérrez, C.; Cornish-Bowden, A.; Soto-Andrade, J. Closure to efficient causation, computability and artificial life. *J. Theor. Biol.* **2010**, *263*, 79–92.

157. Villani, M.; Filisetti, A.; Graudenzi, A.; Damiani, C.; Carletti, T.; Serra, R. Growth and Division in a Dynamic Protocell Model. *Life* **2014**, *4*, 837–864.

158. Shrager, J. The fiction of function. *Bioinformatics* **2003**, *19*, 1934–1936.

159. Brownlee, J. *Complex Adaptive Systems*; CIS Technical Report 070302A; Complex Intelligent Systems Laboratory, Centre for Information Technology Research: Melbourne, Australia, 2007.

160. Wolkenhauer, O. Systems biology: The reincarnation of systems theory applied in biology? *Brief. Bioinform.* **2001**, *2*, 258–270.

161. Szathmáry, E. Founder of systems chemistry and foundational theoretical biologist: Tibor Gánti (1933–2009). *J. Theor. Biol.* **2015**, *381*, 2–5.

162. Lazebnik, Y. Can a biologist fix a radio?—Or, what I learned while studying apoptosis. *Cancer Cell* **2002**, *2*, 179–182.

163. Boogerd, F.; Bruggeman, F.; Jonker, C.; de Jong, H.L.; Tamminga, A.; Treur, J.; Westerhoff, H.; Wijngaards, W. Inter-level relations in computer science, biology, and psychology. *Philos. Psychol.* **2002**, *15*, 463–471.

164. Mayr, E. *This Is Biology: The Science of the Living World*; Belknap Press/Havard University Press: Cambridge, MA, USA, 1997.

165. Kitto, K. A Contextualised General Systems Theory. *Systems* **2014**, *2*, 541–565.

systems

MDPI

Article

Knowledge to Manage the Knowledge Society: The Concept of *Theoretical Incompleteness*

Gianfranco Minati

Italian Systems Society, Milan 20161, Italy; gianfranco.minati@AIRS.it; Tel.: +39-02-6620-2417

Academic Editor: Ockie Bosch
Received: 23 May 2016; Accepted: 11 July 2016; Published: 15 July 2016

Abstract: After having outlined the essential differences between non-complex systems and complex systems we briefly recall the conceptual approaches considered by the pre-complexity General Systems Theory introduced by Von Bertalanffy in 1968 and those of the science of complexity and *post-Bertalanffy* General Systems Theory. In this context, after outlining the concept of completeness, we consider cases of incompleteness in various disciplines to arrive at *theoretical incompleteness*. The latter is clarified through several cases of different natures and by approaches in the literature, such as *logical openness*, the *Dynamic Usage of Models* (DYSAM), and the principle of uncertainty in physics. The treatment and the contrast between completeness and incompleteness are introduced as a conceptual and cultural context, as knowledge to manage the knowledge society in analogy, for example, with the transition from the logic of certainty to that of uncertainty introduced by De Finetti. The conceptual framework of completeness is not appropriate for dealing with complexity. Conversely, the conceptual framework of incompleteness is consistent and appropriate with interdisciplinary complexity.

Keywords: completeness; constructivism; decidability; openness; procedure; system; uniqueness

1. Introduction

This article introduces conceptual specifications regarding the concepts of completeness and incompleteness. The latter concept is treated by distinguishing between (a) incompleteness of a phenomenological nature (given, for instance, by non-proceduralizability; non-completability; incompleteness of constraints, of degrees of freedom or representation; approximation; and by indefiniteness) and (b) theoretical incompleteness (given, for instance, by non-decidability; principles of uncertainty; non-computable uncertainty; non-complete, non-explicit and non-univocal modeling). The theme of theoretical incompleteness has been treated in the literature in correspondence to, or negation of, meanings of completeness in various disciplines [1], such as logic and mathematics, where examples include Russell's paradox [2] and the incompleteness theorems introduced by Gödel [3].

This paper focuses on concepts appropriate for modeling complex systems, such as social systems [4–6]. It attempts to outline conceptual invariants relating to theoretical incompleteness and its disciplinary transversality, therefore, being suitable for generalizations not based on genericity, but being suitable for methodologies, approaches, and representations for problems, typically related to complex systems [7,8], (see Section 2) represented, for instance, as networks [9].

Such problems and representations showing how incompleteness, or better, assumptions of completeness, turn out to be completely or partially ineffective and inadequate [10] are explored. This is done by using procedures and analytical models based on finite representations and *logically closed systems* (see Section 6.1).

We emphasize the search for meaning, cross-usability of the philosophical concept of incompleteness and of theoretical incompleteness. This should really be a cultural process involving all of the concepts of the *post-Bertalanffy* Theory of General Systems [11–13] such as *logical openness* (see Section 6.1) as a multiplicity of modeling; coherence; emergence as continuous and unpredictable, but coherent, acquisition of multiple properties; equivalence/non-equivalence of models and evolutionary paths; theoretical principles of uncertainty and indeterminacy; individuality to break equivalences; induction of properties, rather than prescriptions of solutions; overall conceptual interdependence, non-separability from the environment; irreversibility as the *price* for uniqueness; multiplicity, multiple systems set dynamically by the same components; observed in terms of the observer; networks and properties of networks, such as *small worldness* and their topological properties; simultaneity; and overlapping. These, considering the fact that properties and problems of complex systems, particularly social systems, are still dealt with using the concepts of classical Bertalanffy General Systems Theory [14], such as anticipation; openness as permeability of system barriers to matter and energy; equilibrium; functioning, as for devices possessing and not acquiring properties; optimal and computable organization; optimization always being possible and positive *in principle*; predictability and repeatability; adjustability; feedback; and reversibility.

When dealing with social systems such a distinction, between Bertalanffy and post-Bertalanffy General Systems Theory, is combined with, or even corresponds to, the knowledge to be used to manage. That is, conceive, make emergent, and change properties, and induce processes in business, culture, and economics, within cooperation and assimilation, fashion, food uses, interactions, professions, customs, and general usage.

The post-industrial society [15], or knowledge society [16], is where examples of *sources of complexity* are properties, such as knowledge-intensive products and services; delocalization and globalization; duplicability; general highly-networked interconnections; high virtuality; importance of individuality; coherences rather than equilibrium; interchangeability; online actions; reduced time between design, implementation, and marketing; generally short product lifespan; technological innovations and solutions rapidly creating new problems; non-linearity; and large amounts of data. Conversely, social systems are often managed, i.e., conceived as non-complex systems, by still using the knowledge of pre-complexity [17].

This is not simply to *update* methods, but to vary and structurally adapt the general culture. This article is meant to be a contribution towards social development processes in areas such as education and the management of enterprises and social institutions [18,19]. Here, the word 'management' takes for granted the *directivity* of social systems, in a pre-complexity systemic vision. In reality, the aim is to induce the emergence of properties to be acquired by the Social Systems (as complex systems) and not to adjust or impose them. *Giving orders* to a complex system is inadequate and ineffective. Management of complex systems, which social systems are, requires multiple strategies and approaches, such as action on the environment, adding suitable perturbations, making energetic variations, acting upon the relations with the outside world and on learning processes by interacting with other systems. It is to change constraints and provide perturbations to be processed by the system to *orient* and not to decide. This could include changes in the use of resources, language, organizational approaches, prices, or rates. Probably the word management is now misleading after years of being synonymous with directive and invasive approaches.

The work objectives are based on considering theoretical incompleteness as one of the properties characterizing complex systems amongst others well-known in the literature [20] (see Section 2).

Theoretical incompleteness should be suitably represented, modeled, used, activated or de-activated, graduated, replicated, avoided, and combined in different forms. The assumption of theoretical incompleteness as a *negative* property which must be reduced to *incompleteness to be completed* as the only possible strategy is very ineffective when dealing with complex systems.

We may consider a possible analogy between the mesoscopic level of representation and theoretical incompleteness. In physics, mesoscopic variables, when not relating to the quantum level, relate to an *intermediate* level between the micro and the macro. At this level the micro is not completely neglected as when adopting the macro levels (see, for instance, [21–24]).

The approach based on considering mesoscopic variables relies on the *philosophy of the 'middle way'* as introduced in [25].

Simple examples of mesoscopic variables are:

- In car traffic, when considering the instantaneous number of cars, or clusters, which cannot accelerate. Simultaneously we consider cars blocked in a queue, cars which are decelerating, and cars with constant speed;
- In a building, one can consider the instantaneous number of people, or clusters, using the elevator or stairs, going either up or down; and
- In a flock we may consider the instantaneous number of birds, or clusters, having the same speed, whatever their direction or altitude.

The mesoscopic level is the *place* of *continuous negotiations* between the micro and the macro, where a large variety of mesoscopic representations are possible allowing an undefined number of possible clusters.

On the other hand, the conceptual analogy of the mesoscopic case with theoretical incompleteness lies in the fact that such incompleteness may apply in several different ways as the *place* of those continuous negotiations between completeness and incompleteness. In this case, instead of clusters, there are different levels, graduations, and approximations of completeness and incompleteness referring to different variables describing the case.

Theoretical incompleteness may be intended as the place where phenomena are incomplete enough to allow emergence [26] of dynamical coherences [27], e.g., multiple synchronizations [28], as for the emergence of complex systems (see Section 2) rather than iterations of the same complete rules and properties. This is the case of *logically open systems*, introduced in Section 6, of which their evolution is not driven by the same complete rules, but by dynamically changing *coherent* rules.

One interesting future research issue relates to the possibility of considering theoretical incompleteness as a necessary, or even, in specific cases, a sufficient condition for the establishment of processes of emergence.

The purpose of this paper is to underline and support approaches for the *usages* of incompleteness as a property rather than to avoid it as a negative property, in principle. This relates to complex systems in general, with special attention to social systems.

We conclude by stressing the extreme interdisciplinary nature of this way of understanding theoretical incompleteness to be elaborated by disciplines relevant for dealing with social systems in their post-industrial phase, or knowledge societies intended as those based upon scientific and theoretical knowledge as their main resource (see Section 7). Examples of such disciplines are communication sciences; economics; education; political sciences; psychology; and sociology.

2. Non-Complex and Complex Systems

Non-complex systems are intended as those based on the properties cited in the Introduction and which are related to classical Bertalanffy General Systems Theory. Non-complex systems acquire over time the same *property* due to their functioning (such as the *working* of powered electronic devices when, for instance, they are transformed from a structured set of components to a device acquiring the property of being a radio or some other end-user device). Elementary examples include machinery, electronic devices, and feedback-regulated devices, systems that do not learn, having fixed rules, or otherwise low-parametric variability. It is a matter of *decidable* systems, represented completely, whose evolutionary paths can reach a *finite*, or in any case, a numerable number of states. Their possible openness involves finite and precise modalities to process external environmental input [14]. Into this category fall systems that have probabilistic, but computable, evolutionary behavioral paths (see the difference between computable and non-computable uncertainty [29] as introduced below). Examples of such systems include automata (mechanical and electronic systems), transportation systems, housing systems, telecommunication systems, security systems, energy management systems, and fuzzy systems [30].

Complex systems do not acquire, over time, the same systemic property, but comprise *continuous* processes of structural changes (see Section 5.1, Section 5.4, and Section 5.8), such as phase transition [31,32], with self-organization (when the sequence of new properties acquired in a phase transition-like manner, has regularities and repetitiveness, for example a swarm around a light, the formation of queues, or synchronizations [33–37] and emergence (when the sequence of new properties is not regular, not repetitive, but *coherent*, i.e., having multiple different synchronisations and correlations, for example a flock having multiple and changing shapes, density and direction but maintaining scale invariance) [38–40] taking place, acquiring over time *coherent sequences of new properties* [41,42]. Examples of complex systems without cognitive systems include cellular automata; double pendulum; dissipative structures [43,44], e.g., whirlpools in fluid dynamics and chaotic systems (see below). Examples include the climate system; protein chains and their withdrawal; cells and bacteria; objects on vibrating surfaces that tend to make consistent variations; autonomous lighting networks that tend to adopt coherent variations (such as communities of fireflies); and traffic signals (Internet). We may consider examples of complex systems consisting of coherent communities of living systems provided with cognitive systems, i.e., autonomous systems that are able to *decide* their behavior not only in an algorithmic way. In these cases the decision of behaviors cannot be reduced to compute optimizations, but it is a process of emergence from a wide variety of aspects. In a nutshell, a cognitive system is to be understood as a system of interactions between activities such as those related to attention, perception, language, the affective and emotional sphere, memory and inferential system, logical activity. Examples include swarms; flocks; industrial clusters; industrial district networks; markets; and social systems, such as cities, schools, hospitals, companies, families, and temporary communities, such as passengers, audiences, and telephone networks.

Complex systems have characterizing properties. Some of them are listed below.

- *Scale invariance* when patterns, shapes, and morphological properties are independent of dimensions (e.g., spatial properties, number of components) [45];
- Validity of *power laws* among variables characterizing the system. Power laws are given by a special kind of mathematical relationship between two quantities. When the frequency of an event varies as a power of some attribute of that event, e.g., its size, the frequency is said to follow a power law [46].
- The evolutionary paths of the system fall within the basin of an *attractor*, see Figure 1 and Section 5.7, [47–50]. Evolutionary paths are robust to perturbations. This is the case of *chaotic systems* which are very sensitive to initial conditions, e.g., smoke diffusion and weather [51–53].
- Properties based on *Network Science* [54–58]. We consider properties of a network representing a complex system, such as being scale-free (when the network has a high number of nodes possessing few links, and a small number of nodes possessing a high number of links. In other words, in scale-free networks the probability that a node selected at random will possess a particular number of links follows a power law). Furthermore networks can have the property to be *small-world* when most nodes are not neighbors of one another, but most nodes can be reached from every other node via a small number of intermediate links. Other properties relate to the degree of sequence distribution, cluster coefficient, topology of the network, and fitness (the way the links between nodes change over time depends on the ability of nodes to attract links) [59–62]. Several representations of social systems based on networks are available in the literature (see, for instance, [63,64]).

Other properties which may be considered include their explicit analytical intractability (their behavior cannot be *zipped* nor exhaustively represented with analytical formulae). In such cases systems are modeled using non-ideal models, such as neural networks and cellular automata.

We are considering here theoretical incompleteness as a candidate property of complex systems.

For an overview on properties of complex systems see, for instance, [65–68].

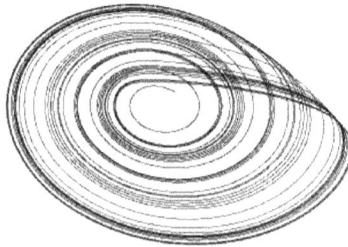

Figure 1. Generic graphic representation of evolutionary paths around an attractor, a single point in this case.

3. Meaning

Definitions are supposed to provide, at the highest levels of precision, formal descriptions of properties and limits with the vocation for completeness or exhaustiveness. That is, we cannot add to it. Examples in mathematics are geometrical axioms. Other generic examples relate to fiscal properties, such as taxation or safety procedures in the workplace.

An incomplete definition is intended to overlook some aspects which would tend to make it only partially usable.

Without going into theoretical specificity treated by semiotics and philosophy, we consider here the meanings of terms as given at different levels of contextuality and sensitivity to usage, suitable for different levels of generalization. The completeness of a definition is in these cases dampened by metaphorical usage, analogies, and extrapolations accepting generic contextuality. For instance, this occurs when one has to specify concepts which need to include all possible negations of a definition (*absolute incompleteness* as opposed to completeness). This is the concept of incompleteness which is not reducible to multiple negations of completeness.

4. Notes on the Concept of Completeness

Considering the topic of this article, there are several ways to indicate the meaning of completeness. For instance, a list may be considered complete when it contains all, and possibly only, the references to the entities under consideration. Of course, the number of elements is usually expected to be finite and limited, even though the definition is conceptually extendable to different cases. The reference is to be exhaustive, with complete information considered and without overlooking any detail that is deemed to be significant.

"Complete" has the meaning of being the maximum, to finish, end, close, reaching the purpose, considering even *equi-finality* starting from different initial conditions in open systems [14]. Completeness can refer to a process which, over time, reaches its final, possibly more than one, state, equivalent or not.

Completeness may be understood as corresponding to the fact that a system or a process has a finite number of degrees of freedom (fully described by a finite number of variables), and a finite number of constraints (values of min and max which may be adopted by variables). An endless completeness may be considered to correspond with incompleteness.

Completeness can also conceptually relate to processes and functionalities assumed as complete. For example, as they are self-consistent and autonomous, able to find resources to maintain a property, such as life, or to perform autopoietic reproduction. This is the case of systems which are incomplete (for example, with a number of states being assumed as being undefined), but autonomous, that is capable of reacting, for example in a stochastic manner or by learning, to specific inputs.

A related concept is that of completability, in general; that is, the possibility of making complete what is not yet complete. In general, in the non-complexity conceptual context, the assumption of completability as potential is assumed valid.

In mathematics, completeness can refer to the properties of a mathematical entity to be contained within another larger entity of the same type. Furthermore, in mathematical logic completeness refers to the fact that a set of axioms is sufficient to prove all the truths of a theory. Based on this it is, therefore, possible to decide the truth or falsity of any statement formulated in the language of the theory. In this context, the *conceptual space of completeness* would be well defined by considering it as consisting of:

1. *Decidability.* As introduced by Alan Turing (1912–1954) a problem is decidable if there is an algorithm that produces the corresponding solution in finite time for each instance of the input data (the equivalent of having a complete, calculable model of the behavior of systems having that problem and subjected to an external intervention). This is the concept of *effective computability* [69].

    A problem is "undecidable" if there is no algorithm which produces the corresponding solution in finite time for each instance of the input data (equivalent to the fact that an algorithm cannot produce the solution in finite time). In mathematical logic, the concept of undecidability refers to the fact that a given formalized theory *T* is not decidable, i.e., there is no algorithm able to mechanically determine for each formula whether or not it is a theorem of *T* (see Sections 5.1 and 5.7). The reference is to the Turing machine, an *ideal finite states machine*. This machine processes, using a predetermined set of rules defined exhaustively by reading and writing symbols, the data contained on an *ideal tape* for input and having potentially infinite length. The process of such an abstract model defines *computability*, the concept of algorithm in general [69].

2. *Deduction,* instead of induction or abduction. Validity of deduction is assumed when, for example, (a) this box contains red balls; (b) these balls are from that box; (c) the balls are all red. If the premises are true, the conclusions cannot be that true. Whereas induction has a probabilistic nature as in the case that (a) the balls are from that box; (b) these balls are red; (c) the more they pull out and if they are all red one might conclude that all the balls in the box are red.

    In the case of abduction, introduced by Sanders Peirce (1839–1914), it is matter of the invention of hypotheses, which can also be understood as a choice amongst the most effective available. For example, when observing facts of type B, a rule such as: if A then B may explain B. So, if at the time no other hypothesis can explain B better, we may assume the validity of A [70]. This is related to creativity, as in second order cybernetics, focusing upon inventing a new game rather than playing an existing game [71].

3. *Certainty.* Essentially this is a matter of reducing, or even cancelling, uncertainty. In the logic of certainty [72,73], it is not conceptually considered to maximize probabilities, but to have computable levels of certainty. However, probabilities relate to *configurations of events,* conditional probabilities as considered by the observer.

    The assumption that events can be considered as isolated, separated by configurations to which they belong should be considered as reductionist simplifications. On the other hand, configurations are not objectivistic but rather depend upon the cognitive approach taken by the observer.

These configurations are inalienably constructivist [74–79], created by the observer, the generator of *cognitive reality* rather than of relativism, as introduced by De Finetti (1906–1985) [80]. The *cognitive strategy* expressed by cognitivism is well-expressed through the distinction between:

- Trying to understand how something *really* is, and
- How it is *more effective* to think of it (which model to adopt).

Note that the former is a particular case of the latter. Objectivism leads to the inevitable phenomenological acceptance of becoming. From the moment we wonder *what* happened..., *how* it happened ... by involving cognitive abilities, constructivism intervenes. We assume then descriptive and interpretative models of the phenomena detected (detection needs a model itself...), generators, in their turn, of new configurations to be modelled. Something happened in what our cognitive system calls *outside* and we adopt a strategy to make models for acting, interacting, and abstracting using the cognitive system we have. Science does this and cognitive science studies this process. There is a vast literature on this subject (see, for instance, [81–84]). For example, is it more effective to think of a social problem as being military *or* political, a mathematical problem as being geometrical *or* algebraic, the problem of corruption as being cultural *or* legal (See Sections 5.2 and 5.3). Effectiveness may increase by suitably considering options in combined and not only exclusive ways (assuming only the "*or*" option), e.g., biological *and* psychological. *Non-computable uncertainty* refers to those contexts having, as a component, the environment with its turbulence, processes of emergence, and autonomous processes [85,86]. Section 5.5 covers incompleteness and non-predictability.

5. Incompleteness

Below is a partial list, in no specific order, illustrating concepts and cases related to incompleteness, considered as constituting examples of dynamic and theoretically-incomplete sets (see Section 6) of cases of phenomenological incompleteness.

The property of incompleteness may refer to one or more specific properties; may have different levels or may be general; that is, valid for *any properties* in general or for *any property of a specific entity*; may be intended as *not yet complete* or *uncompletable* or *chronically* not-completable, in principle. The property of incompleteness should be considered in combination with one or more levels of description and scales.

Section 6 shows how incompleteness refers to aspects of *logical openness* [10,87,88].

5.1. Non-Proceduralizable

The prototype of the concept of procedure is the concept of the algorithm intended as a program for a Turing Machine as an ideal programmable machine (see Section 4 on *decidability*). The general idea is to have approaches well-defined through step-by-step methods indicating what and when to do something mechanically applicable to families of problems, such as those related to construction, health, maintenance, and security. A procedure is intended as that being well-described in *instruction manuals*.

However, the plurality of degrees of freedom of complex systems; their structural change (change of rules rather than of parameters of only the same rule, variability, and unpredictability) make procedural approaches ineffective because they consider fixed and abstract variables in the face of phenomena whose variables change, form, and combine in multiple and unpredictable ways, i.e., they are emergent [89,90]. The procedures may cover areas, aspects of a problem, and provide, for example, the basic behavioral claims in cases of emergency, e.g., health, and ensure legal cover for those who carefully followed the prescribed procedure.

However, the procedures to deal with a problem can increase so much that it becomes impossible or paralyze an appropriate intervention for situations with features of uniqueness (complex emergent systems are made of coherent uniqueness). Such procedures often have as their purpose the prevention of unauthorized actions or personal responsibilities. Examples include actions to tackle environmental disasters or criminal acts, and interventions in the face of irrational, unexpected behavior with predictable, nefarious outcomes.

There are also cases where complexity prevents the adoption of procedures to be considered as exhaustive. For example, the management of safety at work is not reducible to rules, due to the complexity inevitably introduced by the human operator [91].

In mathematics, this issue concerns the non-reducibility of *general computation* to *Turing computability* as for emerging and natural computing [92]).

5.2. Uncompletable

In geometry, an axiom of order introduced by David Hilbert (1862–1943) states that if A and C are two points, then there always exists at least a point B on the line AC, such that C lies between A and B, see Figure 2.

$$A \qquad\qquad C \qquad\qquad\qquad B$$

Figure 2. There always exists at least a point between two points.

It is a theoretical generator of points between two points.

In this case, a hypothetical list of points between two points is *uncompletable* (because of the *non-finite* number).

In a metaphorical or conceptually similar way, the knowledge, i.e., representations and models, of a complex problem is not limited to the formulation of lists having finite numbers of recommended procedures to be followed (the classical instructions for use). Rather, in the case of social systems, it is a question of cognitive, context-sensitive processing, always variable depending, for example, upon experience, or new information or new knowledge, as considered by *logical openness* (see Section 6.1).

Since the classical instructions for use cannot exhaust the possibilities of actual use, modern products and services are presented with appropriate designs to induce adequate but any usage by the generic user.

In the management of public safety any procedure is uncompletable as any type of events can change the context, such as the use of technologies, environmental impact, or unexpected situations, in general.

5.3. Incompleteness and Uniqueness

Completeness can be understood to reflexively possess *repeatability of the action of completing* in the face of the improbable uniqueness of the action of completing, for instance, by applying the same procedure. The action of completing may be assumed to be identically repeatable, for example, because of the sameness of what is to be completed: either a procedure or the mathematical fractal filling curves of Peano and Hilbert [93]. There is conceptual negation of uniqueness, i.e., an assumption of repeatability of the same procedure to deal with the same category of problems. Processes of completion intended as unique, for one-off, non-repeatable, cases contrasts with the iterative, procedural aspect of completion.

In a parallel way, incompleteness can be understood as the non-repeatability of any completion. This may occur, for example, because what is to be completed is considered never equal to itself, e.g., on suitably different scales, or because of the dependence of the process of completing based on different initial conditions. There is conceptual correspondence with the concept of uniqueness.

Strategies and approaches to manage repeatability are inadequate for the possible uniqueness of incompleteness; as for emergent properties, unique and incomplete, not completely definable in an analytical way.

5.4. Undefined and Uncompleted

Indefiniteness can be considered as being due to insufficient levels of accuracy in defining (perhaps as desired or possibly upgradeable until achieving definiteness). The vague lines of the impressionist painters constructively leave the observer to mentally build his/her line at that time. Indefiniteness, imprecision, and incompleteness can be understood to constitute a *constructivist freedom*.

In turn, it is possible to consider indefinability as being given, for example, by structural dynamics for which entities or processes, on a suitable scale and level of description, change fundamental structural properties and identity. For example, structural dynamics apply to the structural autonomy of processes of emergence which cannot be completely, explicitly, and univocally defined (they are *logically open*, see Section 6.1). A flock is always different over time, but it keeps its identity as a flock in that it maintains, for instance, sufficient levels of coherence and scale invariance. Furthermore, appropriate levels of vagueness and incompleteness are *areas of potentiality* for processes which adopt corresponding levels of uniqueness (e.g., oneness of people, fingerprints, medical patients, or snowflakes). Moreover, there are various (even infinite) ways to comply with constraints. Variables considered as representing a specific entity can take on different values while respecting the constraints, e.g., constantly close to the minimum or maximum values of the constraints or oscillating in regular or stochastic ways.

Another example of indefiniteness and incompleteness is given in baroque music by the practice of the *basso continuo*, a historically improvised accompaniment, so that the score does not completely indicate the chords to play along with the low notes, but only numbers (or code words, as the word BACH) which give *clues* on how the instrumentalist should improvise.

5.5. Incompleteness and Non-Predictability

It is possible to consider incompleteness as being combined with non-predictability. We distinguish between at least two types of probabilities which represent predictability. *Certain probability* is intended here as *computable probability* intended as identifying significant possible extremes e.g., maximum-minimum, to the phenomenological becoming of any process, which is free to happen within these extremes, such as, for example, computed through the Bayes' theorem. In short, the theorem states that if $P(A)$ and $P(B)$ are the independent probabilities of A and B, and $P(A \mid B)$ is the probability (conditional probability) of observing event A given that B is true, than $P(B \mid A)$ is the probability of observing event B given that A is true, see also Section 4. *Uncertain probability* is intended here as *non-computable probability* referring to the non-predictability of processes, for example, of emergence which cannot be completely, explicitly, and uniquely defined.

The property of incompleteness of events or processes may be predictable or not, such as the induction of a self-organizing process which could have non-predictable aspects, for example, in producing the Belousov-Zhabotinski reaction [94,95] consisting of an oscillating chemical reaction in which the periodic variation of concentrations is indicated by striking color variations and the formation of convective patterns called Rayleigh-Bénard cells [96] in a liquid evenly heated from below. The occurrence of such processes is predictable but in incomplete ways, e.g., details of patterns and directions of convective patterns are not predictable. On the other hand, aspects of unpredictability, as in processes of emergence, owe their unpredictability to their being constantly incomplete, which makes them unique.

5.6. Multiple and Dynamic Incompleteness

Depending upon the level of description, incompleteness may relate to various aspects of a phenomenon, such as completely owning a property, but at different times (both with and without regularity); with different modes (e.g., intensity); or relating to different properties.

In the scientific literature, multiple and dynamic incompleteness may be intended as being expressed by the concept of *quasiness* where, for example, *quasicrystals* are a particular solid form in which the atoms are arranged in a deterministic but not repetitive structure, which is not periodic, as in the case of normal crystals, honeycombs, and beehives. Quasicrystals have almost periodic patterns for which the local arrangement of the material is fixed and regular, but not periodic throughout the material, i.e., the property is incompletely respected in multiple possible ways [97].

5.7. Incompleteness and Undecidability

A typical problem of undecidability (see Section 4) is the classic *halting problem* for the Turing machine. Turing showed [69] that a Turing machine which decides, i.e., stops in some state, whether another Turing machine processing an input w will stop in some state or will continue indefinitely, cannot exist.

In this regard, in mathematical logic Kurt Gödel (1906–1978) proved two theorems about the *incompleteness of mathematics* [3]. One of these states that, in any consistent formalization of mathematics, sufficiently powerful to allow axiomatization of the elementary theory of natural numbers (that is, define the structure of natural numbers with the operations of sum and product) it is possible to construct a formally-correct proposition that can be neither proven nor disproved within the same system.

This is logically equivalent to the *construction of a logical formula that denies its own provability*. The second states that if one considers a coherent mathematical theory sufficient to contain arithmetic (e.g., the arithmetic based on the Peano axioms: 1-there exists a natural number *0*; 2-every natural number has a natural number successor; 3-different numbers have different successors; 4-zero is not the successor of any natural number; 5-each subset of natural numbers which contains 0 and the successor of each element coincides with the entire set of natural numbers (axiom of induction), it is not possible to prove the coherence of the theory within the theory itself. The result can be extended to any theory that can be represented recursively (by recursive functions and predicates), and for which it is possible to arithmetize the syntax of the theory. A generalization consists in the fact that *it is not possible to use a coherent system to demonstrate its own coherence.*

Incompleteness concerns indecisiveness as a characteristic of a given problem; for example, the non-theoretical availability of an algorithm able to answer questions related to the problem.

As will be shown below when dealing with theoretical incompleteness (see Section 6), incompleteness as undecidability also concerns the conceptual impossibility of identifying solutions and, in particular, optimal solutions to complex problems which require, instead, systems of dynamical and non-equivalent approaches (as with DYSAM and *logical openness* as introduced in Sections 6.1 and 6.2) to intervene effectively on properties acquired continuously rather than being possessed by complex systems.

Of the same nature is the concept of optimization which, for complex systems, does not concern the acquisition of the best value, but of permissible values through the dynamics of variable configurations within dynamic constraints ensuring coherence at least as synchronization (for example, in biology ensuring sustainability and permissible values of vital parameters) or permissible evolutionary trajectories in the vicinity of an attractor. In short sn attractor is a set of numerical values, e.g., a single point or a finite set of points, a curve, and a manifold, toward which a dynamical system, starting from any variety of initial conditions tends to evolve. Evolutionary paths of the system, when close enough to the attractor, remain close even if perturbed (the Lorenz butterfly attractor is a celebrated example, [52,53]).

5.8. Systemic Incompleteness

Systems can be considered as complete or incomplete.

For example, systems can be considered complete when they are fully described by a finite number of variables and models (presumably the same throughout the entire developmental period). Yet a system can be considered complete when all of the evolutionary-reachable states are of a finite and limited number. This is the case of *logical closure* (see Section 6.1) as opposed to *logic openness*.

Systems can be considered incomplete when their complexity in acquiring properties and their structural dynamics are such that a single model is not sufficient for their representation. More non-equivalent models are needed, not only in an indefinite number, but variable in different combinations to be abductively, constructively invented by the observer (as in the case of DYSAM introduced in Section 6.2).

Moreover, a system can be considered incomplete when the set of reachable evolutionary states is undefined or, better, *its finiteness is undecidable*.

5.9. Incompleteness of Constraints

As mentioned in Section 5.9 the incompleteness of the constraints is the fact that nothing is said about how to comply with such constraints and the actual use that a single system makes. Systems with the same constraints may have had different evolutionary paths.

For example, as mentioned above in Section 5.4, the values between the maximum and minimum constraints are all authorized, but they may be acquired by systems in different ways over time, for example, close to the maximum or close to the minimum, or oscillating with regularity, or randomly oscillating.

The mode of use of the constraints (that is, the modes respecting the constraints) allows systems to be considered as structurally equivalent, but actually different, with regard to aspects of their evolutionary process and whereby paths between the constraints represent specific trajectories having specific properties.

The use of the constraints is not prescribable and in this aspect lays the incompleteness.

5.10. Incompleteness and Freedom

We only mention how the concept of freedom refers to the possibility of building new degrees of freedom rather than the extension, both in number and for allowable values, of those already available.

6. Theoretical Incompleteness

The concept of theoretical incompleteness is typically studied in mathematics, as mentioned in Sections 4 and 5.7. In general, the concept of theoretical incompleteness refers to incompleteness in principle, intrinsically, as a possible property possessed by systems, theories, phenomena, or processes, and their properties themselves. Regarding systems, this corresponds, for example, to the concept of *logical openness*.

6.1. The Case of Logical Openness

We recall that closed systems refer to the fact of being isolated, without exchange of neither matter nor energy with the outside world and reaching their final state uniquely determined by the initial conditions. Examples are ideally thermally-insulated machines.

The concept of the *logically closed model* is suitable to describe the evolution of such *thermodynamically closed systems*.

More precisely, we define a model as logically closed when:

1. a full, formal description of the relations between the state variables of the model is available;
2. a complete and explicit, i.e., analytically describable, description of the interaction between the system and its environment is available; and
3. knowledge from the previous two points allows deduction of all possible states which the system can take together with its structural characteristics.

For example, a simple temperature control system with a thermostat can be considered logically closed and complete since the influence of the environment is reduced to its possible variations of temperature to which the thermostat reacts.

A system is intended as *logically open* when there is violation of the three points above (see Table 1). *Logical openness* [10,87,88,98], ([99], pp. 111–112) can be considered as the non-depleting or infinite number of degrees of freedom when the system includes the environment (in principle independent), thus making the system incomplete as regards the environment and its influence. In complex systems, as introduced in Sections 1 and 2, the degrees of freedom (system variables) are not only in an imprecise and variable number, but can be continuously acquired. In complex systems processes of emergence occur for which the system acquires n-sequences of new non-equivalent properties to be dealt with by adopting n-different levels of description to be treated with corresponding n-different models and combinations (as with DYSAM, introduced below). The *n*-levels of modeling are based on *n*-levels of representation by:

- constructively creating suitable possible correspondences between levels,
- adopting a strategy to *move* between levels, and
- considering, simultaneously, more than one level.

The incompleteness of *logical openness* is given both by usage of a variable number of models and by the indefiniteness of *n* due to the fact that models and representations are abductively and constructively generated by the observer [99].

Table 1. A schematic comparison between aspects of logically closed systems and logically open systems.

Logically Closed Systems	Logically Open Systems
Passive	Active
Insensitive to the context	Context-sensitive
Do not learn	Learn
Object-oriented	Process-oriented
Not flexible	Flexible
Do not change the rules, at most the parameters	Change the rules
Avoid contradictions	They use the contradictions
Operate on the basis of mono-strategies	Multi-use strategies, such as DYSAM
Deductive	Deductive, inductive and abductive
Objectivist conceptual framework	Use of objectivism and constructivism
Observer considered external, generator of relativism	Observer is an integral part of the system and generator of its cognitive existence

It is possible to consider *logical openness* for autonomous systems provided with cognitive systems which process input and decide case by case, each in a possibly different way from the other (a) the behavior to be taken; (b) the degrees of freedom in unknown number, variable and continually acquired. This is the case for human behavior where assumptions of rationality and optimization are simplistic and inadequate (for example, to model the behavior of market customers by adopting decisions based on optimizing the trade-off between price and quality) as is the assumption of complete respect for procedures or repeatability over time.

In the case of *logical openness* for systems with cognitive systems, one may consider the following examples of levels of logical openness.

1. A first level can be considered as being given by thermodynamic openness to which we have referred, for which matter and energy are able to cross the boundaries of the system. For example, the system is able to send and receive signals, but nothing is said about their processing or attribution or the processing of their meaning. This is the case for two computers physically exchanging "strings of bits" and for the moves of two opponents who are playing against each other.

2. A second level can be considered where received and transmitted signals are processed in the hypothesis of an absolute semantics, predetermined, and equal for all. In this case there is the assumption that the meaning of the messages is identical and constant between transmitter and receiver, as with the formal language of operating systems for computers: in the interactions with the user it is the latter which must adapt and understand the pre-established meanings.

3. A third level can be considered where one system generates a model of the other, and communication takes place between the respective models. For example, two systems exchange messages whose meanings are constructed by using each other's models. The problem of *user modeling* in computer science involves these issues. Moreover, mutual modeling takes place through learning activities and it is refined over time, such as relations between teacher and pupil, or between companions in a sports team and within families.

4. A fourth level can be considered where a system in the communication process sends not only the message but also the context (as an extension, a completion, of the message itself) in which it takes the meaning which the sender wants to transmit, providing the receiver with the ability

to generate the context, inducing it (for example the relationship between two or more subjects during a negotiation). In this process, the use of examples and redundancy can be expected precisely to induce the generation of a certain meaning of that message.

5. A fifth level can be considered where a system can use the previous levels of openness and decide how to act: either as a closed system or an open one at various levels. The ability to decide on the level of openness can be understood as the maximum expression of openness. For example, during a conversation, the system can decide to refuse to understand or pay attention, or be active interacting with interlocutors; sending examples even without the guarantee they will be used; or sending confusing, ambiguous examples, on purpose.

Logical closure is complete by definition (see above). Logical openness is incomplete on principle, as negation of logical closure.

6.2. The Case of the DYnamic uSAge of Models (DYSAM)

Logical openness has resulted in approaches such as the use of Dynamic Usage of Models (DYSAM) [99] (pp. 64–75), based on established approaches in the literature, such as ensemble learning [100,101] and Evolutionary Game Theory [102].

From an objectivist conception the dynamics consist of the temporal representation through a model which is expected to be the best existing one, in principle, and which should only be implemented with an appropriate research activity. We distinguish here, however, between dynamic models having time in their equations and the dynamics of models as methods and strategies to use more models over time. This is to cope with the multitude of properties acquired by complex systems over time.

Examples are given by considering and using non-equivalent models simultaneously as the biochemical and psychological aspects in medicine; the economic, cultural, and religious aspect in sociology; in childhood, where a child uses the five senses, not with the purpose of choosing the one which feels best, but to learn how to use them simultaneously in a coherent manner; in physics, to decide when it is more appropriate to use classical and quantum models.

However, the problem is not only to model dynamic systems which change their evolution over time, but complex systems which change themselves by acquiring new properties. The dynamics refer not only to how the same system changes parameters, but also to how the system acquires several aspects simultaneously and non-equivalent properties (not linearly convertible one to another), and continually transforms itself coherently, requiring models based on *logical openness*. This is the case, for example, for Multiple Systems and Collective Beings [99] (pp. 97–134) constituted by the same elements interacting in different ways to accommodate the acquisition of various diverse properties.

In Multiple Systems the same component elements (a) play interchangeable, multiple and overlapping roles (that is, an action, a position, the value of a variable have different meanings depending on the respective systems they establish); and (b) simultaneously or sequentially interact in various ways, dynamically constituting sequences of different systems. Examples of Multiple Systems are given, for example, by co-operative and multiple roles of programs and nodes of the Internet, in electrical networks, by the values of sensors constituting, at one and the same time, safety and regulatory systems.

Multiple Systems are called Collective Beings when the constituent elements are autonomous, possessing cognitive systems which allow them to decide their behavior and mode of interaction. Examples of Collective Beings are given, for example, by people and families who can simultaneously form corporate systems, traffic, markets, and users of telephone networks.

DYSAM refers to the logically open use of multiple strategies and approaches not to solve, but to induce, for instance, by using redundancies to confirm the same meaning of the intervention, but in multiple coherent ways.

This can be done, for example, by acting on the environment, inserting suitable perturbations, making energetic variations, acting on external relations, acting on communication processes, learning by interacting with other systems, and by inserting *attractors of meaning* with repetitions and reformulations. Thus, perturbations are processed by the system to orient it and not to decide its destiny.

DYSAM makes sense in cases of incompleteness; otherwise there would be instructions for use.

6.3. Uncertainty Principles

Theoretical incompleteness is closely related to the uncertainty principles in physics for which, in certain phenomena, the search for increasing accuracy in knowing the value of one variable correspondingly involves reduction in knowing the value taken by another. It is the measurement of *homologous components*, such as position and momentum (the product of the mass of an object and its speed). This is the well-known Uncertainty Principle [103] introduced in 1927 by Werner Heisenberg (1901–1976). One must also consider the related Complementarity Principle introduced by Neils Bohr (1885–1962) in 1928 [104] for which the corpuscular and wave aspects of a physical phenomenon will never occur simultaneously. The experimental observation of one prevents observation of the other.

Finally, from a generic point of view, the property of incompleteness can be general or refer to a specific property or several properties; have different levels; mean not yet complete; mean uncompletable or chronically uncompletable; and theoretically non-completable as for procedures (see Section 5.1).

7. Handling Incompleteness

All of this should help us to recognize incompleteness situations against which approaches based on negative extensions, or iterations, or combinations of those used for completeness are inadequate, both theoretically and practically.

As outlined above, when it is not about abstract problems, as in mathematics, incompleteness matches complexity as a property. This is true across disciplines.

In general, the issue concerns the inadequacy of dealing with problems and complex phenomena using classical General Systems Theory, as mentioned in the introduction. An example of such inadequacy is given by the management of properties of the post-industrial society (knowledge society) using criteria and approaches typical of the industrial society instead of using the principles and approaches of post-Bertalanffy General Systems Theory.

In particular, as outlined above, multiplicity, non-equivalence, concurrent and dynamic contextualities, cognitive skills, such as abduction, coherence, and the ability to induce, have to be used. Learning abilities are not considered for obtaining repeatability or the equilibrium of situations, but for selecting and creating strategies, models, and approaches in the face of unpredictability and non-repeatability of complexity. Examples include *logical openness* and DYSAM.

In social systems, this regards their formation, emergence, and management processes. This requires modeling based on principles beyond classical ones moving, for example, from stratified differentiation to functional differentiation [105,106], which leads to ineffectiveness when considering roles and structures to be replaced with other functionals. One has to rethink the concept of management and of value, for example, with respect to the independence of economics and finance. The management of companies modeled as autopoietic, emerging, distributed, and incomplete has nothing to do with that of companies considered as structurally stable, in which the non-emergent aspect is considered to be absolutely dominant.

In social systems, it also involves their processes of education previously considered as dispensing knowledge, divided into disciplines and focusing on events and places (school terms and places like schools) when the environment was stable, as was the role of education. In the knowledge society, education processes are multiple, interdisciplinary, non-explicit, non-complete, non-unique, and non-equivalent, even with different and dynamic coherences (a typical situation of *logical openness*).

8. Conclusions

The transition from the logic of certainty to the logic of uncertainty is a matter of transition from research with the assumption of completeness to that of incompleteness; from equilibrium and regulation to coherence; from the search for the solution to the inducement of the acquisition of properties, as considered by post-Bertalanffy General Systems Theory. Incompleteness is not formulated as a possibly temporary theoretical limit, but as the intrinsic foundations of *logical openness*.

We must face the balance between completeness and incompleteness, as limit and opportunity.

It is not simply to *replace* previously-accepted concepts and approaches relating to completeness with others relating to incompleteness, but to use *both of them* properly, depending on the context, and also simultaneously using a DYSAM-like approach. In the first case, related to replacing completeness, there seems to be a reassuring relativism, able to make algorithmic processes of choice for the best approach. In the second case, however, the possible simultaneous use of approaches of different natures is dynamic, braided with incompleteness and completeness, suitable for complexity. Dealing with complexity, the usage of algorithms, removing responsibilities and always tending to guarantee the optimal objective choice, is inconceivable. Incompleteness represents this situation by making explicit the responsibility of choosing between various currently *equivalent possibilities*, with particular regard to various kinds of problems, such as social, medical and scientific for which choices are irreducible to optimizations. The focus on incompleteness not only makes explicit the impossibility to choose algorithmically, it also opens the way for various cultural reflections of a different nature, such as the role of *unintended effects* [107] in social emergence [108] of social dynamics [109]. Such reflections should be useful to move from considering complexity as "inaccurate machinery", not fully understood (but this is precisely the perspective), to carry out research, to represent the multiplicity of incompleteness and dynamic coherence as an irreducible conceptual space (with its own properties), for which it is theoretically impossible to zip them into complete, analytical representations.

Conflicts of Interest: The authors declare no conflict of interest.

References

1. Carrier, M. *The Completeness of Scientific Theories*; Kluwer Academic Publisher: Dordrecht, The Netherlands, 1994.
2. Godehard, L. *One Hundred Years of Russell's Paradox: Mathematics, Logic, Philosophy*; Gruyter: Berlin, Germany, 2004.
3. Gödel, K. *On Formally Undecidable Propositions of Principia Mathematica and Related Systems*; Dover Publications Inc.: Mineola, NY, USA, 1962.
4. Castellani, B.; Hafferty, F.W. *Sociology and Complexity Science: A New Field of Inquiry*; Springer-Verlag: Berlin/Heidelberg, Germany, 2009.
5. Estrada, E. *The Structure of Complex Networks: Theory and Applications*; Oxford University Press: Oxford, UK, 2016.
6. Rouse, W.B. *Modeling and Visualization of Complex Systems and Enterprises*; Wiley: Hoboken, NJ, USA, 2015.
7. Hooker, C. (Ed.) *Philosophy of Complex Systems*; Elsevier: Oxford, UK, 2011.
8. Bellomo, N.; Ajmone Marsan, G.; Tosin, A. *Complex Systems and Society: Modeling and Simulation*; Springer: New York, NY, USA, 2013.
9. Barabási, A.L. *Linked: The New Science of Networks*; Perseus Publishing: Cambridge, MA, USA, 2002.
10. Licata, I. Logical openness in cognitive models. *Epistemologia* **2008**, *31*, 177–191.
11. Minati, G.; Abram, M.; Pessa, E. (Eds.) *Towards a Post-Bertalanffy Systemics*; Springer: New York, NY, USA, 2016.
12. Minati, G. General System(s) Theory 2.0: A brief outline. In *Towards a Post-Bertalanffy Systemics, Proceedings of the Sixth National Conference of the Italian Systems Society, Rome, Italy, 21–22 November 2014*; Minati, G., Abram, M., Pessa, E., Eds.; Springer: New York, NY, USA, 2016; pp. 211–219.
13. Minati, G.; Pessa, E. Special Issue on Second Generation General System Theory. *Systems* **2014**. Available online: http://www.mdpi.com/journal/systems/special_issues/second-generationgeneral-system-theory (accessed on 13 July 2016).
14. Von Bertalanffy, L. *General System Theory: Foundations, Development, Applications*; George Braziller: New York, NY, USA, 1968.

15. Kumar, K. *From Post-Industrial to Post-Modern Society: New Theories of the Contemporary World*; Blackwell Publishers: Oxford, UK, 2004.
16. Haunss, S. *Conflicts in the Knowledge Society*; Cambridge University Press: Cambridge, UK, 2015.
17. Minati, G. Some new theoretical issues in Systems Thinking relevant for modelling corporate learning. *Learn. Organ.* **2007**, *14*, 480–488.
18. Minati, G. Knowledge to manage the Knowledge Society. *Learn. Organ.* **2012**, *19*, 352–370. [CrossRef]
19. Minati, G. Special Issue: Knowledge to manage the Knowledge Society. *Learn. Organ.* **2012**, *19*, 296–382. [CrossRef]
20. Boccara, N. *Modeling Complex Systems*; Springer: New York, NY, USA, 2010.
21. Altshuler, B.L., Lee, P.A., Webb, R.A. (Eds.) *Mesoscopic Phenomena in Solids*; North Holland: Amsterdam, The Netherlands, 1991.
22. Haken, H. *Information and Self-Organization. A macroscopic approach to complex systems*; Springer: Berlin, Germany, 1988.
23. Imry, Y. Physics of Mesoscopic Systems. In *Directions in Condensed Matter Physics*; Grinstein, G., Mazenko, G., Eds.; World Scientific: Singapore, 1986; pp. 101–163.
24. Liljenstrom, H., Svedin, U. (Eds.) *Micro, Meso, Macro: Addressing Complex Systems Couplings*; World Scientific: Singapore, 2005.
25. Laughlin, R.B.; Pines, D.; Schmalian, J.; Stojkovic, B.P.; Wolynes, P. The Middle Way. *PNAS* **2000**, *97*, 32–37. [CrossRef] [PubMed]
26. Goldstein, J. Emergence as a Construct: History and Issues. *Emergence* **1999**, *1*, 49–72. [CrossRef]
27. Mikhailov, A.S.; Calenbuhr, V. *From Cells to Societies. Models of Complex Coherent Actions*; Springer: Berlin, Germany, 2002.
28. Boccaletti, S. *The Synchronized Dynamics of Complex Systems*; Elsevier: Oxford, UK, 2008.
29. Licata, I. Living with Radical Uncertainty. The Exemplary case of Folding Protein. In *Crossing in Complexity: Interdisciplinary Application of Physics in Biological and Social Systems*; Licata, I., Sakaji, A., Eds.; Nova Science: New York, NY, USA, 2010; pp. 1–9.
30. Zadeh, L.A., Klir, G.J., Yuan, B. (Eds.) *Fuzzy Sets, Fuzzy Logic, and Fuzzy Systems: Selected Papers by Lotfi A. Zadeh*; World Scientific: Singapore, 1996.
31. Pessa, E. Phase Transitions in Biological Matter. In *Physics of Emergence and Organization*; Licata, I., Sakaji, A., Eds.; World Scientific: Singapore, 2008; pp. 165–228.
32. Nishimori, H. *Elements of Phase Transitions and Critical Phenomena*; Oxford University Press: New York, NY, USA, 2015.
33. Nagaev, R.F. *Dynamics of Synchronising Systems*; Springer-Verlag: Berlin, Germany, 2002.
34. Ashby, W.R. Principles of the Self-Organizing Dynamic System. *J. Gen. Psychol.* **1947**, *37*, 125–128. [CrossRef] [PubMed]
35. De Wolf, T.; Holvoet, T. Emergence Versus Self Organisation: Different Concepts but Promising when Combined. In *Engineering Self-Organising Systems: Methodologies and Applications*; Brueckner, S.A., Di Marzo Serugendo, G., Karageorgos, A., Eds.; Springer: New York, NY, USA, 2005; pp. 1–15.
36. Boccaletti, S.; Kurths, J.; Osipov, G.; Valladares, D.L.; Zhouc, C.S. The synchronization of chaotic systems. *Phys. Rep.* **2002**, *366*, 1–98. [CrossRef]
37. Buck, J.; Buck, E. Biology of synchronous flashing of fireflies. *Nature* **1966**, *211*, 562–564. [CrossRef]
38. Baas, N.A. Emergence, Hierarchies, and Hyperstructures. In *Alife III, Santa Fe Studies in the Sciences of Complexity*; Langton, C.G., Ed.; Addison-Wesley: Redwood City, CA, USA, 1994; pp. 515–537.
39. Ryan, A.J. Emergence is Coupled to Scope, not Level. *Complexity* **2006**, *67*, 67–77. [CrossRef]
40. Emmeche, C.; Koppe, S.; Stjernfelt, F. Explaining Emergence: Towards an Ontology of Levels. *J. Gen. Philos. Sci.* **1997**, *28*, 83–119. [CrossRef]
41. Minati, G.; Licata, I. Meta-Structural properties in Collective Behaviours. *Int. J. Gen. Syst.* **2012**, *41*, 289–311. [CrossRef]
42. Minati, G.; Licata, I. Emergence as Mesoscopic Coherence. *Systems* **2013**, *1*, 50–65. [CrossRef]
43. Nicolis, G.; Prigogine, I. *Self-Organization in Nonequilibrium Systems: From Dissipative Structures to Order through Fluctuations*; Wiley: New York, NY, USA, 1977.
44. Prigogine, I. *From Being to Becoming: Time and Complexity in the Physical Sciences*; W.H. Freeman & Co.: New York, NY, USA, 1981.

45. Stanley, H.E.; Amaral, L.A.N.; Gopikrishnan, P.; Ivanov, P.C.; Keitt, T.H.; Plerou, V. Scale invariance and universality: Organizing principles in complex systems. *Phys. A Stat. Mech. Its Appl.* **2000**, *281*, 60–68. [CrossRef]
46. Schroeder, M. *Fractals, Chaos, Power Laws: Minutes from an Infinite Paradise*; Dover Publications Inc.: New York, NY, USA, 2009.
47. Bunde, A.; Havlin, S. *Fractals and Disordered Systems*; Springer: New York, NY, USA, 2012.
48. Ruelle, D. *Chaotic Evolution and Attractors*; Cambridge University Press: Cambridge, UK, 2008.
49. Grassberger, P.; Procaccia, I. Measuring the Strangeness of Strange Attractors. *Phys. D* **1983**, *9*, 189–208. [CrossRef]
50. Liu, C.; Liu, T.; Liu, L.; Liu, K. A new chaotic attractor. *Chaos Solitons Fractals* **2004**, *22*, 1031–1038. [CrossRef]
51. Giuliani, A.; Zbilut, J. *The Latent Order of Complexity*; NovaScience: New York, NY, USA, 2008.
52. Lorenz, E. Deterministic Non Period Flow. *J. Atmos. Sci.* **1963**, *20*, 130–141. [CrossRef]
53. Sparrow, C. *The Lorenz Equations: Bifurcations, Chaos, and Strange Attractors*; Springer-Verlag: New York, NY, USA, 2013.
54. Lewis, T.G. *Network Science: Theory and Applications*; Wiley: Hoboken, NJ, USA, 2009.
55. Motter, A.E.; Albert, R. Networks in motion. *Phys. Today* **2012**, *65*, 43–48. [CrossRef]
56. Valente, T.W. Network Interventions. *Science* **2012**, *337*, 49–53. [CrossRef] [PubMed]
57. Dorogovtsev, S.N.; Goltsev, A.V.; Mendes, J.F.F. Critical phenomena in complex networks. *Rev. Mod. Phys.* **2008**, *80*, 1275–1335. [CrossRef]
58. Baker, A. Complexity, Networks, and Non-Uniqueness. *Found. Sci.* **2013**, *18*, 687–705. [CrossRef]
59. Newman, M. *Networks: An Introduction*; Oxford University Press: New York, NY, USA, 2010.
60. Cohen, R.; Havlin, S. *Complex Networks: Structure, Robustness and Function*; Cambridge University Press: Cambridge, UK, 2010.
61. Giuliani, A. Networks as a Privileged Way to Develop Mesoscopic Level Approaches in Systems Biology. *Systems* **2014**, *2*, 237–242. [CrossRef]
62. Kohestani, H.; Giuliani, A. Organization principles of biological networks: An explorative study. *Biosystems* **2016**, *141*, 31–39. [CrossRef] [PubMed]
63. Kadushin, C. *Understanding Social Networks: Theories, Concepts, and Findings*; Oxford University Press: New York, NY, USA, 2011.
64. Missaoui, R.; Sarr, I. *Social Network Analysis—Community Detection and Evolution*; Springer: New York, NY, USA, 2015.
65. Johnson, S. *Emergence: The Connected Lives of Ants, Brains, Cities and Software*; Touchstone: New York, NY, USA, 2002.
66. Kauffman, S. *At Home in the Universe: The Search for the Laws of Self-Organization and Complexity*; Oxford University Press: New York, NY, USA, 1993.
67. Gros, C. *Complex and Adaptive Dynamical Systems*; Springer: New York, NY, USA, 2013.
68. Hong, H.; Park, H.; Choi, M.Y. Collective synchronization in spatially extended systems of coupled oscillators with random frequencies. *Phys. Rev. E* **2005**, *72*, 036217. [CrossRef] [PubMed]
69. Turing, A. On computable numbers, with an application to the Entscheidungs problem. *Proc. Lond. Math. Soc.* **1936**, *42*, 230–265.
70. Magnani, L. *Abduction, Reason and Science: Processes of Discovery and Explanation*; Springer: New York, NY, USA, 2001.
71. Von Foerster, H. Cybernetics of Cybernetics. In *Communication and Control in Society*; Krippendorff, K., Ed.; Gordon and Breach: New York, NY, USA, 1979; pp. 5–8.
72. Hacking, I. *The Emergence of Probability: A Philosophical Study of Early Ideas about Probability Induction and Statistical Inference*; Cambridge University Press: Cambridge, UK, 2006.
73. Liu, B. *Uncertainty Theory*; Springer-Verlag: Berlin, Germany, 2014.
74. Gash, H. Constructing constructivism. *Constr. Found.* **2014**, *9*, 302–327.
75. Gergen, K.J. *An Invitation to Social Construction*; Sage Publications: London, UK, 2015.
76. Gash, H. Systems and Beliefs. *Found. Sci.* **2016**, *21*, 177–187. [CrossRef]
77. Butts, R.; Brown, J. (Eds.) *Constructivism and Science*; Kluwer: Dordrecht, The Netherlands, 1989.
78. Segal, L. *The Dream of Reality: Heinz Von Foerster's Constructivism*; Springer-Verlag: New York, NY, USA, 2013.
79. Von Glasersfeld, E. *Radical Constructivism: A Way of Knowing and Learning*; Falmer Press: London, UK, 1995.
80. Galavotti, M.C. (Ed.) *Bruno de Finetti Radical Probabilist*; College Publications: London, UK, 2008.

81. Friedenberg, J.D.; Silverman, G.W. *Cognitive Science: An Introduction to the Study of Mind*; SAGE Publications Inc.: Los Angeles, CA, USA, 2015.
82. Cecconi, F. (Ed.) *New Frontiers in the Study of Social Phenomena: Cognition, Complexity, Adaptation*; Springer: New York, NY, USA, 2016.
83. Nescolarde-Selva, J.A.; Usó-Doménech, J.-L. Gash A theorical point of view of reality, perception, and language. *Complexity* **2014**, *20*, 27–37. [CrossRef]
84. Usó-Doménech, J.L.; Nescolarde-Selva, J.; Gash, H. Guest editorial: Belief systems and science. *Cybern. Syst.* **2015**, *46*, 379–389. [CrossRef]
85. Lee, M.D. *Bayesian Cognitive Modeling*; Cambridge University Press: Cambridge, UK, 2014.
86. De Finetti, B. *Theory of Probability—A Critical Introductory Treatment*; John Wiley & Sons: London, UK, 1975.
87. Licata, I. Seeing by models: Vision as adaptive epistemology. In *Methods, Models, Simulations and Approaches towards a General Theory of Change*; Minati, G., Abram, M., Pessa, E., Eds.; World Scientific: Singapore, 2012; pp. 385–400.
88. Minati, G.; Penna, M.P.; Pessa, E. Thermodynamic and Logical Openness in General Systems. *Syst. Res. Behav. Sci.* **1998**, *15*, 131–145. [CrossRef]
89. Manrubia, S.C.; Mikhailov, A.S. *Emergence of Dynamical Order: Synchronization Phenomena in Complex Systems*; World Scientific: Singapore, 2004.
90. Pikovsky, A.; Rosenblum, M.; Kurths, J. *Synchronization: A Universal Concept in Nonlinear Sciences*; Cambridge University Press: Cambridge, UK, 2001.
91. Bonometti, P. Improving safety, quality and efficiency through the management of emerging processes: The TenarisDalmine experience. *Learn. Organ.* **2012**, *19*, 299–310.
92. Mac Lennan, B.J. Natural computation and non-Turing models of computation. *Theor. Comput. Sci.* **2004**, *317*, 115–145. [CrossRef]
93. Bader, M. *Space-Filling Curves: An Introduction with Applications in Scientific Computing*; Springer-Verlag: Berlin, Germany, 2013.
94. Tyson, J.J. *The Belousov-Zhabotinskii Reaction*; Springer: Berlin, Germany, 1976.
95. Kinoshita, S. (Ed.) *Pattern Formations and Oscillatory Phenomena &Belousov-Zhabotinsky Reaction*; Elsevier: Amsterdam, The Netherlands, 2013.
96. Getling, A.V. *Rayleigh-Bénard Convection: Structures and Dynamics*; World Scientific: Singapore, 1998.
97. Janot, C. *Quasicrystals: A Primer*; Oxford University Press: Oxford, UK, 2012.
98. Licata, I.; Minati, G. Creativity as Cognitive design-The case of mesoscopic variables in Meta-Structures. In *Creativity: Fostering, Measuring and Contexts*; Corrigan, A.M., Ed.; Nova Publishers: New York, NY, USA, 2010.
99. Minati, G.; Pessa, E. *Collective Beings*; Springer: New York, NY, USA, 2006.
100. Hinton, G.E.; Van Camp, D. Keeping neural networks simple by minimizing the description length of the weights. In *Proceedings of the Sixth Annual Conference on Computational Learning Theory*; Pitt, L., Ed.; ACM Press: New York, NY, USA, 1993; pp. 5–13.
101. Zhou, Z.-H. *Ensemble Methods: Foundations and Algorithms*; CRC Press: Boca Raton, FL, USA, 2012.
102. Vincent, T.L. *Evolutionary Game Theory, Natural Selection, and Darwinian Dynamics*; Cambridge University Press: Cambridge, UK, 2012.
103. Heisenberg, W. *Physics and Beyond*; Harper & Row: New York, NY, USA, 1971.
104. Bohr, N. The Quantum Postulate and the Recent Development of Atomic Theory. *Nature* **1928**, *121*, 580590. [CrossRef]
105. Luhmann, N. How Can the Mind Participate in Communication? In *Theories of Distinction: Redescribing the Descriptions of Modernity*; Rasch, W., Ed.; Stanford University Press: Stanford, CA, USA, 2002; pp. 169–184.
106. Luhmann, N. Limits of Steering in Theory. *Cult. Soc.* **1997**, *14*, 41–57. [CrossRef]
107. Usó-Doménech, J.L.; Nescolarde-Selva, J.; Lloret-Climent, M. "Unintended effects": A theorem for complex systems. *Complexity* **2015**, *21*, 342–354. [CrossRef]
108. Sawyer, R.K. *Social Emergence: Societies as Complex Systems*; Cambridge University Press: Cambridge, UK, 2005.
109. Skyrms, B. *Social Dynamics*; Oxford University Press: Oxford, UK, 2014.

systems

MDPI

Article

System-of-Systems Design Thinking on Behavior

Christian Stary

Department of Business Information Systems, Communications Engineering,
Johannes Kepler University of Linz, 4040 Linz, Austria; Christian.Stary@jku.at;
Tel.: +43-732-2468-4320

Academic Editors: Gianfranco Minati, Eliano Pessa and Ignazio Licata
Received: 31 October 2016; Accepted: 6 January 2017; Published: 13 January 2017

Abstract: Due to the increasing digitalization of all societal systems, informed design of services and systems becomes pertinent for various stakeholders. This paper discusses the design of digital systems in a user-centered way with the help of subject-oriented design. The approach follows a communication-driven and network-centric perspective on a System-of-Systems, whereby system specifications encapsulate behavior and exchange messages, including relevant data, such as business objects. Systems can represent activities of human actors, as well as artefacts. Stakeholders can be actively involved in their roles in the design of a System-of-Systems. In the course of design, they identify and refine role-specific behavior, based on communication to other actors or systems. A System-of-Systems specification evolves as a network of cooperating behavior entities. It develops according to communication needs and system-specific capabilities, on the level of synchronized execution agents, or as an overlay mechanism on existing applications or sub networks. Since certain behavior sequences, such as decision-making procedures, are re-occurring in organizations or eco-systems, the design of complex systems can be facilitated by behavior patterns stemming from existing modeling experiences.

Keywords: System of Systems; design; subject-orientation; communication structures; collaboration network; interaction; diagrammatic specification; design patterns; choreography

1. Introduction

Digitalization means the continuous penetration of IT applications into domains and areas of economic and lifestyle concern. A typical example are calendars. Once becoming digital they have been connected to other systems, like room and meeting management systems, even though there exist diverse personal ecologies of calendar artifacts. As Dittmar et al. [1] could show "the changing demands in daily life, the availability of new tools, and the participants' knowledge about the costs and benefits of their calendar work and about the consequences of potential failures influence their tendency to explore and possibly integrate new calendar artifacts and appear implicated in the deliberate non-use of new technology". Such findings indicate that design of these artefacts need to be reconsidered for the sake of their applicability and social acceptance once being embodied in digital infrastructures.

One opportunity of digital artifacts to put potential users in control of design is their capability to be increasingly editable, interactive, reprogrammable, and distributable [2]. This capability goes beyond adding or visually arranging apps on a tablet or smartphone interfaces. Rather, it refers to creating interactive digital environments involving various stakeholders, such as developers, consumers, facilitators, brokers, etc. (cf. [3]). When looking for methodological support of stakeholder involvement in system design, first inputs in terms of theories of design pop up (cf. [4]), while structured guidance seems still to be lacking [5]. Hence, more attention needs to be drawn in investigating, creating the design of digital systems in a stakeholder-sensitive way. It needs to go beyond user involvement as commonly pursued in software development, as already existing services or applications need to be configured in a context-sensitive and adaptive way. In addition, different stakeholders may be involved, as they may need to be represented by interaction or system features, e.g., when specifying individualized alarm systems in healthcare (in the context of this paper, stakeholders denote persons having interest or share in an active community, e.g., a societal sector or organization. They are involved in, or affected by, some course of action of this community).

Co-creation of services and products is not only of relevance in consumer settings, but also in industrial settings (cf. [6–8]). For instance, Tang et al. [7] address the co-creation of digital services and applications in the Web 2.0 digital ecosystem where companies can co-create business with their customers. However, the focus here is in combining product and technological capabilities, rather than the process and method of co-creation. In a distributed product and service setting, the process of design needs to be supported in a constructive and methodologically grounded way, as digital innovations in vehicle maintenance reveal, at least following a layered architecture of digital technology [9]. Feature or service layers provide the opportunity to various stakeholders to create, manipulate, and store different systems, either as apps, functions, or services. For stakeholder-centered development execution capabilities must exist independently, but still intertwined. The latter support implementing designed artefacts and putting systems into operation [10].

According to these requirements we follow an 'assemblage of systems' approach in System-of-Systems (SoS) design. This is one of the timely identified perspectives on SoS [11]. Thereby, we put the stakeholders in control of the development process, aiming to integrate (existing) system behaviors to achieve capabilities that cannot be achieved by the constituent systems. The resulting behavior forms a collaborative network system. Methodologically, we follow Maier [12], as the definition of SoS as "a collaborative assemblage captures a grouping distinctively different in terms of developing best practices" (p. 3149). We leverage socio-technical equilibria according to the stakeholder perspective without requesting upper layers in the hierarchy, as originally being superimposed for architecting SoS (cf. [13]). However, in case a network node represents a complex system, e.g., providing access to a module of an enterprise resource planning system, it can be considered as a hierarchical overlay to existing system functions. In order to allow the execution of system models, we follow the idea of providing communicating structures as inherent part of SoS specifications (ibid.). The proposed subject-oriented SoS architecture encapsulates communication tasks with functional ones as elementary specification structure. The triggered send- and receive-patterns involve the exchange of messages and establish a choreographic flow of control in the collaborative assemblage.

The next section of this paper provides a review on the design of System-of-Systems (SoS), in order to detail the selection of SoS type according to the objective of the work. Then, a framework, corresponding methodological steps for SoS, and diagrammatic notational elements, stemming from subject-oriented business process management, are introduced. We detail behavior modeling capabilities designed for SoS stakeholders. We use those diagrammatic modeling capabilities to introduce subject-oriented SoS thinking for designing and re-designing SoS from a stakeholder perspective. Hence, both the diagrammatical manifestation of SoS design, and the automated execution of SoS model representations can be considered a step towards interactive and dynamic SoS development. The final section concludes the paper referring to these achievements.

2. System-of-Systems (SoS) Design

This section introduces System-of-Systems (SoS) as a design entity and reviews methodological inputs capturing the process of SoS design. The first subsection deals with structural foundations, the second subsection with behavior issues, including the emergence of SoS behavior.

2.1. System-of-Systems

System-of-Systems (SoS) has been conceptualized for engineering addressing the construction and development of complex artefacts [14]. Due to the variety of application domains, a variety of definitions and explanations exists (cf. [15]). However, they consider a system as "a group of interacting elements (or subsystems) having an internal structure which links them into a unified whole. The boundary of a system is to be defined, as well as the nature of the internal structure linking its elements (physical, logical, etc.). Its essential properties are autonomy, coherence, permanence, and organization" (ibid., p. 1). Complex systems are constituted "by many components interacting in a network structure", with most often physically and functionally heterogeneous components, and organized in a hierarchy of subsystems that contributes to the system function [13]. As Jaradat et al. [11] have shown, several structures and categorization schemes have been used in the history of complex systems interpreted as system-of systems, ranging from closed coupling (systems within systems) to loosely coupling (assemblage of systems). In architectural terms, these properties correspond to embodied systems cooperating in an interoperable way (cf. [16]), allowing for autonomous behavior of systems or components while being part of a network collaborating with other systems and, thus, contributing to the objective of the network [12].

Referring to structural and dynamic complexity, 'structural complexity derives from (i) heterogeneity of components across different technological domains due to increased integration among systems; and (ii) scale and dimensionality of connectivity through a large number of components (nodes) highly interconnected by dependences and interdependences. Dynamic complexity manifests through the emergence of (unexpected) system behavior in response to changes in the environmental and operational conditions of its components' ([15], p. 1). The review of the Technical Committee of the IEEE-Reliability Society concludes with considering a System-of-Systems (SoS) as a system that involves several systems "that are operated independently but have to share the same space and somehow cooperate" (ibid., p. 2). As such, they have several properties in common: operational and managerial independence, geographical distribution, emergent behavior, evolutionary development, and heterogeneity of constituent systems (ibid.). As Jaradat et al. [11] pointed out (p. 206), these properties affect setting the boundaries of SoS and the internal behavior of SoS and, thus, influences methodological SoS developments. More concrete, according to Jaradat et al. ([11], p. 206) SoS are distinct with respect to:

(1) *autonomy*, where constituent systems within the SoS can operate and function independently and the capabilities of the SoS depends on this autonomy;
(2) *belonging* (integration), which implies that the constituent systems and their parts have the option to integrate to enable SoS capabilities;
(3) *connectivity* between components and their environment;
(4) *diversity* (different perspectives and functions); and
(5) *emergence* (foreseen or unexpected).

A typical example are apps being available on a smartphone. They can be considered as systems. When adjusting them along a workflow, e.g., to raise alerts and guide a patient to the doctor, in case certain thresholds with respect to medical conditions are reached for a user, several systems, such as a blood pressure app, calendar app, and navigation app, need to be coordinated and aligned for personal healthcare. In this case, the smartphone serves as a SoS carrier, eventually supporting patient-oriented redesign of the workflow and, thus, the SoS structure. The smartphone can still be used as a device to talk to other people while serving as a communication infrastructure of medically relevant systems. The latter identifies the smartphone a SoS component.

2.2. SoS Design as an Informed Process

As SoS thinking is grounded in recognizing the network-centric and knowledge-based nature of systems [17], a realist perspective on developing them seems to be appropriate (cf. [18]). Its focus is on adaptation and flexibility and, thus, on "local context and expressing findings as broad principles of action and contingent approaches" (ibid., p. 424f.). Hence, any abstraction to describe and specify needs to be adequate to the situation of use for its stakeholders (cf. [19] for process representations). A system's situatedness is awareness about its world, such as the organization, society, or other contingent systems, and its capability to induce changes in it (cf. [20]). 'The essence of situation awareness lies in the monitoring of various entities and the relations that occur among them. Since the properties of relations, unlike the properties of objects, are not directly measurable, one needs to have some background knowledge (such as ontologies and rules) to specify how to derive the existence and meaning of particular relations' [21].

Consequently, SoS development should lead to architectures allowing dynamic changes (cf. [22]). Situatedness of behavior is a key issue in engineering support of communities (cf. [23]). Prescriptions need to be adapted to the situation at hand, allowing for systems dynamics in the course of development. Most important, we need to recognize that stakeholders and support systems, in particular when considered from a system perspective, are an integral part of situations, as termed by cognitive scientists: actors are considered as embodied and interactively situated in worlds [24]. When analyzing the meanings attached to these terms a set of conditions for situatedness and embodiment can be derived, based on the conclusive assumption that external representational schemas are required for adaptation. While virtual actors in virtual worlds are neither considered situated nor embodied, awareness of evolving goals, various modalities for interaction and task accomplishment procedures could lead to a rich repertoire of interactions. Embedded actors could develop individual points of view, relative to their starting position workspaces, and have a capacity to develop a dedicated interaction space. None of these capabilities are possible without representational capacities, such as diagrammatic or formal notations.

Hereby, system thinking plays a crucial role, as it is a way of looking at situations as an ecosystem. A situation is analyzed in terms of how the different parts influence and relate to each other rather than decomposing it into parts that are studied in isolation [11,25]. System thinking focuses on actors in a mutually dependent setting. Their concern is how they work together and respond to each other even following complex paths of behavior. Design is focused on development work of systems while keeping the whole in mind. According to Frank [26] stakeholders creating systems through system thinking need a variety of cognitive competencies. They need to "understand the whole system beyond its elements, sub-systems, assemblies and components, and recognize how each element/sub-system/assembly/component functions as part of the entire system. They are multifaceted, able to consider issues from a wide range of perspectives and points of view and possess a generalist's perspective" (p. 276). They also need to "understand the interconnections and the mutual influences and interrelations among system elements. Systems thinking involves thinking about the system's interactions, interrelationships, and interdependencies of a technical, social, socio-technical or multi-level nature" (ibid.).

In doing so, developers lay ground for emergent properties of systems, effecting perspectives beyond engineering an isolated system, e.g., a navigation app. A systemic representation, such as a SoS specification, enables one not to get stuck on details, and tolerate ambiguity and uncertainty. From a methodological perspective, the "good (the right) questions" ([26], p. 277) need to be asked, and given information needs to be questioned constantly. Moreover, these questions enable curiosity and open-mindedness, in order to consider a system beyond the limited area of a certain expertise. Without a sense of vision, novel system behavior might not emerge, neither when optimizing, nor re-engineering existing systems.

3. Subject-Oriented Design of SoS

Subject-oriented SoS design seeks to assist system development by providing a methodology that presents behavior-relevant information in a manner analogous to natural language features, namely, subject, object, and predicate constructs from the stakeholder's perspective [27]. These characteristics enable stakeholders to be engaged more effectively via a simple and intuitive behavior representation and via human-centered elicitation.

As informed design has the intention of reducing the time spent for, and complexity of, modeling, development support should also include the execution of models. Having an easy to learn, and deployable, behavior modeling tool intertwined with execution also enables other stakeholders to have a platform for probing with existing reference patterns or models when specifying their mental representation. They could directly specify their behavior, e.g., in terms of a flow of activities, to achieve a certain objective. In case behavior knowledge is not available in explicit representations so far, probing supports stakeholders expressing themselves successively in terms of executable actions or communication acts (cf. [28]). In contrast to explicit elicitation techniques, such as interviews, stakeholders need not have to rely on information provided by analysts. In settings involving external people, such as for interviewing, it cannot be assumed that analysts are familiar with the domain at hand [29]. Moreover, stakeholders—in particular, experts—forget tasks to mention they assume to be widely known, or have difficulties explaining what they do without actually doing it [30], as knowledge is inseparable from doing (cf. [31]).

Putting situated cognition theory in the context of system modeling, generated models in a natural and intuitive way should potentially have greater accuracy than what could traditionally be achieved with common acquisition and analysis techniques (cf. [32]). Reducing the requirement of involving external people enables a wider scope of application, as more stakeholders could participate in organizational change and development. Subject-oriented design is focusing on parallel processes; thus, stakeholders can detail their behavior specification individually and concurrently, after agreeing on respective communication interfaces.

An underlying concept of behavior-based SoS design is agency: according to Himma [33] the "idea of agency is conceptually associated with the idea of being capable of doing something that counts as an act or action. As a conceptual matter, X is an agent if and only if X is capable of performing action; breathing is something we do, but it does not count as an action. Typing these words is an action, and it is in virtue of my ability to do this kind of thing that, as a conceptual matter, I am an agent. ...Agents are not merely capable of performing acts; they inevitably perform them (in the relevant sense). ...The very concept of agency presupposes that agents are conscious." (p. 19)

Reflecting this understanding reveals the manner of involvement in a situation when humans are acting or interacting. It underpins the requirement to devote design effort to human-centered behavior. Moreover, it enables paradigmatic shifts towards communication and interaction, in addition to functional task accomplishment or functional role fulfilment. While traditional approaches to modeling mostly rely primarily on an exclusive functional perspective on task accomplishment, subject-oriented SoS design focuses initially on communication links between active components and, thus, interactions between systems.

3.1. Specification Constituents and Models

In subject-oriented SoS design, systems are viewed as emerging from both the interaction between systems (termed subjects) and their specific behaviors encapsulated within the individual subjects. Like in reality, subjects (systems) operate in parallel and can exchange messages asynchronously or synchronously. It is a view of SoS with operating systems as autonomous, concurrent behaviors of distributed entities. A system (subject) is a behavioral role assumed by some "actor", i.e., an entity that is capable of performing actions. The entity can be a human, a piece of software, a machine (e.g., a robot), a device (e.g., a sensor), or a combination of these. (SoS) Subjects can execute local actions that do not involve interacting with other subjects (e.g., calculating a price and storing a postal address), and communicative actions that are concerned with exchanging messages between SoS subjects, i.e., sending and receiving messages.

SoS subjects are one of five core symbols used in specifying designs. Based on these symbols, two types of diagrams can be produced to conjointly represent a system: subject interaction diagrams (SIDs) and subject behavior diagrams (SBDs). SIDs provide a global view of a SoS, comprising the subjects involved and the messages they exchange. The SID of a simple ordering process is shown in Figure 1. Subject behavior diagrams (SBDs) provide a local view of the process from the perspective of individual subjects. They include sequences of states representing local actions and communicative actions, including sending messages and receiving messages. State transitions are represented as arrows, with labels indicating the outcome of the preceding state (see Figures 2 and 3).

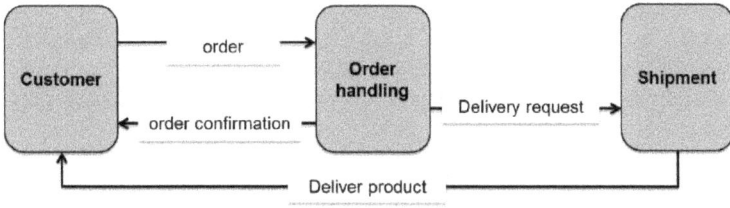

Figure 1. Order handling—subject interaction diagram.

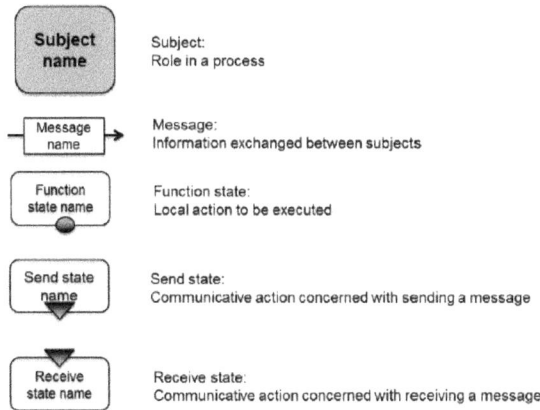

Figure 2. Diagrammatic elements in subject-oriented SoS design.

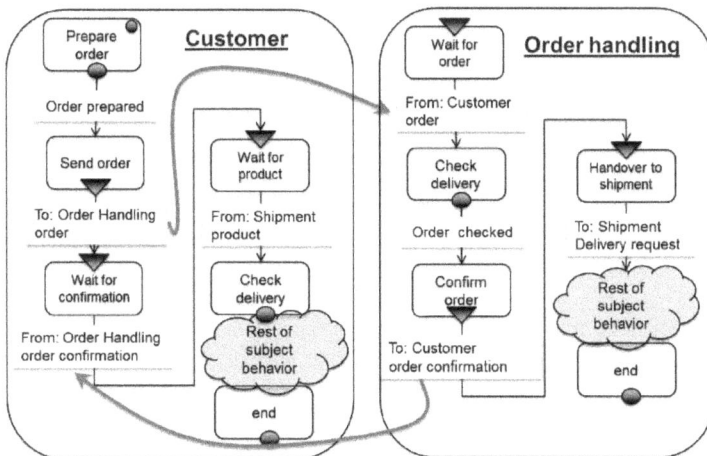

Figure 3. Order handling—Subject behavior diagrams "customer" and "order handling".

Given these capabilities SoS designs are characterized by:

- A simple communication protocol (using SIDs) and, thus,

- standardized behavior structures (enabled by SBDs), and
- scaling in terms of complexity and scope.

The approach allows meeting ad hoc, non-deterministic, and domain-specific requirements. The ultimate stage of scalability could be reached through dynamic and situation-sensitive formation of systems and their architecture beyond domains, referring to adaptability. As validated behavior specifications can be executed without further transformation (see Figure 4), stakeholders guide the implementation of their SoS specifications.

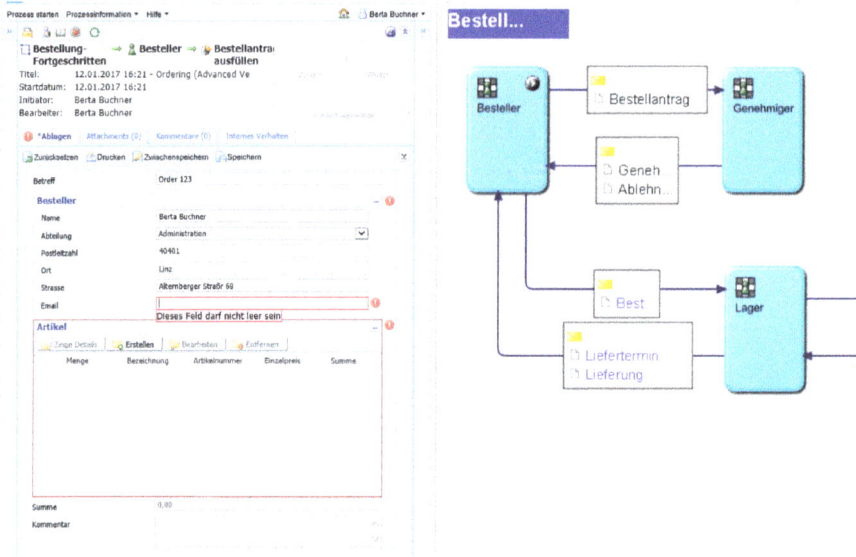

Figure 4. Execution of subject-oriented SoS behavior sequences.

3.2. Supporting Adaptive SoS Design

In design thinking workshops stakeholders work on subject-oriented models representing SoS from a cognitive, social, and organizational/domain perspective. The workshops include ad hoc models, presentations of pre-fabricated models, or user stories by stakeholders on their (work) situation, their task implementations, and their mental models of a domain. In the course of the workshops, both types of subject-oriented models, subject interaction diagrams (SIDs), and subject behavior diagrams (SBDs), are developed, in order to achieve an interactional understanding and corresponding SoS of the concerned organizational setting. Facilitators care about encapsulating system behavior and the required SoS communication structures. In addition, they also support probing SoS models through prototypical implementations. Since a group of stakeholders is usually affected by subject-oriented SoS representations—each subject can be represented by a person or system—the models need to be discussed and aligned according to the interaction patterns. In this way, the SoS models get validated. Starting point for all activities is the SoS scope (universe of discourse), e.g., getting help in case of an individual homecare emergency, handling customer orders once being processed, or revealing decision-making on challenging production processes.

As subject-orientation has been created by Business Process Management (BPM) practitioners for practitioners the approach has been applied in various (re-)design settings (cf. [34,35]). In the course of the performed case studies behavior patterns, or even reference models, evolved, sometimes guided by knowledge management methods (cf. [36]). In the tradition of the design thinking approach these behavior patterns need to undergo application and review cycles to allow learning for further refinements and developments. They are offered not as a template, but rather as an opportunity for thought simplifying reflection of experienced/ envisioned situations and a corresponding model for SoS construction. In contrast to traditional pattern development, subject-oriented patterns focus on the communication structures and, thus, likewise representation of functional and interactional behavior. In the following we report on the patterns that could be revealed so far. We continue following the order processing scenario aiming to capture dynamic customer and supplier behavior, as well as product/service development.

Figure 5 shows patterns for proactive and reactive behaviors in subject-oriented notation. Reactive behavior is based on temporal sequences of loosely coupled receive—do—send states. Proactive behavior requires asking for inputs, namely sending requests, while performing regular tasks, and awaiting response—see Figure 6 for the proactive order handling system. Utilizing such a pattern, e.g., encapsulated by a system component "ensure re-confirmation", proactive behavior along a customer or supplier relation can be modelled in an effective way. The system component can be activated in different contexts and activated at runtime, given running system components by the instantiated subject. In one case the process request can be performed by an automated decision-making component, e.g., checking the availability of goods, in another case process request requires human-computer interaction.

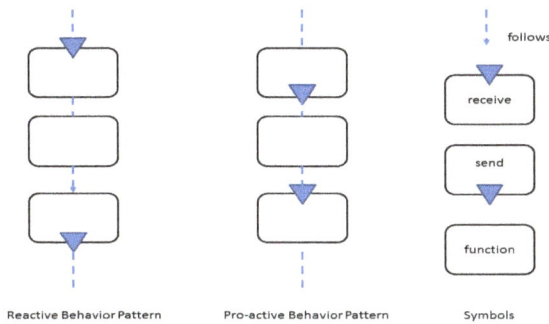

Figure 5. Subject-oriented representation of proactive and reactive behavior patterns.

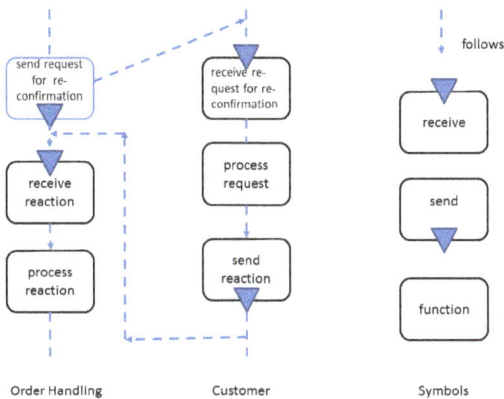

Figure 6. Proactive behavior pattern of order handling.

Of particular interest is the capability to combine pro- and reactive patterns. Consider a customer relation that is driven by incomplete orders over a longer period of time. In such cases, a proactive reminder to complete the order could be effective for business operation, while the default could be a reactive SoS, waiting for the customer to become active. In that context, further examples utilizing that pattern are payment and supplier: consider traditional late providers and customers. They could become more reliable with respect to timely delivery when the proactive SoS component is activated.

A pattern-based approach of this kind can be a first step towards representing adaptive behavior of actors in self-organizing SoS. According to Sterling et al. [37], such systems should provide several qualities:

- Purposefulness, pursuing a goal and determining paths of action accordingly;
- Controlled autonomy, as a system component is able to pursue its own goals seemingly independently; and
- Situatedness of system components, i.e., being aware of the surrounding environment, being capable of perceiving changes, and responding appropriately.

Subject-oriented models allow meeting these qualities when explicitly representing the goal-relevant part of the SoS environment as subjects:

- A system component (subject) is a behavior abstraction for a specific purpose. It can stem from a functional role, type of application, or intention to capture behavior on an abstract level;
- A subject as a system component has controlled autonomy, as it encapsulates behavior to pursue a specific goal independently (while being interrelated through communication structures with other system components); and
- A subject is situated in an environment of subjects, and, most importantly, it is aware of this environment of the other subjects as it is exchanging messages within this environment (SoS). This mechanism allows not only perceiving changes and responding appropriately, but acquiring information about behavior of other subjects of the environment.

In particular, the latter property helps to proactively collect information of relevance to change behavior. Each SBD captures all possible local states a subject can be in, namely in terms of sending, receiving, and acting. They represent the actions it can perform, as well as the interfaces to other subjects. The local protocol is given by triggering actions internally (do), or externally (send). In this way the protocol for subject interaction is defined: receiving a message triggers an internal action of the addressed subject. The inputs to subjects trigger the evolution procedure in terms of proceeding from one local state to another depending on its own activities and the actions of other subjects (i.e., incoming messages from other subjects).

For further explicating SoS intelligence, North et al. [38] propose rules based on theories of individual rational behavior. Hereby, individuals collaborate with others when it is in their best (economic) interest to do so. Decision-making should be based on simple rules, in order to let rule structures evolve. In this way, bounded rationality can be taken into account. Reasoning regarding goals is progressively refined by means of procedures accounting for the limited knowledge and abilities of individual decision-makers (ibid.). In line with modular decision-making, patterns can be introduced, as shown in Figure 7. Such structures help mapping a system component's behavior effectively to executable patterns. Since subject-oriented SoS models can be executed after validation and in a concurrent way they allow for simulating behavior in complex situations.

A typical example is the aforementioned decision making on selecting the pro- or reactive pattern. Checking the reliability of a customer paying the bill, or of a supplier delivering a part in a certain period of time, requires applying a criteria-based selection. The pattern allows multiple criteria to be checked which in turn triggers subject behavior, either through activating other subjects or immediate acting (cf. considering a customer delaying payment of a good of high value which requires payment of suppliers in due time). In this way, behavior in SoS can be controlled on multiple system properties in a structured way.

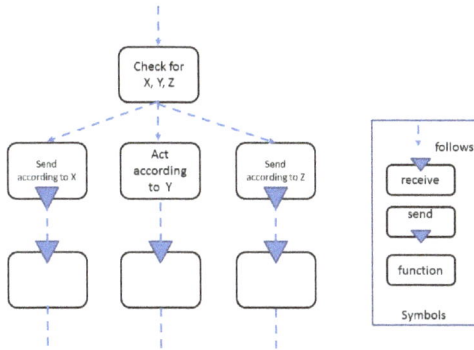

Figure 7. Decision-making patterns.

For adaptation subjects need to (i) become informed; and (ii) be selective with respect to their behavior. Following Gero et al. [39,40], the subject's "sensors" are monitor subjects that (actively) search the environment for situation-relevant data and produce direct input for an acting subject. In addition, a decision support subject can be invoked by an acting subject. It is provided with monitored information and situation-sensitive data to identify mismatches between the current and desired situation. Based on the results of the decision support process, the acting subject can decide which sequence of operations to execute.

Figure 8 provides the corresponding interaction scheme. It reveals that an <acting subject> can ask for both monitoring of the situation and decision support based on the monitored information. An <acting subject> is any subject in a situation that requires some action to accomplish a task. The monitor subject embodied in the environment either accepts requests to collect data on the current situation or does automatically receive data as a sensor, before processing and delivering the monitored data to the <acting subject>. The decision support subject is available for consultancy with respect to selecting the next action. This requires all available information on a certain situation.

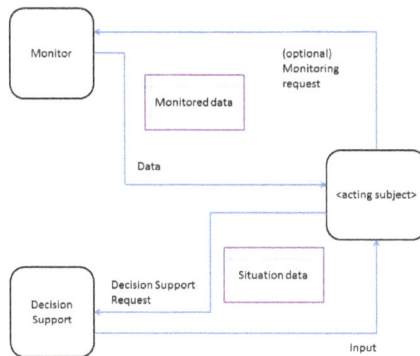

Figure 8. An <acting subject> can ask for monitoring a situation and decision support based on the monitored information.

Figure 9 demonstrates the possible impact for processing orders. Customer service asks the supply chain monitor subject for monitoring producing Part A. The monitored data of the supply chain are sent to customer service which in turn sends them together with order processing data to the decision support subject. Based on the results further handling of the order is triggered.

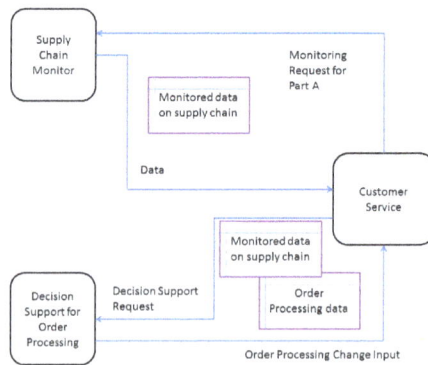

Figure 9. A customer service subject asks for monitoring the supply chain and decision support based on the monitored supply chain for changing order processing.

A monitoring subject can either receive and process environment data automatically or monitor the behavior of other subjects. Figure 10 shows both types of monitors. The environment monitor is a signal receiver processing them to deliver data, e.g., an event sensor indicating a delay in traffic due to changing weather conditions. The social monitor actively requests data from other subjects, e.g., whether certain data values have changed. Both could iterate for refining results.

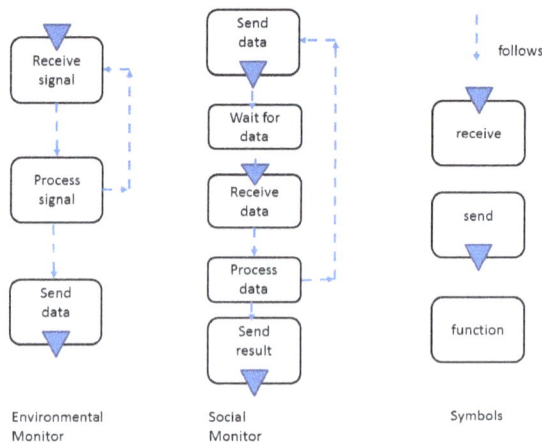

Figure 10. Push- and pull monitoring of a situation.

The communication of the <acting subject> can be conceptualized accordingly. Figure 11 details an <acting subject> releasing a request and waiting for the result (case of social monitoring). The request for monitoring can be modeled before any function state (action), thus providing for each critical function a preprocessing sequence. It considers environmental data beyond subject interaction and can also be captured in behavior descriptions. Once an <acting subject> has received monitor data it can act in response to the situation it is part of. The input data from the monitor are then processed along the subject's behavior. In case a subject acts requires decision support, it needs to activate the decision support subject.

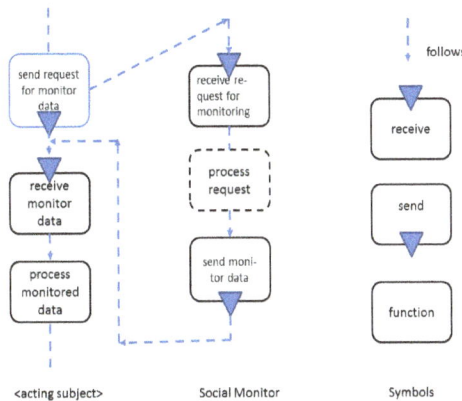

Figure 11. A <subject-in-charge> interacts with a monitor (encapsulating how the request is processed).

Referring to the running example, the supply chain is monitored by a social monitor. Once the request is completed they can be transferred to customer service for further processing, in our case asking for a decision based on order processing data.

The initial step hereby is requesting inference on situation-specific data, not only using the monitored data, but also data created or processed by the <acting subject> itself. The input data are sent to decision support as they describe the current situation of the <acting subject>. Processing the request for decision support requires a business rule engine that holds for the situation and environment (e.g., an organization) at hand. This concept can be cascaded for situation-relevant decision-making according to the scope of the SoS. The received results are processed based on available sets of rules and the situation data provided by the <acting subject>. The results are finally delivered to the <acting subject>—see Figure 12.

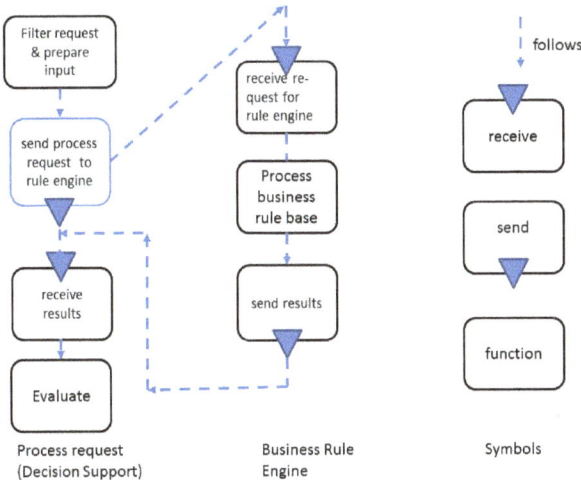

Figure 12. Rule processing.

The evaluated results are checked and a decision pattern can be triggered to complete a work task. In Figure 13 we exemplify a customer subject reflecting its order and deciding to opt for another product while ordering. The monitor subject is activated to find out whether the other product is in stock. In order to avoid frustration in case it is not in stock, an alternative product is looked up—a decision support subject with respect to consumer behavior is activated and checked whether it is available, until the customer can be satisfied or informed about a lack of further alternatives.

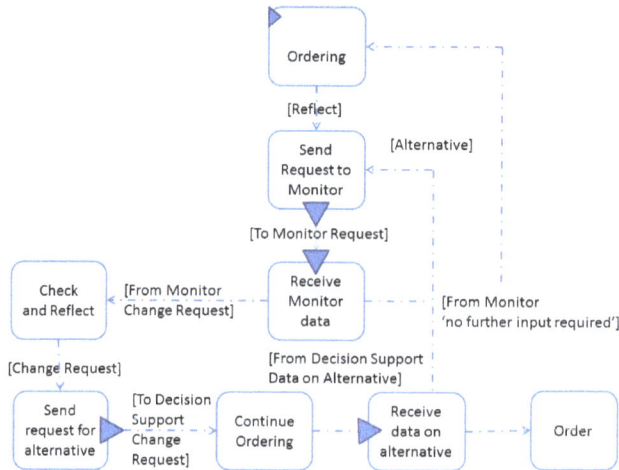

Figure 13. Case involving product monitoring and decision support.

This example shows a developed network as a SoS for a given scope. The utilized patterns allow composing a representation according to stakeholder needs and his/her understanding of a certain situation. Although it contains all relevant functions required for ordering, it can be implemented in a variety of ways, e.g., either by automated monitoring or manual observation.

Given monitoring and decision support, an acting subject (system) can perform in a reflexive way, using input data from the environment, either preprocessed from interaction behavior from other subjects or from observer components, such as sensors. Moreover, an acting subject can also activate a decision support subject. In this case, monitored data from the environment can be enriched with situation-sensitive data by the subject for further processing, e.g., according to general rules of encoding behavior. The results lead to informed decision taking of the acting subject at hand.

4. Conclusions

The penetration of digitalization into the variety of societal systems requires informed design of services and systems and involves stakeholders in the role of designers. Rather than looking for purely task or functional support, digital system design could follow a communication-driven perspective. In the presented work, system specifications encapsulate behavior of active entities (in terms of doing) and their exchange of messages (sending, receiving). A System-of-Systems specification is a network of self-contained behavior entities, developing according to communication needs and system-specific capabilities. Reoccurring behavior patterns can be used for complex system design, as they form the backbone of intelligent systems.

Once model elements, as well as contextual parameters, can be determined, a situation model can be constructed that can be used to learn from previous modeling steps to predict future activities. Such a model is also of particular use, in case reference or re-occurring behaviors are to be represented. It could guide stakeholders when describing a situation. Dai et al. [41] reveal opportunities to predict human activity and social links with greater accuracy, given human mobility data. As stakeholders move, they might be tagged or their device can be tracked location-wise. Stakeholders, when moving through their workspace, provide inputs to pattern mining and prediction opportunities according to their digital footprints, including fixing an object to represent it as a subject or business object, annotating information, such as marking a piece of work incomplete, and digital media consumption (video or audio).

This rich data can be used to analyze human behavior patterns, build networks of actors, and predict future activities of individuals using their activity space. In particular, the predictions of an individual's future activity allow more relevant recommendations to be provided for modeling individual behaviors. While previous research mainly focused on location prediction—location prediction algorithms are based on the prior knowledge of the probability distribution of the mobile velocity and/or direction of movement—recent work investigates human behavior prediction based on only locations and GPS data obtained from smart phones [41].

Such prediction frameworks allow categorically quantifying upcoming target activity frequency such as no activity, normal user activity, and high-frequency activity, or other more refined categorization based on user (or business) data, e.g., referential role behavior or preferences. Hence, not only the existence of an activity, but also frequency-related quantities, can be taken into account. Recent prediction frameworks, as Dai et al.'s, tend to utilize the more general concept of partial repetitive behavior (instead of the stronger periodicity condition), and work with landmark behaviors in terms of representative user activity temporal patterns, where reference representations can be derived from probing certain patterns. Future research will go beyond predicting series of mobile phone features to modeling tool chain features, such as identifying an object, tagging it as a subject or context, storing it, and relating it to the already stored model elements.

Chen et al. [42] have explored parameters like the duration of task accomplishment, the amount of resources spent on tasks, etc., from cloud users to identify patterns for prediction. This mechanism can be used in the context of stakeholder (user) modeling to capture the dynamics of the series of incoming modeling data, in order provide an effective prediction towards the cognitive workload (i.e., the interarrival time between objects to be processed), including resources/tools that can be requested in terms of processing engines for pattern matching, validating, and processing models based on the historical data.

In summary, besides content- or domain-specific issues, research questions with respect to upcoming stakeholder behavior will be studied in future research.

Conflicts of Interest: The author declares no conflict of interest.

References

1. Dittmar, A.; Dardar, L. Personal ecologies of calendar artifacts. *J. Interact. Sci.* **2015**, *3*, 2. [CrossRef]
2. Kallinikos, J.; Aaltonen, A.; Marton, A. The Ambivalent Ontology of Digital Artifacts. *MIS Q.* **2013**, *37*, 357–370.
3. Gruber, M.; De Leon, N.; George, G.; Thompson, P. Managing by design. *Acad. Manag. J.* **2015**, *58*, 1–7. [CrossRef]
4. Storni, C.; Binder, T.; Linde, P.; Stuedahl, D. Designing things together: Intersections of co-design and actor–network theory. *CoDesign* **2015**, *11*, 149–151. [CrossRef]
5. Acheson, P.; Daglo, C.; Kilicay-Ergin, N. Model Based Systems Engineering for System of Systems Using Agent-Based Modeling. *Procedia Comput. Sci.* **2013**, *16*, 11–19. [CrossRef]
6. Alam, I.I. Moving Beyond the Stage Gate Models for Service Innovation: The Trend and the Future. *Int. J. Econ. Pract. Theor.* **2014**, *4*, 637–646.
7. Tang, T.; Wu, Z.; Karhu, K.; Hämäläinen, M.; Ji, Y. Internationally distributed living labs and digital ecosystems for fostering local innovations in everyday life. *J. Emerg. Technol. Web Intell.* **2012**, *4*, 106–115. [CrossRef]
8. Zheng, M.; Song, W.; Ming, X. A Framework for Integrating Industrial Product-Service Systems and Cyber-Physical Systems. In *Proceedings of the 8th International Conference on Cross-Cultural Design, Toronto, ON, Canada, 17–22 July 2016*; Lecture Notes in Computer Science; Springer International Publishing: Berlin, Germany, 2016; Volume 9741, pp. 628–637.

9. Yoo, Y.; Boland, R.J., Jr.; Lyytinen, K.; Majchrzak, A. Organizing for innovation in the digitized world. *Organ. Sci.* **2012**, *23*, 1398–1408. [CrossRef]

10. Kim, K.; Altmann, J.; Baek, S. *Role of Platform Providers in Software Ecosystems*; Technology Management, Economics, and Policy Program (TEMEP). Discussion Paper No. 2015:120; Seoul National University: Seoul, Korea, 2015.

11. Jaradat, R.M.; Keating, C.B.; Bradley, J.M. A histogram analysis for system of systems. *Int. J. Syst. Syst. Eng.* **2014**, *5*, 193–227. [CrossRef]

12. Maier, M.W. Research challenges for systems-of-systems. In *Proceedings of the International Conference on Systems, Man and Cybernetics, Waikoloa, HI, USA, 10–12 October 2005*; IEEE: New York, NY, USA, 2005; Volume 4, pp. 3149–3154.

13. Kotov, V. *Systems of Systems as Communicating Structures*; Hewlett Packard: Bristol, UK, 1997.

14. Jamshidi, M. (Ed.) *System of Systems Engineering: Innovations for the Twenty-First Century*; John Wiley & Sons: New York, NY, USA, 2011; Volume 58.

15. IEEE-Reliability Society. Technical Committee on 'Systems of Systems'. In *Systems-of-Systems*; White Paper; IEEE: New York, NY, USA, 2014; p. 5.

16. Stary, C.; Wachholder, D. System-of-systems support—A bigraph approach to interoperability and emergent behavior. *Data Knowl. Eng.* **2016**, *105*, 155–172. [CrossRef]

17. Lane, J.A.; Epstein, D. *What Is a System of Systems and Why Should I Care?* Report USC-CSSE-2013-001; University of Southern California: Los Angeles, CA, USA, 2013.

18. Best, A.; Greenhalgh, T.; Lewis, S.; Saul, J.E.; Carroll, S.; Bitz, J. Large-system transformation in health care: A realist review. *Milbank Q.* **2012**, *90*, 421–456. [CrossRef] [PubMed]

19. Margaria, T.; Boelmann, S.; Doedt, M.; Floyd, B.; Steffen, B. Customer-oriented business process management: Vision and obstacles. In *Conquering Complexity*; Hinchey, M., Coyle, L., Eds.; Springer: London, UK, 2012; pp. 407–429.

20. Campos, J.; Lopez-Sanchez, M.; Rodriguez-Aguilar, J.A.; Esteva, M. Formalizing Situatedness and Adaption in Electronic Institutions. In *Proceedings of the COIN 2008*; LNAI 5428; Springer: Berlin, Germany, 2009; pp. 126–139.

21. Matheus, C.J.; Baclawski, K.; Kokar, M.M.; Letkowski, J.J. Using SWRL and OWL to capture domain knowledge for a situation awareness application applied to a supply logistics scenario. In *Proceedings of the International Workshop on Rules and Rule Markup Languages for the Semantic Web, Galway, Ireland, 10–12 November 2005*; Springer: Berlin/Heidelberg, Germany, 2005; pp. 130–144.

22. Rolland, C.; Prakash, N.; Benjamen, A. A Multi-Model View of Process Modelling. *Requir. Eng.* **1999**, *4*, 169–187. [CrossRef]

23. Barwise, J.; Perry, J. Situations and attitudes. *J. Philos.* **1981**, *78*, 668–691. [CrossRef]

24. Dobbyn, C.; Stuart, S. The Self as Embedded Agent. *Minds Mach.* **2003**, *13*, 187–201. [CrossRef]

25. Senge, P. *The Fifth Discipline: The Art and Practice of the Learning Organization*; Doubleday: New York, NY, USA, 2005.

26. Frank, M. Engineering Systems Thinking: Cognitive Competencies of Successful Systems Engineers. *Procedia Comput. Sci.* **2012**, *8*, 273–278. [CrossRef]

27. Fleischmann, A.; Schmidt, W.; Stary, C.; Obermeier, S.; Börger, E. *Subject-Oriented Business Process Management*; Springer: Berlin, Germany, 2012.

28. Herrgard, T. Difficulties in diffusion of tacit knowledge in organizations. *J. Intell. Cap.* **2000**, *1*, 357–365. [CrossRef]

29. Parsaye, K.; Chignell, M. *Expert Systems for Experts*; Wiley: New York, NY, USA, 1988.

30. Grosskopf, A.; Edelman, J.; Weske, M. Tangible business process modeling methodology and experiment design. In *Business Process Management Workshops*; Lecture Notes in Business Information Processing; Rinderle-Ma, S., Sadiq, S., Leymann, F., Eds.; Springer: Berlin/Heidelberg, Germany, 2010; Volume 43, pp. 489–500.

31. Brey, P. The Epistemology and Ontology of Human-Computer Interaction. *Mind Mach.* **2005**, *15*, 383–398. [CrossRef]

32. Harman, J.; Brown, R.; Kannengiesser, U.; Meyer, N.; Rothschädl, T. Model while you do. Engaging an S-BPM vendor on process modeling in 3D worlds. In *S-BPM in the Wild—Value Creating Practice in the Field*; Fleischmann, A., Schmidt, W., Stary, C., Eds.; Springer: Berlin, Germany, 2015.

33. Himma, K.E. Artificial Agency, Consciousness, and the Criteria for Moral Agency. *Ethics Inf. Technol.* **2009**, *11*, 19–29. [CrossRef]

34. Fleischmann, A.; Schmidt, W.; Stary, C. *S-BPM in the Wild: Practical Value Creation*; Springer Publishing Company, Incorporated: Berlin, Germany, 2015.

35. Neubauer, M.; Stary, C. *S-BPM in the Production Industry. A Stakeholder Approach*; Springer Publishing Company, Incorporated: Berlin, Germany, 2017.

36. Stary, C. Non-disruptive knowledge and business processing in knowledge life cycles-aligning value network analysis to process management. *J. Knowl. Manag.* **2014**, *18*, 651–686. [CrossRef]

37. Sterling, L.; Taveter, K. *The Art of Agent-Oriented Modeling*; MIT Press: Cambridge, MA, USA, 2009.

38. North, M.J.; Macal, C.M. *Managing Business Complexity: Discovering Strategic Solutions with Agent-Based Modeling and Simulation*; Oxford University Press: Oxford, UK, 2007.

39. Gero, J.S.; Fujii, H. A computational framework for concept formation for a situated design agent. *Knowl. Based Syst.* **2000**, *13*, 361–368. [CrossRef]

40. Gero, J.S.; Kannengiesser, U. The Situated-Function-Behaviour-Structure framework. In *Artificial Intelligence in Design'02*; Gero, J.S., Ed.; Kluwer: Dordrecht, The Netherlands, 2002; pp. 89–104.

41. Dai, P.; Ho, S.S. A smartphone user activity prediction framework utilizing partial repetitive and landmark behaviors. In *Proceedings Mobile Data Management*; IEEE: New York, NY, USA, 2014; Volume 1, pp. 205–210.

42. Chen, S.; Ghorbani, M.; Wang, Y.; Bogdan, P.; Pedram, M. Trace-Based Analysis and Prediction of Cloud Computing User Behavior Using the Fractal Modeling Technique. In *Proceedings Big Data Congress*; IEEE: New York, NY, USA, 2014; pp. 733–739.

Article

Reaction Networks as a Language for Systemic Modeling: On the Study of Structural Changes

Tomas Veloz [1,2,*] and Pablo Razeto-Barry [2,3]

[1] Center Leo Apostel, Brussels Free University, 1050 Brussels, Belgium
[2] Insituto de Filosofía y Ciencias de la Complejidad, Los Alerces 3024, Ñuñoa, Chile; prazeto@ificc.cl
[3] Vicerrectoría Acadḿica, Universidad Diego Portales, Manuel Rodŕguez Sur 415, Santiago, Chile
* Correspondence: tveloz@gmail.com; Tel.: +56-2-27276403

Academic Editors: Gianfranco Minati, Eliano Pessa and Ignazio Licata
Received: 18 November 2016; Accepted: 23 March 2017; Published: 31 March 2017

Abstract: Reaction Networks have been recently proposed as a framework for systems modeling due to its capability to describe many entities interacting in contextual ways and leading to the emergence of meta-structures. Since systems can be subjected to structural changes that not only alter their inner functioning, but also their underlying ontological features, a crucial issue is how to address these structural changes within a formal representational framework. When modeling systems using reaction networks, we find that three fundamentally different types of structural change are possible. The first corresponds to the usual notion of perturbation in dynamical systems, i.e., change in system's state. The second corresponds to behavioral changes, i.e., changes not in the state of the system but on the properties of its behavioral rules. The third corresponds to radical structural changes, i.e., changes in the state-set structure and/or in reaction-set structure. In this article, we describe in detail the three types of structural changes that can occur in a reaction network, and how these changes relate to changes in the systems observable within this reaction network. In particular, we develop a decomposition theorem to partition a reaction network as a collection of dynamically independent modules, and show how such decomposition allows for precisely identifying the parts of the reaction network that are affected by a structural change.

Keywords: reaction networks; chemical organization theory; system theory; emergence; structural change

1. Introduction

The dynamics of a system, either modeled with a reaction network or any other representational language (such as agent-based models, rule-based modeling, or systems of equations; for an overview, see [1]) is generally examined by combining analytic and computational tools, in the so-called paradigm of dynamical systems [2]. The dynamical systems approach aims at explaining the time evolution of a system and how such evolution can be influenced by small but sudden changes in the system's state, known as perturbations. The theory of dynamical systems has produced various notions that describe non-trivial aspects of a dynamical system (e.g., attractor), and has produced powerful mathematical results to explain the evolution of a system after a perturbation (e.g., Lyapunov exponent), or subjected to frequent perturbations in time [3].

However, since perturbations are conceived to occur *within* the state space (phase space) of the system, external influences that *qualitatively* modify the state space of a system can hardly be represented in this paradigm. In fact, little is known about how to represent external influences that lead to structural transformations of the state space. The concept of *qualitative* change is a fundamental notion for system theory and is at the core of important notions developed by system theorists such as structural coupling [4] and change of code [5]. Therefore, a formal notion of qualitative changes is crucial for improving our understanding of what constitutes the identity of a system, as well as for developing notions that might go beyond the traditional notion of perturbation such as resilience, robustness, and adaptivity.

In [6], we have proposed using reaction networks (RN) as a language for modeling systems. In particular, we showed that, by using Chemical Organization Theory (COT), we can compute the set of possible observable systems from a reaction network universe. In COT, the notion of *organization* is introduced in order to identify structurally closed and self-maintaining subnetworks of the reaction network. The set of organizations of a reaction network can be proven to contain all of the stationary regimes of the dynamics of the reaction network [7,8]. In the present article, we will extend the formalism of COT and propose it as a framework for representing the qualitative changes mentioned above.

Although most theoretical research in COT has been focused on the stoichiometric level of representation (with the exception of a few attempts such as [8,9]), which lacks dynamical equations for time-evolution, and thus is insufficient for the dynamical systems paradigm, COT is suited to model qualitative changes by simply modifying the way in which species are transformed in the reaction network universe. In particular, we identify two types of changes that go beyond the dynamical system's notion of perturbation. The first is the modification of the way in which reactions occur in the network, and the second is the addition and/or elimination of species and reactions in the reaction network. These changes will be called *process-structure perturbation* and *topological-structure perturbation*.

Since a change in the reaction network universe leads to changes in the set of organizations of the reaction network, we have that a qualitative change in the reaction network shapes a new landscape of observable systems. However, we will show that a deeper analysis in the organization's structure and in their self-maintaining processes allows for precisely identifying the influence of qualitative changes in a system.

In Section 2, we recall the basics of the reaction network formalism and COT. In Section 3, we introduce the notions of process-structure and topology-structure perturbation using the language of reaction networks. In Section 4, we elaborate a decomposition Theorem that, later in Section 5, is applied to formalize the structural perturbations and precisely identify their impact in the reaction network. In Section 6, we present an example to illustrate the Definitions and results of this work, and, finally, in Section 7, we present some general conclusions.

2. COT Summary

In COT, we consider three increasingly complex ways to represent a reaction network. In the first, so-called *relational* description, reactions include information about which species are consumed and produced, but no quantitative information about the production and consumption of reactions is given. In the second, so-called *stoichiometric* description, quantitative information about the number of species consumed and produced by each reaction is included. In the third, so-called *kinetic* level of description, quantitative information about the number or concentration of species as well as rules for the ways in which reactions occur are included. The core ideas of this work lie at the stoichiometric level of representation. For a comprehensive introduction to COT, we refer to [6,7].

Let $\mathcal{M} = \{s_1, \ldots, s_m\}$ be a finite set of m *species* reacting with each other according to a finite set $\mathcal{R} = \{r_1, \ldots, r_n\}$ of n *reactions*. Together, the set of species and the set of reactions is called the *reaction network* $(\mathcal{M}, \mathcal{R})$. A reaction r_i is represented by

$$r_i = a_{i1}s_1 + \ldots a_{im}s_m \rightarrow b_{i1}s_1 + \ldots b_{im}s_m, \tag{1}$$

with $a_{ij}, b_{ij} \in \mathbb{N}_0$, for $i = 1, \ldots, n$.

Reactions describe what collections of species transform into what new collections. For a given reaction $r_i \in \mathcal{R}$, the species s_j to be transformed, i.e., such that $a_{ij} > 0$, are called *reactants* of r, and the species to be created, such that $b_{ij} > 0$, are called *products*.

In COT, we study subsets of species $X \subseteq \mathcal{M}$. Note that, for all X, there is a unique maximal set of reactions $\mathcal{R}_X \subseteq \mathcal{R}$ defined as the set of all reactions whose reactants are in X. Thus, each set X induces a *sub-network* $(X, \mathcal{R}_X)$.

Definition 1. *X is structurally closed iff the products of every reaction in $\mathcal{R}_X$ are in X* [6].

A structurally closed set X entail a sub-network of the reaction network whose reactions do not produce species outside X.

Definition 2. *Two species $s_j, s_k \in X$ are directly-connected in X if and only if there exist a reaction $r_i \in \mathcal{R}_X$ such that both species are active in the reaction, i.e., s_j and s_k are either reactants or products of r_i. We say s_j and s_k are connected in X if and only if there exists a sequence of species $s_0, ..., s_p \in X$ such that $s_0 = s_j$, $s_p = s_k$ and for all $l = 0, ..., p - 1$, we have that s_l and s_{l+1} are directly-connected in X.*

Connected species in X can be seen as *potentially co-dependent* species in the reaction network because the consumption of one of them might affect the production of all the species connected to it. In general, X can be decomposed into connected modules whose reactions are *independent*. Identifying independent behavioral modules of a reaction network is useful from both computational and mathematical perspectives because it provides resources for an algorithmic *divide-and-conquer* strategy, and also can deepen the understanding of the structure of the reaction network.

The dynamics of the reaction network is determined by how often reactions occur. A particular specification of the occurrence of reactions within the reaction network is called *reaction process*, or simply *process*, and we denoted it by **v**. In reaction network modeling, **v** is usually called flux vector. We are introducing a slightly more general notion because our aim lies beyond the modeling of biochemical systems. Thus, a process corresponds to a non-negative vector $\mathbf{v} = (\mathbf{v}[1], ..., \mathbf{v}[n])$, whose components are natural or real numbers for representing discrete and continuous dynamics, respectively. We say a process **v** *can be applied to* X if all the reactions in the process can be triggered by the species in the set X. Hence, **v** *can be applied* to X only if $\mathbf{v}[i] > 0$ implies $r_i \in \mathcal{R}_X$, for $i = 1, ..., n$.

Definition 3. *Let $\mathcal{R}^* \subseteq \mathcal{R}_X$ and $\Pi(\mathcal{R}^*)$ be a set of processes such that $\mathbf{v} \in \Pi(\mathcal{R}^*)$ implies $\mathbf{v}[i] > 0$ for $r_i \in \mathcal{R}^*$, and $\mathbf{v}[i] = 0$. Otherwise, $i = 1, ..., n$.*

Lemma 1. *A process **v** can be applied to X if and only if there exist $\mathcal{R}^* \subseteq \mathcal{R}_X$ such that $\mathbf{v} \in \mathcal{R}^*$.*

Corollary 1. *$X \subset X'$ implies $\mathcal{R}_X \subseteq \mathcal{R}_{X'}$, which, in turn, implies that $\Pi(\mathcal{R}_X) \subseteq \Pi(\mathcal{R}_{X'})$.*

In order to represent how species are globally transformed in the reaction network by the application of a process, let us represent the state of a reaction network by a vector **x** of non-negative coordinates such that $\mathbf{x}[j]$ corresponds to the number (or concentration) of species of type s_j in the reaction network, $j = 1, ..., m$. In addition, note that the numbers a_{ij} and b_{ij} in Equation (1) can be used to encode the way in which species are consumed and produced by the reactions. Namely, we can build a *stoichiometric matrix* $\mathbf{S} \in \mathbb{N}_{0\geq}^{m \times n}$ such that $\mathbf{S}[j, i] = b_{ij} - a_{ij}$.

From here, we can compute the state $\mathbf{x_v}$ of the reaction network associated to a state **x** and a process **v** by the following equation:

$$\mathbf{x_v} = \mathbf{x} + \mathbf{Sv}. \tag{2}$$

For simplicity, we have implicitly assumed that the coordinates of **x** are sufficiently large to consume the species required by the reactions in **v** in any order. For a study of order effects in reaction networks we refer to [10].

Definition 4. *X is semi-self-maintaining with respect to a set of reactions $\mathcal{R}^*$ if for each reactant $s \in X$ of a reaction $r \in \mathcal{R}^*$, there is a reaction $\bar{r} \in \mathcal{R}^*$ such that s is a product of $\bar{r}$. We say X is semi-self-maintaining if and only if X is semi-self-maintaining with respect to $\mathcal{R}_X$.*

A set of species that is semi-self-maintaining can recreate the species consumed by its associated set of reactions. However, such recreation might not be quantitatively balanced. For quantitatively balanced processes of recreations, we introduce self-maintaining sets.

Definition 5. *X is weak-self-maintaining with respect to $\mathcal{R}^*$ if there exists $\mathbf{v} \in \mathbf{\Pi}(\mathcal{R}^*)$ such that $\mathbf{x_v}[j] \geq \mathbf{x}[j]$, $j = 1, ..., m$. We say X is self-maintaining if and only if X is weak-self-maintaining with respect to $\mathcal{R}_X$.*

A set of species that is self-maintaining has processes such that all the reactions of a reaction network into consideration have a positive rate and, when applied, all species of the set are quantitatively recreated.

Definition 6. *Let $X \subseteq \mathcal{M}$, X be an organization if and only if X is structurally closed and self-maintaining.*

Remarkably, all stable dynamical regimes of a reaction network correspond to organizations [7,8]. This fact was used to propose the reaction network formalism as a language for modeling systems, by noting that the set of organizations of a reaction network corresponds to the observable systems in the reaction network universe [6].

From now on, we will assume that X is structurally closed and connected (both properties can be verified at low computational cost). Note that, since the species in X are connected, for all species $s_j \in X$, there is at least one reaction $r_i \in \mathcal{R}_X$ such that either $a_{ij} > 0$ or $b_{ij} > 0$.

Definition 7. *A species $s_j \in X$ is a catalyst w.r.t X if and only if $r_i \in \mathcal{R}_X$ implies $a_{ij} = b_{ij}$, for $i = 1, ..., n$. The maximal set of catalysts w.r.t X will be denoted by E.*

Definition 8. *Let $\mathbf{x}$ be any initial state, $\mathcal{R}^* \subseteq \mathcal{R}_X$, and $\mathbf{v} \in \mathbf{\Pi}(\mathcal{R}^*)$ be a non-null process vector such that $\mathbf{x_v}$ in Equation (2) is a non-negative vector. If $\mathbf{x_v}[j] > \mathbf{x}[j]$, we say that s_j is an overproduced species by $\mathbf{v}$ in X, or simply that s is overproducible in X. Furthermore, if all the species of a set $G \subseteq X$ are simultaneously overproduced by a process vector $\mathbf{v}_G$, we say that G is overproduced by $\mathbf{v}_G$ in X, or that G is overproducible in X.*

Lemma 2. *Let s be overproducible in X. Then, s is overproducible in every $X' \supset X$.*

Proof. Direct consequence of Corollary 1. □

Lemma 3. *The maximal set of overproducible species F in X is unique.*

Proof. Since for every species $s \in X$ we have that s is either overproducible or not overproducible, we can build the set $\{s_{j_1}, ..., s_{j_k}\} = G$ of all the overproducible species in X, and the set of processes $\{\mathbf{v}_1, ..., \mathbf{v}_k\}$ that overproduces the species in G. Note that each $\mathbf{v}_i \in \mathbf{\Pi}(\mathcal{R}_i^*)$, with $\mathcal{R}_i^* \subseteq \mathcal{R}_X$. Thus, the process $\mathbf{v}_F = \mathbf{v}_1 + ... + \mathbf{v}_k \in \mathbf{\Pi}(\mathcal{R}^*)$, with $\mathcal{R}^* = \cup_{l=1}^k \mathcal{R}_l^*$, overproduces all the species in G simultaneously. Since $\mathcal{R}^* \subseteq \mathcal{R}_X$, we conclude $G = F$. □

Species in F can be unlimitedly overproduced by the repeated application of the process $\mathbf{v}_F$. Hence, for any process $\mathbf{v}$, we can create a new process $\mathbf{v} + \alpha \mathbf{v}_F$, with sufficiently large α, such that species in F (and thus in $E \cup F$) are ensured to have non-negative production.

Definition 9. *Let $\mathbf{v}$ be a process. We define $\mathbf{v}(F) = \mathbf{v} + \alpha \mathbf{v}_F$ by choosing sufficiently large α so that all species in F are overproduced by $\mathbf{v}(F)$.*

The following Lemma shows that computing overproducible species can be very simple in some cases.

Lemma 4. *Let $G \subseteq F$. The products of the reactions in $\mathcal{R}_{E \cup G}$ are either catalysts or overproducible.*

Proof. Let $r_i \in \mathcal{R}_{E \cup G}$ and s be a product of r_i. If $s \in E \cup G$, the Lemma follows directly. If $s \notin E \cup G$, let a process $\mathbf{v}$ such that $\mathbf{v}[i] = 1$ and $\mathbf{v}[l] = 0$ if $l \neq i$, and $\mathbf{v}_G$ be a process that overproduces all species in G. Since $\mathbf{v}$ consumes species in $E \cup G$ only, we have that for sufficiently large α, the process $\mathbf{v}(G) = \mathbf{v} + \alpha \mathbf{v}_G$ overproduces all the species in G. Now, since $\mathbf{v}$ produces s, and consumes species in $E \cup G$ only, we have that s is overproduced by $\mathbf{v}(G)$. $\square$

3. On the Types of Change

In dynamical systems, the evolution equation of a system being in state $\mathbf{x}$ is given by

$$0 = \mathbf{F}(t, \mathbf{k}, \mathbf{x}, \dot{\mathbf{x}}, \ddot{\mathbf{x}}, \ldots), \quad (\mathbf{x}, \dot{\mathbf{x}}, \ddot{\mathbf{x}}, \ldots)_{t=0} = \mathbf{c}_0, \tag{3}$$

where $\mathbf{F}$ is the evolution operator, t represents time, $\mathbf{k}$ is a vector of parameters describing the particular dynamical process, and $\mathbf{c}_0$ are the initial conditions. A perturbation is a sudden change from $\mathbf{x}$ to $\bar{\mathbf{x}} = \mathbf{x} + \vec{e}$ where $\vec{e}$ is a small vector. The goal of dynamical systems is to study how these perturbations affect the dynamical evolution of the system. For various cases, the influence of $\mathbf{k}$ on such evolution is also investigated, and more abstract approaches determine classes of evolution operators $\mathbf{F}$ where analytic results concerning the dynamics of the system can be obtained [2].

In traditional dynamical systems theory, there is no clear way to represent a structural change. Namely, such change would correspond to a *change* on $\mathbf{F}$. However, there are two problems for qualifying or quantifying a change on $\mathbf{F}$. First, the algebraic structure of two such operators might be completely different, so no algebraic comparison can be established in a sensible way. This problem can be sorted out relying on abstract measures of such as distances induced by norms in abstract operator spaces. However, a second problem emerges here. Namely, it is not clear how to relate such abstract distance with a structural change in the system into consideration.

In the language of reaction networks, the analogous of Equation (3) lies at the kinetic level of representation, and is described by the equation

$$\dot{\mathbf{x}} = \mathbf{S} \mathbf{v},$$

where $\mathbf{S}$ is the stoichiometric matrix introduced in Section 2, and $\mathbf{v}$ usually is a function (whose functional form depends on the kinetic law) of $\mathbf{x}$, t, and a vector of parameters $\mathbf{k}$. Unfortunately, the analysis of the dynamics of a reaction network at the kinetic level is as complex as the analysis of general dynamical systems. Therefore, at this level of representation, there are no major advantages in using reaction networks over other languages for representing systems.

However, since COT allows for *lift* information gathered at the stoichiometric level to the kinetic level (at a reasonable computational cost, see Theorem 1 in [6]), we can focus on exploiting stoichiometric properties that provide a better understanding at the kinetic level.

At the stoichiometric level, we consider a state vector $\mathbf{x}$ and a process $\mathbf{v}$ that is simply a non-negative vector that represents the occurrence of reactions within a certain time frame. Hence, the change to a new state $\mathbf{x}_\mathbf{v}$ produced by the process $\mathbf{v}$ is given by Equation (2):

$$\mathbf{x}_\mathbf{v} = \mathbf{x} + \mathbf{S} \mathbf{v}.$$

Remarkably, the right hand side of Equation (2) provides evidence of the different aspects of the dynamics of the reaction network universe that can be modified. Namely, $\mathbf{x}$, $\mathbf{v}$, and $\mathbf{S}$. In the first case, $\mathbf{x}$ changes to a new state $\bar{\mathbf{x}} \neq \mathbf{x}$. Such a change can be quantified by the geometrical distance of the two vectors (as it is done in dynamical systems). In the second case, $\mathbf{v}$ is changed to another process $\tilde{\mathbf{v}} \neq \mathbf{v}$. This is similar to changing the parameters of a dynamical system's equation. More generally, we can extend this change by considering a change on the set of permitted (or forbidden) processes Λ. The set Λ of permitted processes is the analogous of a kinetic law at the stoichiometric level. Indeed, since the stoichiometric level lacks a kinetic law, we cannot determine what process will occur, but we can constrain what processes might occur in the reaction network universe by, for example, limiting the ratios between certain reactions, setting the permitted processes within a convex set, limiting the minimum/maximum value of particular coordinates of the process vector, etc.

From here, since systems are structurally closed and self-maintaining sets (i.e., organizations), we have that the set of observable systems depends on the set Λ. A change to the set of permitted processes will be referred to as a *process-structure perturbation*. In the third case, $\mathbf{S}$ is changed to $\tilde{\mathbf{S}} \neq \mathbf{S}$. Therefore, at least one reaction has been modified, added to, or eliminated from the reaction network universe. In case $\tilde{\mathbf{S}}$ contains new (or eliminates certain) species, we must update the state vector dimensionality, and thus such change induces a change of state, and in case new reactions are added (or removed), such change induces a process-structure change. This change, so called *topology-structure perturbation*, operationalizes the notion of qualitative change of a reaction network universe.

The process-structure and topology-structure perturbations are both non-trivial ways to change the structure of a system and both go beyond the notion of perturbation in dynamical systems. In order to characterize these new types of perturbation, we need to identify a theoretically convenient way to quantify them so that dynamical concepts related to the notion of structural change (such as resilience, robustness, adaptivity, etc.) can be put forward.

In what follows, we will advance the structural notions of COT presented in [6]. In particular, we will develop a Theorem that allows for decomposing the inner structure of a system into a collection of dynamically independent self-maintaining sets. From here, we can precisely identify the influence of a structural change in a system.

4. Modularizing the Behavior of a Reaction Network

Let E be the set of catalysts in X, and F its maximal overproducible set. The latter assumption will simplify the analysis of this section. However, the formal elaboration presented here can be done for a set X without any prior knowledge about its inner structure. Recall from Definition 9 that, for every process $\mathbf{v}$ applied to X, we can build a process $\mathbf{v}(F)$ such that the species in F (and thus in $E \cup F$) have non-negative production. Thus, in order to comprehend the structure of self-maintaining processes in X, we must focus on the structure of $X - (E \cup F)$.

Definition 10. *Let $C = X - (E \cup F)$. We call C the potential fragile-circuit of X.*

The potential fragile-circuit C of X entails *the most sensitive* part in the dynamics of X. In fact, note that, for a species s in C, we can infer that (i) its maximal overproduction is zero ($s \notin F$); and (ii) it cannot be equally consumed and produced in all of the reactions, it participates in ($s \notin E$). Therefore, s must be consumed more than produced by at least one reaction. Hence, if s is not produced more than consumed by another reaction, then X is not semi-self-maintaining, and thus X is not self-maintaining.

Lemma 5. *If C is not semi-self-maintaining with respect to $\mathcal{R}_X$, then X is not self-maintaining.*

Analyzing the inner structure of potential fragile-circuits reveals interesting features that contribute to the understanding of organizations.

Suppose that X is self-maintaining, and let $s \in C$, $\mathbf{v}$ a self-maintaining process, and $r \in \mathcal{R}_X$ such that s is a reactant of r. Since X is self-maintaining, s must be the product of another reaction $\bar{r}$. Note, however, that we can deduce that at least one reactant $\bar{s}$ of $\bar{r}$ is in C. Indeed, suppose no reactant of $\bar{r}$ is in C. Then, all the reactants of $\bar{r}$ are in $E \cup F$, and, by Lemma 4, we have that all the products of $\bar{r}$ are in $E \cup F$. Thus, if no reactant of $\bar{r}$ is in C, we have that all the products of $\bar{r}$, and in particular s, would be in $E \cup F$. This contradicts our assumption that $s \in C$. Therefore, there is at least one reactant $\bar{s}$ of $\bar{r}$ in C. Subsequently, every reaction that produces $\bar{s}$ has a reactant in C. Hence, we can deduce that every reaction that is used to produce a species in C has a reactant in C. Since the set of species C is finite, we have that for any two species $s, \bar{s} \in C$, we have that either both species are needed to produce each other, or there are two disjoint sets of reactions $\mathcal{R}_s^*$ and $\mathcal{R}_{\bar{s}}^*$ such that self-maintainance of s and $\bar{s}$ are verified by process vectors triggering reactions in $\mathcal{R}_s^*$ and $\mathcal{R}_{\bar{s}}^*$, respectively.

The previous deduction entails a powerful structural property of self-maintaining reaction networks that, to the knowledge of the authors, has not been widely acknowledged in the biochemical literature (with the exception of [11]). In Figure 1, we illustrate the two most simple cases where such structural property can be used to decompose a reaction network into *dynamically-connected* subnetworks.

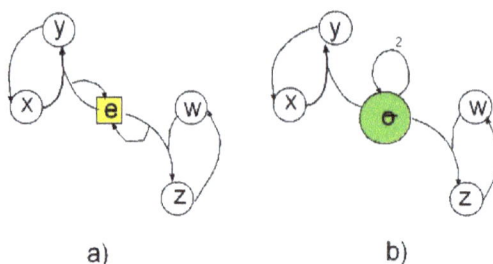

a)   b)

Figure 1. Note that in (**a**), the self-maintainance of $C_1 = \{x, y, e\}$ and $C_2 = \{z, w, e\}$ are independent because e is a catalyst; The same situation occurs in (**b**) with sets $C_1 = \{x, y, o\}$ and $C_2 = \{z, w, o\}$ with o an overproducible species. Hence, although X is connected (see Definition 2), the potential fragile-circuit $C = \{x, y, w, z\}$ can be partitioned into two *dynamically-independent* sets $\{x, y\}$ and $\{w, z\}$.

Definition 11. *Two species s and $\bar{s}$ in C are dynamically-connected in X if and only if there exists a sequence of species $s_0, ..., s_p \in C$ such that $s_0 = s$, $s_p = \bar{s}$ and for all $k = 0, ..., p-1$, we have that s_k and s_{k+1} are directly-connected in X (see Definition 2).*

Dynamical-connection entails connection through reactions which have reactants in C. Note that $s, s' \in C$ can be dynamically-connected in X but not connected in C. For example, consider the reactions:

$$\varnothing \rightarrow o,$$
$$s + o \rightarrow s'. \tag{4}$$

Note that s is dynamically-connected to s' in $\{o, s, s'\}$, but s is not connected to s' in $\{s, s'\}$. For simplicity, when s and s' are dynamically-connected in X, we will simply say that s and s' are dynamically-connected.

Definition 12. *We define $dyn(s)$ as the maximal set of species dynamically-connected to s.*

Corollary 2. *Let $s, \bar{s} \in C$. $\bar{s} \in dyn(s)$ if and only if $dyn(s) = dyn(\bar{s})$.*

Corollary 3. *Let $s, \bar{s} \in C$. If $dyn(s) \neq dyn(\bar{s})$, then $dyn(s) \cap dyn(\bar{s}) = \varnothing$.*

Definition 13. *Any set $C' \subseteq C$ s.t $C = \bigcup_{s \in C'} dyn(s)$ is called a generating set of C. Any minimal cardinality generating set of C is called a base of C.*

Since any generating set can be reduced up to a base, we will concentrate on the bases of C. Note that C can have several bases. However, the following Lemma shows that such bases are all equivalent.

Technically speaking, we will show that $\text{dyn}(\cdot)$ is an equivalence class of the dynamically-connected equivalence relation over C. An equivalence relation is a reflexive, symmetric, and transitive binary relation that allows to partition a set into equivalence classes and simplify the study of a set in terms of its quotient space [12]. For simplicity, we will not elaborate on more mathematical details in this line.

Lemma 6. *Let D_1, D_2 be two bases of C. Then, each species in D_1 is dynamically-connected to one and only one species of D_2.*

Proof. Let $s \in D_1$ and suppose that s is not dynamically-connected to any species in D_2. By Corollary 2, we have that $\text{dyn}(s)$ is not contained in $\bigcup_{\bar{s} \in D_2} \text{dyn}(\bar{s}) = C$, which entails a contradiction. Then, s is dynamically-connected to at least one species in D_2. Now, suppose there are two species $s_1, s_2 \in D_2$ dynamically-connected to s. Since s_1 and s_2 are dynamically-connected to s, we have that $s_1, s_2 \in \text{dyn}(s)$. Then, by Corollary 2, we have $\text{dyn}(s_1) = \text{dyn}(s_2)$. Thus, D_2 is not a base of C, which entails a contradiction. $\square$

From now on, let $D = \{\bar{s}_1, ..., \bar{s}_d\}$ be a base of C

Definition 14. *We define for $j = 1, ..., d$:*

$$D_j = \text{dyn}(\bar{s}_j), \text{ and}$$

$$\mathcal{R}_j^* = \mathcal{R}_{D_j \cup EUF} - \mathcal{R}_{EUF}.$$

The set D_j is called the j-th minimal fragile-circuit of C, and $\mathcal{R}_j^*$ the path of D_j.

Note that minimal fragile-circuits decompose the fragile-circuit into dynamically-connected sets. For example, in both networks in Figure 1, we have that the minimal fragile-circuits are $D_1 = \{x, y\}$, and $D_2 = \{w, z\}$, and the bases are $\{x, w\}, \{x, z\}, \{y, w\}$, and $\{y, z\}$.

Lemma 7. *For all $i, j \in \{1, ..., d\}$, $i \neq j$ implies $\mathcal{R}_i^* \cap \mathcal{R}_j^* = \varnothing$*

Proof. Suppose that $\mathcal{R}_i^* \cap \mathcal{R}_j^* \neq \varnothing$. Thus, there is one reaction $r \in \mathcal{R}_i^* \cap \mathcal{R}_j^*$ such that one of its reactants is in D_i or D_j. Without loss of generality, assume a species $s \in D_i$ is a reactant of r. Since $r \in \mathcal{R}_j^*$, we have that $s \in D_j$. Hence, by Corollary 2, we have that $D_i = D_j$, which entails a contradiction. $\square$

Corollary 4.

$$\mathcal{R}_X - \mathcal{R}_{EUF} = \bigcup_{i=1}^{d} \mathcal{R}_{\bar{s}_i}^*.$$

Lemma 7 proves that the paths of the minimal fragile-circuits do not overlap. This implies that a process $\mathbf{v}$ applied to X can be represented as a sum of non-overlapping processes (orthogonal vectors) applied to the different minimal fragile-circuit D_j, i.e., containing only reactions in the path $\mathcal{R}_j^*, j = 1, ..., d$, plus a process $\mathbf{v}_{EUF}$ (orthogonal to all other processes) applied to $E \cup F$.

Corollary 5. *Let $\mathbf{v}$ be a process applied to X. Then, there exist processes $\mathbf{v}_1, ..., \mathbf{v}_d$ such that*

$$\mathbf{v} = \mathbf{v}_D + \mathbf{v}_{EUF}, \text{with}$$
$$\mathbf{v}_D = \mathbf{v}_1 + ... + \mathbf{v}_d,$$

$$\tag{5}$$

where $|\mathbf{v}_D \cdot \mathbf{v}_{EUF}| = 0$, *and for all* $k, l \in \{1, ..., d\}$, *it holds* $|\mathbf{v}_k \cdot \mathbf{v}_l| = |\mathbf{v}_k|^2 \delta_{kl}$, *where* $\cdot$ *is the component-wise product and* $|\mathbf{v}|$ *is the Euclidean norm of* $\mathbf{v}$.

Proof. The proof works by simple construction. We build first $\mathbf{v}_{EUF}$ by setting $\mathbf{v}_{EUF}[i] = \mathbf{v}[i]$ if $r_i \in \mathcal{R}_{EUF}$, and $\mathbf{v}_{EUF}[i] = 0$ otherwise, for $i = 1, ..., n$. Since we know by Lemma 7 that for species $\bar{s}_1, \bar{s}_2 \in D$, we have that $\mathcal{R}_1^* \cap \mathcal{R}_2^* = \varnothing$, we build $\mathbf{v}_j$ by setting $\mathbf{v}_j[i] = \mathbf{v}[i]$ if $r_i \in \mathcal{R}_j^*$, and $\mathbf{v}_j[i] = 0$ otherwise, for $j = 1, ..., d$, $i = 1, ..., n$. $\square$

We can now present a decomposition Theorem for reaction networks.

Theorem 1. *The reaction network* $(X, \mathcal{R}_X)$ *can be partitioned as follows:*

$$X = (E \cup F) \cup D_1 \cup \cdots \cup D_d,$$
$$\mathcal{R}_X = \mathcal{R}_{EUF} \cup \mathcal{R}_1^* \cup \cdots \cup \mathcal{R}_d^*. \tag{6}$$

Moreover, X is self-maintaining if and only if D_j *is weakly-self-maintaining with respect to* $\mathcal{R}_j^*$ *for* $j = 1, ..., d$.

Proof. The first statement is a trivial consequence of Definitions 10 and 13 for partitioning X, and of Corollary 4 for partitioning $\mathcal{R}_X$. For proving the second statement:
$\Rightarrow$: Let $\mathbf{v}$ be a vector which verifies the self-maintainance of X. By using Corollary 5, we decompose the process $\mathbf{v} = \mathbf{v}_1 + ... + \mathbf{v}_d + \mathbf{v}_{EUF}$. Since $\mathbf{v}_j$ contains all the reactions in $\mathcal{R}_j^*$, we have $\mathbf{v}_j(F)$ as in Definition 9 proves that D_j is weak-self-maintaining w.r.t $\mathcal{R}_j^*$, for $j = 1, ..., d$.
$\Leftarrow$: Let $\mathbf{v}_1, ..., \mathbf{v}_d$ be the processes that verify the weak-self-maintainance of D_i, for $i = 1, ..., d$. Then, the vector $\mathbf{v} = \mathbf{v}_1 + ... + \mathbf{v}_d$ verifies the weak-self-maintainance of C w.r.t to $\cup_{j=1}^d \mathcal{R}_j^*$. Finally, we have that $\mathbf{v}(F)$ as in Definition 9 proves the self-maintenance of X. $\square$

5. Revisiting the Types of Change of a System

Theorem 1 provides a way to identify independent dynamical modules in a system and decompose the action of a process in the reaction network into independent actions in these modules. From here, we can target the modules of the systems that become influenced by a structural change in the reaction network. For the sake of simplicity, we will not present a fully-mathematically detailed formulation of the types of change of a system, but show some interesting notions and results that advance in this direction. From now on, we assume that $X = E \cup F \cup D_1 \cup \cdots \cup D_d$ is self-maintaining.

In order to proceed with the analysis, we must formalize some notions.

Definition 15. *Let* $\hat{\mathbf{v}}_{EUF}[i] = 1$ *if* $r_i \in \mathcal{R}_{EUF}$, *and* $\hat{\mathbf{v}}_{EUF}[i] = 0$ *otherwise. Similarly, for each* $j = 1, ..., d$ *we define* $\hat{\mathbf{v}}_j[i] = 1$ *if* $r_i \in \mathcal{R}_j^*$, *and* $\hat{\mathbf{v}}_j[i] = 0$ *otherwise. We say* $\hat{\mathbf{v}}_{EUF}$ *is the unitary process for* $E \cup F$, $\hat{\mathbf{v}}_i$ *is the unitary process for* D_i, *and* $\hat{\mathbf{v}}_X = \hat{\mathbf{v}}_{EUF} + \hat{\mathbf{v}}_1 + ... + \hat{\mathbf{v}}_d$ *is the unitary process in* X.

Definition 16. *Let* $\Lambda \subset \mathbb{R}_{\geq 0}^n$. *We say that* Λ *is a set of allowed processes, or process-structure, of* $(X, \mathcal{R}_X)$ *if and only if* $\mathbf{v} \in \Lambda$ *implies* $\mathbf{v} = \mathbf{v} \cdot \hat{\mathbf{v}}_X$. *For each process-structure* Λ, *we define*

$$\Lambda_{EUF} = \{\bar{\mathbf{v}} = \hat{\mathbf{v}}_{EUF} \cdot \mathbf{v}, \text{ with } \mathbf{v} \in \Lambda\}, \text{ and}$$
$$\Lambda_j = \{\bar{\mathbf{v}} = \hat{\mathbf{v}}_j \cdot \mathbf{v}, \text{ with } \mathbf{v} \in \Lambda\}. \tag{7}$$

We call Λ_{EUF} *the process-structure restricted to* $E \cup F$, *and* Λ_j *the process structure restricted to* D_j.

The process-structure Λ represents all the processes that can be applied to X. The restricted process-structures represent the process-structure when we observe the application of the process in the decomposition modules (obtained in Theorem 1) only.

Note that in the same way that the process represents the stoichiometric counterpart of the time-evolution of the reaction network at the dynamical level of representation, the process-structure is the stiochiometric counterpart of the kinetic law, where a dynamical law is usually a function $\mathbf{K} : \mathbb{R}_{\geq 0}^m \times \mathbb{R}_{\geq 0}^n \to \mathbb{R}_{\geq 0}^n$ that maps a state $\mathbf{x} \in \mathbb{R}_{\geq 0}^n$ of the reaction network and a set of kinetic parameters $\mathbf{k} \in \mathbb{R}_{\geq 0}^m$ to a set of possible processes $\mathbf{K}(\mathbf{x}, \mathbf{k})$. In particular, if the set $\mathbf{K}(\mathbf{x}, \mathbf{k})$ is a singleton (point) for all pair $(\mathbf{x}, \mathbf{k})$, then $\mathbf{K}$ is a deterministic kinetic law at the dynamical level of representation. Since in principle every process is possible, the largest possible process-structure is the set $\mathbb{R}_{\geq 0}^n$. However, we might have to consider a more restricted process-structure in some cases. For example, in order to verify that X is self-maintaining, we must focus on the smaller set of processes $\Pi(\mathcal{R}_X)$. Moreover, dynamical constraints of diverse nature might forbid certain process, where, for example, some reactions occur too much with respect to others.

Therefore, the decomposition of the process, first noticed in Corollary 5 introduced in Definition 16 can be used to trace the influence of each of the behavioral modules in the dynamics of the reaction network, and also to target the modules of the reaction network where a process-structure perturbation produces an impact.

A perturbation to a process-structure Λ leads to a new process-structure Λ' that is similar to Λ according to some geometric criteria. We will not give details on how to formally define a perturbation in the space of process-structures (a simple example of such criteria can be the distance function induced by the sup norm). Applying Definition 16, we can also decompose the process-structure Λ', and study the influence of the perturbation in the decomposition modules. As an example, we discuss two examples of process-structure perturbation and summarize them in Table 1.

Table 1. Table of properties associated with a reaction network analysis for a process-structure perturbation.

Λ'	Change	Might Create	Possible Consequence
Ex. 1	$\Lambda'_{EUF} \subset \Lambda_{EUF}$	$F' \subset F$	Larger minimal fragile-circuits
Ex. 2	$\Lambda'_i \subset \Lambda_i$	D_i is not weak-SM w.r.t $\mathcal{R}_i^*$	X is not self-maintaining

In the first case, suppose Λ' is such that $\Lambda'_{EUF} \subset \Lambda_{EUF}$. In this case, some species $s \in F$ that were overproducible within Λ might not be overproducible within Λ' anymore. Therefore, the set of overproducible species F might change to a smaller set $F' \subset F$. This, in turn, will modify what species are dynamically-connected. In particular, some pairs of species s_1, s_2 that were not dynamically-connected, i.e., they were connected by species $s \in (F - F')$, would become dynamically-connected. Therefore, some dynamically-connected sets will merge, and some other new dynamically-connected sets might be created. Thus, the new decomposition of X will have larger minimal fragile-circuits.

In the second case, suppose Λ' is such that $\Lambda'_i \subset \Lambda_i$. Here, it can be that the self-maintaining processes of D_i become unavailable. Thus, D_i becomes not weak-self-maintaining with respect to $\mathcal{R}_i^*$ anymore. By Theorem 1, we have that this implies that X is not self-maintaining.

For the case of a topology-structure perturbation, we identify four basic types of perturbation and explain some of the possible consequences of such perturbations:

(i) Exclusion of a species $s \in X$: the new set $X' = X - \{s\}$ induces a new set of reactions $\mathcal{R}_{X'} = \mathcal{R}_{X-s}$, and the product s in the reactions $\mathcal{R}_{X'}$ is also eliminated. If s is a catalyst or overproducible species in X, the decomposition of X' might eventually have less minimal fragile-circuits than the decomposition of X. However, it is also possible that some species that were not catalysts or overproducible in X might become a catalyst or overproducible in X', creating more minimal fragile-circuits in the decomposition of X'. Furthermore, if s is in the potential fragile-circuit of X, some species in C might dynamically-disconnect in X', creating more minimal fragile-circuits in the decomposition of X'.

(ii) Exclusion of a reaction $r \in \mathcal{R}_X$: note that, in this case, the set of overproducible species can only become smaller or remain the same, while the set of catalysts can only become larger or remain the same. Moreover, it can also be that r is crucial for dynamically-connecting certain species. In such a case, the decomposition can produce more minimal fragile-circuits.

(iii)   Inclusion of a reaction with existing species: here, we include a new reaction $r \in \mathcal{R}_X$. We identify three basic cases. First, if r becomes part of $\mathcal{R}_{E \cup F}$, then some species in the minimal fragile-circuits might become overproducible (this, in turn, might produce more modules in the decomposition of X). Second, it can be that r consumes (or produces) a catalyst. This will reduce the set of catalysts. Third, r might dynamically-connect species from C that were in different minimal fragile-circuits, which, in turn, will merge such minimal fragile-circuits.

(iv)   Inclusion of a reaction r with a new species s: in this case, the perturbation leads to a new set of species $X \cup \{s\}$ and a new set of reactions $\mathcal{R}_{X \cup \{s\}} = \mathcal{R}_X \cup \{r\}$. We must first identify whether s is a catalyst or an overproducible species. If $s \in C$, we must identify the minimal fragile-circuits that are dynamically-connected. Then, the analysis of this case is equivalent to case (iii).

We summarize these changes and their possible influences in the decomposition of a reaction network in Table 2. In the light of this analysis, we can see that these four basic types of topology-structure perturbation provide an adequate but modest categorization with respect to the influence of a perturbation in the overproducible species and catalyst sets. In order to fully understand the changes in E and F, and to precisely understand the changes in C, and on the minimal fragile-circuits D_i, a much deeper categorization is needed. For example, we can deepen in the categorization of type (i), by separating the cases where the species s excluded from the reaction network is either an overproducible species, a catalyst, or a species from a particular fragile-circuit. Moreover, we can determine to what minimal fragile-circuits s is connected and dynamically-connected. With this information. we can understand precisely how such perturbation would affect the decomposition of the reaction network, and thus its self-maintenance.

Table 2. Table of properties associated with reaction network analysis for a topology-structure perturbation. The symbols $\subset$ ($\subseteq$) and $\supset$ ($\supseteq$) symbols means that the set referred by the column will become smaller (or equal) and larger (or equal), respectively, after a perturbation. # Min Cyc. stands for number of minimal cycles, and SM stands for self-maintainance.

Type	X	$\mathcal{R}_X$	F	E	C	# Min Cyc.	SM
(i)	$\subset$	$\subset$	$\subseteq$ or $\supseteq$	$\subseteq$ or $\supseteq$	$\subseteq$ or $\supseteq$	More/Less	?
(ii)	$=$	$\subset$	$\subseteq$	$\supseteq$	$\subseteq$ or $\supseteq$	More/Less	?
(iii)	$=$	$\supset$	$\supseteq$	$\subseteq$	$\subseteq$ or $\supseteq$	More/Less	?
(iv)	$\supset$	$\supset$	$\supseteq$	$\subseteq$ or $\supseteq$	$\subseteq$ or $\supseteq$	More/Less	?

Remarkably, the information necessary to provide a fine-grained analysis of the topology-structure perturbation, and its relation with the process-structure perturbation, is already confined within the formal notions introduced to develop the decomposition Theorem 1. Thus, we have a framework to analyze modern systemic dynamical notions that depend on these types of perturbations such as resilience, adaptivity, robustness, etc.

For simplicity, we do not elaborate on such fine-grained categorization of perturbations here, but provide an example that illustrates various possible perturbations in what follows.

6. Example

In Figure 2, we show an example of a reaction network with species $\mathcal{M} = \{s, e, f, s_1, ..., s_6\}$ and the reactions:

$$r_1 = s_1 \rightarrow s_2,$$
$$r_2 = s_2 + e \rightarrow e + s_1,$$
$$r_3 = e + s_4 \rightarrow e + s_3,$$
$$r_4 = f + s_3 \rightarrow s_4,$$
$$r_5 = \emptyset \rightarrow f, \tag{8}$$
$$r_6 = f + s_6 \rightarrow s_5,$$
$$r_7 = s_5 \rightarrow s_6,$$
$$r_8 = s \rightarrow e,$$
$$r_9 = e + s \rightarrow 2s.$$

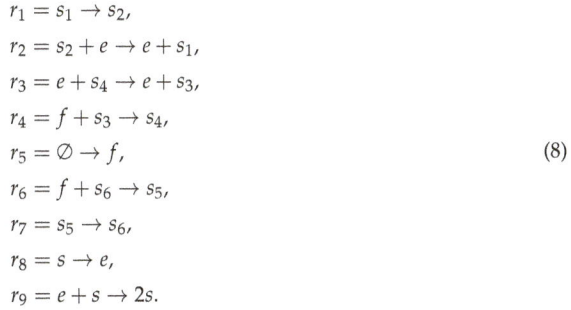

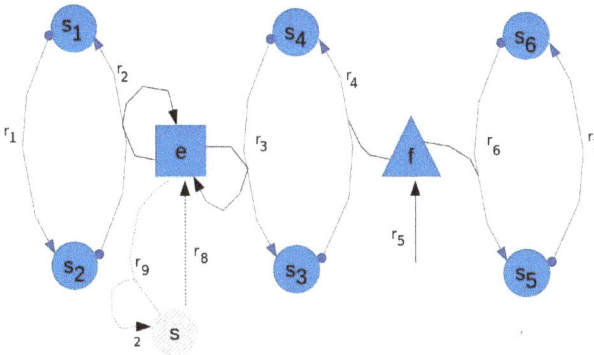

Figure 2. Example of a reaction network for showing notions of decomposition, process-structure and topology-structure perturbation. Arrows are labelled by the number of the reaction that they represent, and, at the end of the arrow, the number of produced species by the reaction is shown only when such number is larger than one.

Let us first consider the set of species $X = \mathcal{M} - \{s\}$. Note that $E = \{e\}$ and $F = \{f\}$ in X. Moreover, we have $C = D_1 \cup D_2 \cup D_3$ with $D_1 = \{s_1, s_2\}$, $D_2 = \{s_3, s_4\}$, and $D_3 = \{s_5, s_6\}$. Hence, we have that a process in $\mathbf{v} \in \Lambda$ is decomposed as $\mathbf{v} = \mathbf{v}_1 + \mathbf{v}_2 + \mathbf{v}_3 + \mathbf{v}_{EUF}$, where

The non-zero components of $\mathbf{v}_1 \in \Lambda_1$ correspond to the reactions $\mathcal{R}_1^* = \{r_1, r_2\}$,

The non-zero components of $\mathbf{v}_2 \in \Lambda_2$ correspond to the reactions $\mathcal{R}_2^* = \{r_3, r_4\}$,

The non-zero components of $\mathbf{v}_3 \in \Lambda_3$ correspond to the reactions $\mathcal{R}_3^* = \{r_6, r_7\}$, (9)

The non-zero components of $\mathbf{v}_{EUF} \in \Lambda_{EUF}$ correspond to the reactions $\mathcal{R}_{EUF} = \{r_5\}$.

For $i = 1, 2, 3$, the minimal fragile-circuit D_i, is weak-self-maintaining if and only if there exists a process $\mathbf{v}_i$ such that its non-zero coordinates are all equal. Note that since f is consumed by the weak-self-maintaining processes of the minimal fragile-circuits D_2 and D_3, the process $\mathbf{v}$ that self-maintains X requires that $\mathbf{v}_{EUF}[5] \geq v_2 + v_3$. By Theorem 1, we have that if such processes $\mathbf{v}_1, \mathbf{v}_2, \mathbf{v}_3$, and $\mathbf{v}_{EUF}$ exist in the respective restricted process structures $\Lambda_i, i = 1, 2, 3$, and Λ_{EUF}, then X is self-maintaining.

In summary, X is self-maintaining if and only if

$$\mathbf{v}_1[1] = \mathbf{v}_1[2] = v_1, \quad \mathbf{v}_1[i] = 0, \text{ otherwise,} \tag{10}$$

$$\mathbf{v}_2[3] = \mathbf{v}_2[4] = v_2, \quad \mathbf{v}_2[i] = 0, \text{ otherwise,} \tag{11}$$

$$\mathbf{v}_3[6] = \mathbf{v}_3[7] = v_3, \quad \mathbf{v}_3[i] = 0, \text{ otherwise,} \tag{12}$$

$$\mathbf{v}_{EUF}[5] \geq v_2 + v_3, \quad \mathbf{v}_{EUF}[i] = 0, \text{ otherwise, } i = 1, ..., n. \tag{13}$$

Consider a process-structure perturbation that leads to a new set of processes $\Lambda' \neq \Lambda$, and the possible cases:

- If Λ' does not allow Equation (10), X is not self-maintaining, and the set $X'_1 = X - D_1$ becomes the largest self-maintaining set.
- If Λ' does not allow Equation (11), X is not self-maintaining, and the set $X'_2 = X - D_2$ becomes the largest self-maintaining set. Also, note that X'_2 is a disconnected set.
- If Λ' does not allow Equation (12), X is not self-maintaining, and the set $X'_3 = X - D_3$ becomes the largest self-maintaining set.
- If Λ' does not allow Equation (13), X is not self-maintaining. Moreover, in case $\mathbf{v}_{EUF}[5] \geq v_2$, the set $X'_3 = X - D_3$ becomes the largest self-maintaining set, and in case $\mathbf{v}_{EUF}[5] < v_2$ $X'_4 = X - (D_3 \cup D_2)$ becomes the largest self-maintaining set inside X, respectively.

We now consider the inclusion of the species s and reactions r_8 and r_9 as a topology-structure perturbation. Since r_8 and r_9 consumes and produces the species e, respectively, we have that e is not a catalyst in $\mathcal{M}$, and thus the minimal fragile-circuits D_1 and D_2 merge into a new minimal fragile-circuit $D_4 = D_1 \cup D_2 \cup \{s\}$ in the decomposition of $\mathcal{M}$. Thus, $\mathcal{M} = F \cup D_3 \cup D_4$.

The analysis of self-maintainance for $\mathcal{M}$ can be done in an analogous way to the analysis for X by defining the restricted process structure Λ_4 of the minimal fragile-circuit D_4. Note, however, that if Λ_4 does not allow processes such that D_4 is not weak-self-maintaining, e.g., when the process vector $\mathbf{v}_4$ cannot have equal values for the coordinates representing r_8 and r_9, we have that the largest self-maintaining set inside $\mathcal{M}$ is $D_3 \cup F$.

Therefore, we can propose that $\mathcal{M}$ is more sensitive than X to a process-structure perturbation because the decomposition of X has more minimal fragile-circuits than the decomposition of $\mathcal{M}$.

7. Conclusions

We have extended the application of reaction networks, and its COT implementation, as a representational framework for systemic modeling [6]. In this framework, a system corresponds to a sub-network that holds structural properties that ensure its qualitative identity (structurally closed) and observability (self-maintaining), i.e., organizations in the COT sense. Technically speaking, organizations characterize the global invariants of the local dynamics and can be computed at a computationally tractable cost. In particular, we developed a formal framework to study structural changes of a system. A structural change goes beyond the change of state, and encompasses changes that modify the inner functioning rules of the system or the very entities defining the system.

We have introduced the notion of dynamical-connectivity, which explains when two species s_1 and s_2 are co-dependent for verifying the self-maintenance of a set X that contains them. This notion, which strongly depends on the overproducible species and catalysts, can be applied to maximally decompose the set X into a collection of minimal fragile-circuits, whose weak-self-maintenance implies the self-maintenance of X (Theorem 1). The decomposition Theorem allows for formalizing two types of structural perturbation that extend the notion of perturbation usually applied in the field of dynamical systems: process-structure and topology-structure perturbation. A process-structure perturbation represents a change of the possible processes in the reaction network universe. The topology-structure perturbation represents a change in the reaction network universe itself, i.e., more/less species and/or more/less reactions. We showed how the decomposition allows for precisely identifying the impact of

Systems **2017**, 5, 30

a perturbation. In particular, a perturbation can either affect a minimal fragile-circuit D_i and/or the set of catalysts and overproducible species $E \cup F$. When the perturbation affects a minimal fragile-circuit, the analysis of its impact can be studied within the minimal fragile-circuit, while when the perturbation affects $E \cup F$, the overall decomposition structure becomes affected. A general perturbation combines effects in some minimal fragile-circuits as well as in $E \cup F$. This, in turn, modifies the set of organizations in the perturbed reaction network. Hence, by comparing the sets of organizations before and after the perturbation, we obtain a picture of the effects of the perturbation in the long-term dynamical properties of the system.

We propose that the structural perturbations introduced in this paper, and the analysis of such perturbations under the light of the decomposition Theorem, can be used as a starting point for the formal study of modern systemic dynamical notions such as resilience, adaptivity, robustness, etc. For example, resilience is generally understood as *the ability of a system to cope with change*. However, this statement is hard to formalize because, since the functioning of a system is understood as a complex process, it is not clear what is a change, or how we identify the part of the system affected by a change, and thus there is no formal framework for defining resilient response-mechanisms. In our framework, we define the notion of structural perturbation that encompass the idea of 'change', vaguely defined in resilience literature. Moreover, since we are able to precisely identify the impact of a structural perturbation, we can characterize response mechanisms within the process-structure.

In order to fully develop the mathematical framework to study modern systemic notions, various aspects must be considered. In particular, a deeper characterization of the process-structure and topology-structure perturbations is needed. Such characterization requires combining geometrical properties of the process-structure Λ with structural information of the reaction network. To do so, not only are more theoretical notions needed, but also an algorithmic framework for computing these theoretical notions for moderately large reaction networks.

As a final comment, we believe that the language of reaction networks could provide a powerful formal framework to open up systemic modeling and systemic thinking to the scientific domain.

Acknowledgments: The FONDECYT postdoctoral scholarship 3170122 has supported the first author and the FONDECYT grant 1150229 has supported the second author.

Author Contributions: Both authors contributed equally.

Conflicts of Interest: The authors declare no conflict of interest.

References

1. Vemuri, V. *Modeling of Complex Systems: An Introduction*; Academic Press: New York, NY, USA, 2014.
2. Strogatz, S.H. *Nonlinear Dynamics and Chaos: With Applications to Physics, Biology, Chemistry, and Engineering*; Westview Press: Boulder, CO, USA, 2014.
3. Boccaletti, S.; Grebogi, C.; Lai, Y.C.; Mancini, H.; Maza, D. The control of chaos: Theory and applications. *Phys. Rep.* **2000**, *329*, 103–197.
4. Maturana, H.R. The organization of the living: A theory of the living organization. *Int. J. Man-Mach. Stud.* **1975**, *7*, 313–332.
5. Licata, I.; Minati, G. Emergence, computation and the freedom degree loss information principle in complex systems. *Found. Sci.* **2016**, doi:10.1007/s10699-016-9503-x.
6. Veloz, T.; Razeto-Barry, P. Reaction networks as a language for systemic modeling: Fundamentals and examples, *Systems* **2017**, *5*, 11.
7. Dittrich, P.; Di Fenizio, P.S. Chemical organisation theory. *Bull. Math. Biol.* **2007**, *69*, 1199–1231.
8. Peter, S.; Dittrich, P. On the relation between organizations and limit sets in chemical reaction systems. *Adv. Complex Syst.* **2011**, *14*, 77–96.
9. Peter, S.; Veloz, T.; Dittrich, P. Feasibility of organizations-a refinement of chemical organization theory with application to p systems. In Proceedings of the Eleventh International Conference on Membrane Computing (CMC11), Jena, Germany, 24–27 August 2010; p. 369.

10. Kreyssig, P.; Wozar, C.; Peter, S.; Veloz, T.; Ibrahim, B.; Dittrich, P. Effects of small particle numbers on long-term behaviour in discrete biochemical systems. *Bioinformatics* **2014**, *30*, i475–i481.
11. Veloz, T.; Reynaert, B.; Rojas, D.; Dittrich, P. A decomposition Theorem in chemical organizations. In Proceedings of the European Conference in Artificial Life—ECAL, Paris, France, 8–12 August 2011.
12. Strang, G. *Introduction to Linear Algebra*; Wellesley-Cambridge Press: Wellesley, MA, USA, 2011.

systems

MDPI

Article

Reaction Networks as a Language for Systemic Modeling: Fundamentals and Examples

Tomas Veloz [1,2,]* **and Pablo Razeto-Barry** [2,3,]*

[1]   Center Leo Apostel for Interdisciplinary Studies, Brussels Free University, 1050 Brussels, Belgium
[2]   Insituto de Filosofía y Ciencias de la Complejidad, Los Alerces 3024, Ñuñoa, Santiago, Chile
[3]   Universidad Diego Portales, Vicerrectoría Académica, Manuel Rodríguez Sur 415, Santiago, Chile
[*]   Correspondence: tveloz@gmail.com (T.V.); prazeto@ificc.cl (P.R.B.)

Academic Editors: Gianfranco Minati, Eliano Pessa and Ignazio Licata
Received: 4 November 2016; Accepted: 18 January 2017; Published: 8 February 2017

Abstract: The basic processes that bring about living systems are conventionally represented in the framework of chemical reaction networks. Recently, it has been proposed that this framework can be exploited for studying various other phenomena. Reaction networks are specially suited for representing situations where different types of entities interact in contextual ways leading to the emergence of meta-structures. At an abstract level, a reaction network represents a universe whose evolution corresponds to the transformation of collections of entities into other collections of entities. Hence, we propose that systems correspond to the sub-networks that are stable enough to be observed. In this article, we discuss how to use reaction networks for representing systems. Namely, we introduce the different representational levels available (relational, stoichiometric, and kinetic), we show how to identify observable systems in the reaction network, discuss some relevant systemic notions such as context, emergence, and meta-system, and present some examples.

Keywords: Reaction Networks; System Theory; Chemical Organization Theory; emergence; context; meta-structure

---

1. Introduction

Systems theory (ST) focuses on the properties, laws, principles and phenomena that different kinds of systems share. Particularly, the founders of systems theory emphasized the importance of studying the structural isomorphism between systems of different domains of reality, which are studied by different scientific disciplines. Consistently, mathematical modeling was proposed as the main interdisciplinary tool of systems theory due to its suitability to represent and handle the formal structure of systems independently of the nature of their components [1,2].

The mathematical modeling of systems has considered a variety of frameworks, usually different when changing from one area of knowledge to another. In general, there is a tradeoff between how precise is the description of the system and its properties, and the number of entities and types of interaction. For example, when an ecological system of only a few species is considered, differential equations are used and exact knowledge of the system can be gathered. However, this approach is not scalable to large ecological systems because the equations involved are too complex and thus the dynamics cannot be computed. In such cases, alternative frameworks such as network or agent-based models are used. These frameworks can simulate some aspects of the dynamics of large systems, but important features that can be studied using differential equations such as the sensitivity to perturbations and the dependence on the system's parameters are lost [3].

Moreover, although well-grounded philosophical and structural principles have been developed for systemic thinking, and although the importance of integrating different areas of knowledge in the mathematical representations has been constantly stressed [4,5], interdisciplinary scientists have had a hard time trying to develop formal representations of systems that integrate diverse areas of knowledge. As a consequence of this, we can conclude that systemic thinking lacks a formal language to express the full scope of systemic thoughts [6].

In conclusion, one of the big challenges in ST is developing a language that allows for putting multiple perspectives into play, but at the same time is mathematically well-grounded so large scale interdisciplinary models can be developed and tested.

In various biochemistry-related areas such as systems biology, bioinformatics, and chemical computing, reaction networks are the mainstream language of representation [7–9]. Interestingly, the language of reaction networks allows for three levels of representation: relational, stoichiometric, and kinetic, respectively. These three levels are increasingly richer in their mathematical structure. On the relational level, one can represent simple structural properties such as connectivity and cycles, and is mainly used for visualization purposes [10,11]. On the stoichiometric level, one can analyze quantitative structural properties of the processes occurring in a reaction network such as elementary flux-modes and self-maintainance [12], and, on the kinetic level, one can compute the time-evolution of the reaction network and perform a detailed study of it [13].

A recent advance of the reaction network formalism is Chemical Organization Theory (COT) [14]. In COT, a reaction network can be associated with a set of *organizations*, which represent the sustainable subnetworks of the reaction network, and can be computed relatively easily. In fact, the study of biochemical reaction networks can hardly be developed using traditional system dynamical tools for large reaction networks [15]. However, organizations have been proven to provide a landscape of the long-term dynamics, and thus characterize the *observable systems* emerging from a reaction network. Hence, COT is a language that helps to bridge the gap between precision of the representation and size of the system. COT has been applied to study several metabolic and other biochemical systems [16–18], and has been proposed as a framework for chemical computation [19] and model checking [20].

Formalisms mathematically equivalent to the language of reaction networks have emerged, rather unexpectedly, in areas outside biology. Namely, formalisms in the early times of parallel computation such as Vector Addition Systems [21] and Petri Nets [22], and, in Linguistics, a formalism known as commutative grammars [23], have been proven to be mathematically equivalent to reaction networks [24,25]. The discovery of these equivalences has led to important cross-fertilizations between biology and computer science [26]. Moreover, since traditional networks are a special type of reaction networks, where only one-to-one relations are allowed, the network-based models in areas such as ecology and social science [27,28] can also be understood as mathematical representations using a simplified version of the language of reaction networks.

Why have reaction networks emerged as a representational language in seemingly different areas of knowledge? The reason is simple but profound. Reaction networks (or any language mathematically equivalent to it) entail a natural way to represent *universes where the interactions among entities are of transformational nature*.

Hence, thinking of reaction networks not as a framework for representing biochemical interactions, but as a language for representing processes of transformation, proposes an interesting way to understand and represent systems as processes that are self-maintaining, and thus stable enough to be observed in time [29,30]. Representing systems by reaction networks not only permits the incorporation of different perspectives into play but also the possibility to represent and study the long-term dynamics of systems with a large number of entities and interactions. Attempts of modeling systems using reaction networks beyond the biochemical domain have been developed in political, decision-making, and economical systems [31–33]. Remarkably, these applications have been carried out combining scholars in various fields including social science, bioinformatics, and mathematics. Hence, these works entail a truly formal interdisciplinary dialogue.

In this paper, we discuss how reaction networks can serve as both *a framework-for-thinking-about* and *a language-for-modeling* systems. In Section 2, we introduce the three levels of description available to represent reaction networks and introduce some relevant properties. In Section 3, we present COT, in Section 4, we discuss how to model systems using reaction networks and present some systemic relevant notions in our framework, in Section 5, we present examples of non biochemical systems modeled using reaction networks, and we present our conclusions and future perspectives in Section 6.

2. Reaction Networks

We introduce the notions of the reaction network formalism that are necessary to understand the relational, stiochiometric and dynamical levels of representation, and to introduce COT. Most of the material of this section, with the exception of the identification of different layers of representation, is standard material in the reaction networks literature. For a comprehensive treatment of the reaction network formalism, we refer to [9,34]. From now on, let $\mathcal{M} = \{s_1, \ldots, s_m\}$ be a set of m *species* that can react with each other according to a set $\mathcal{R} = \{r_1, \ldots, r_n\}$ of n *reactions*. Reactions describe how certain collections of species transform into other collections of species. For a given reaction $r \in \mathcal{R}$, the species to be transformed, i.e., consumed by r, are called *reactants* of r, and the species to be created by this transformation are called *products*. Together, the set of species and the set of reactions are called *reaction network* $(\mathcal{M}, \mathcal{R})$.

In general, some reactions in $\mathcal{R}$ might occur more often than others. A particular specification of the occurrence of reactions within the reaction network is called *reaction process*, or simply *process* (In reaction network modeling, $\mathbf{v}$ is usually called flux vector. We are introducing a slightly more general notion because our aim lies beyond the modeling of biochemical systems) and denoted by $\mathbf{v}$.

Before proceeding with a more detailed description of how reaction networks and processes are represented, note that, for any set of species $X \subseteq \mathcal{M}$, we find a unique maximal set of reactions $\mathcal{R}_X$, defined as the set of all reactions whose reactants are in X. Thus, each set X induces a *sub-network* $(X, \mathcal{R}_X)$. Hence, a process $\mathbf{v}$ *applied to* X can only contain reactions from $\mathcal{R}_X$.

Note that $X \subset X'$ implies $\mathcal{R}_X \subset \mathcal{R}_{X'}$. Thus, a process $\mathbf{v}$ applied to X can always be applied to X'. On the contrary, if $\mathbf{v}$ is a process applied to X', $\mathbf{v}$ can be applied to X only if $\mathbf{v}$ considers reactions in $\mathcal{R}_X$ only.

In this section, we will be concerned with the structure of a set $X \subseteq \mathcal{M}$. To do so, we will introduce three (increasingly more complex) ways to represent a process, and define some properties related to the consumption and production of species. Since we focus on a set $X \subseteq \mathcal{M}$, we assume that $\mathbf{v}$ contain reactions in $\mathcal{R}_X$ only.

2.1. Relational Descriptions

Relational descriptions are the simplest form of representation of reactions and processes. Reactions $r \in \mathcal{R}$ is specified by a set r_C of consumed species, and a set r_P of produced species, and denoted by $r = (r_C, r_P)$, and a process $\mathbf{v}$ is simply a set of reactions.

Definition 1. *X is closed w.r.t.* $\mathbf{v} = \{(r_C^1, r_P^1), \cdots, (r_C^k, r_P^k)\}$ *iff* $\cup_{i=1}^{k} r_P^i \subseteq X$. *If X closed w.r.t. the process* $\mathbf{v} = \mathcal{R}_X$*, we say that X is structurally closed.*

The notion of closure formalizes the fact that no new species are created by a process. For structurally closed sets, no process can create new species. Note that, although X can be closed for certain processes, such as the (trivial) empty process, if X is not structurally closed, the processes for which X is not closed might change its structure. Indeed, when X is not structurally closed, some reactions $(r_C^i, r_P^i) \in \mathcal{R}_X$ are such that r_P^i is not a subset of X. Therefore, at least one species $s \in r_P^i$ is not in X. Hence, whenever such reactions occur in a process applied to X, new species are added. As a consequence, reactions that are in $\mathcal{R}_{X \cup \{s\}}$ but not in $\mathcal{R}_X$ might become available for further processes that, in turn, might add new species. This mechanism can continue until no new species can be added by any process, i.e., when a structurally closed set has been reached.

Lemma 1. *For all X, the structurally closed set $G_{CL}(X)$ of minimal cardinality that contains X is unique [14].*

Definition 2. *X is semi-self-maintaining w.r.t.* $\mathbf{v} = \{(r_C^1, r_P^1), \cdots, (r_C^k, r_P^k)\}$ *if and only if* $\cup_{i=1}^k r_C^i \subseteq \cup_{i=1}^k r_P^i$. *If X is semi-self-maintaining w.r.t.* $\mathbf{v} = \mathcal{R}_X$, *we say that X is structurally semi-self-maintaining.*

Similarly to structural closure, the evolution of a reaction network generally leads to a structural semi-self-maintaining network. Indeed, if a set X is not structurally semi-self-maintaining, then, for some processes, we have that the reactants consumed by the reactions in the process are not being produced by any reaction in the process. Therefore, such species that are not produced are going to be consumed by the processes occurring in the reaction network, until a structurally semi-self-maintaining set is obtained.

Definition 3. *Two species* $s, s' \in X$ *are directly-connected w.r.t. X if and only if there exist a reaction* $r \in \mathcal{R}_X$ *such that* $s, s' \in r_C \cup r_P$. *We say* $s, s' \in X$ *are connected w.r.t. X if and only if there is a sequence of species* $\{s_1, ..., s_k\}$ *such that* $s_1 = s$, s_i *is directly-connected to* s_{i+1} *and* $s_k = s'$.

The connected relation is a generalization of connectivity for traditional networks. In particular, it allows for decomposing a reaction network into a collection of non-interacting subnetworks. In fact, note that every reaction consumes/produces species that are directly-connected to the reactants, and when two species s_1 and s_2 are not connected, we have that none of the species connected to s_1 are connected to any of the species connected to s_2 and vice-versa. From here, it is easy to deduce that any process applied to X can be partitioned into a list of disjoint sub-processes, and that each of these sub-processes in turn correspond to disjoint subsets of species of X. From now on, we will assume that all species in X are connected.

2.2. Stoichiometric Description

Note that, in a relational description, reactions provide information about the type of species transformed only, but not about how many species each type are transformed by the reaction. A stoichiometric description provides a new level of information on how reactions and processes are represented. A reaction r_i is represented by

$$r_i = a_{i1}s_1 + ...a_{im}s_m \rightarrow b_{i1}s_1 + ...b_{im}s_m \tag{1}$$

with a_{ij}, and $b_{ij} \in \mathbb{N}_0$, and $i = 1, ..., n$.

The number $a_{ij} \in \mathbb{N}_0$ denotes the number of reactants of type s_j of the i-th reaction. Together, these numbers form a *reactant matrix* $\mathbf{A} \in \mathbb{N}_0^{n \times m}$. Analogously, the number b_{ij} denominates the number of products of type s_j of the i-th reaction. Together, these numbers form a *product matrix* $\mathbf{B} \in \mathbb{N}_0^{n \times m}$. From here, we can encode the way in which species are consumed and produced by the reactions in the stoichiometric matrix $\mathbf{S} = \mathbf{B} - \mathbf{A}$.

Since the stoichiometric description counts the amount of each type of species involved in the reactions, processes can be extended to specify the number $v_i \in \mathbb{N}_0$ of times that each reaction r_i occurs. Thus, a process corresponds to a vector $\mathbf{v} = (\mathbf{v}[1], ..., \mathbf{v}[n])$.

We can also represent the state of a reaction network by a vector $\mathbf{x}$ of non-negative coordinates such that $\mathbf{x}[j]$ corresponds to the number of species of type s_j in the reaction network, $j = 1, ..., m$. Moreover, we can compute the state $\mathbf{x_v}$ of the reaction network associated to a state $\mathbf{x}$ and a process $\mathbf{v}$ by the following equation:

$$\mathbf{x_v} = \mathbf{x} + \mathbf{Sv}. \tag{2}$$

For simplicity, we have assumed that the coordinates of $\mathbf{x}$ are large enough for the reactions in $\mathbf{v}$ to take place in any order (The study of processes where the number of species is small has been profoundly studied in the context of Petri Nets using the notion of deadlock state [22]. See also [35].).

From here, we can define some relevant quantitative roles for the species participating in a process. For simplicity, we will assume that X is a closed set (When this condition is not satisfied, the definitions below require minor modifications that are not relevant for the purposes of this article.), and that a process $\mathbf{v}$ applied to X is such that if $r_i \notin \mathcal{R}_X$ then $\mathbf{v}[i] = 0$.

Definition 4. *A species $s_j \in X$ is a catalizer w.r.t. X if and only if $r_i \in \mathcal{R}_X$ implies $a_{ij} = b_{ij}$.*

A catalizer is not affected by the action of a process in the reaction network. There are two interesting facts about catalizers. First, if s is a catalizer w.r.t. X, then s is a catalizer with respect to all $X' \subseteq X$, but not necessarily with respect to $X' \supset X$. Second, catalizers w.r.t. X correspond to a row of zeroes in the stoichiometric matrix associated to $\mathcal{R}_X$, and thus are easy to compute.

Lemma 2. *A species $s_j \in X$ is a catalizer w.r.t. X if and only if for all process $\mathbf{v}$ applied to X, we have that $\mathbf{x_v}[j] = \mathbf{x}[j]$.*

Definition 5. *Let $\mathbf{v}$ be a non-null process vector applied to X such that $\mathbf{x_v}$ does not have negative coordinates. If $\mathbf{x_v}[j] > \mathbf{x}[j]$, we say that s_j is overproduced by $\mathbf{v}$ in X.*

Overproduced species have the potential to unlimitedly increase their amount by repeating the process $\mathbf{v}$ which overproduces them. Therefore, overproduced species can be considered as an *unlimited resource* in the reaction network. Interestingly, an overproduced species by $\mathbf{v}$ in X is also overproduced by $\mathbf{v}$ in any $X' \supset X$. Moreover, if two species s_j and s_k in X are overproduced by the processes $\mathbf{v}_j$ and $\mathbf{v}_k$, respectively, it is trivial that $\mathbf{v}_j + \mathbf{v}_k$ overproduces the set $\{s_j, s_k\}$ in X.

Both overproduced species and catalizers can be used to refine the notion of connectivity (see Definition 3), and thus help to provide a much more elegant decomposition of the reaction network into dynamically independent sub-networks. For the sake of simplicity, we will not elaborate on this issue here, but refer the reader to [36] for an elaborated exposition of these results.

Moreover, the notion of (structural) semi-self-maintaining networks can be extended in the stoichiometric description, leading to a quantitative definition of sustainable reaction network.

Definition 6. *Let $\mathbf{v}$ be a non-null process. X is weak-self-maintaining with respect to $\mathbf{v}$ if and only if $\mathbf{x_v}[j] \geq \mathbf{x}[j]$, $j = 1, ..., m$. If, additionally, such process satisfies $\mathbf{v}[i] > 0$ if and only if $r_i \in \mathcal{R}_X$, we say X self-maintaining.*

Lemma 3. *If X is self-maintaining, then X is structurally semi-self-maintaining.*

For a weak-self-maintaining set X, there are processes that lead to a non-negative production of all the species involved in the process. These processes, however, might not use all the reactions in $\mathcal{R}_X$. For self-maintaining sets, we can find processes such that every reaction in $\mathcal{R}_X$ occurs, and the result of the process does not lead to the consumption of any species. Therefore, self-maintaining sets entail the parts of the reaction network where self-sustainable processes, at a quantitative level of description, can occur.

2.3. Kinetic Description

In order to quantify the overall transformation of species derived from a process $\mathbf{v}$ occurring in time, we represent the state vector as a function of time $\mathbf{x}(t) = (x_1(t), ..., x_m(t))$, where $x_j(t)$ encodes the number of species s_j at time t.

Suppose that the process $\mathbf{v}$ occurs between times t_0 and t_1. Therefore, we can obtain $\mathbf{x}(t_1)$ from $\mathbf{x}(t_0)$, $\mathbf{S}$, and $\mathbf{v}$ as follows:

$$\mathbf{x}(t_1) = \mathbf{x}(t_0) + \mathbf{Sv}. \tag{3}$$

This equation provides a formal description for the change of the number of species driven by a process **v** [37].

By setting the diffenence between t_1 and t_2 infinitely small, Equation (3) becomes the differential equation

$$\dot{x} = Sv, \qquad (4)$$

with initial conditions specified by $x(t_0)$.

In this case, the process vector is a function of time. Usually, **v** is conceived as a function of time t, the state vector $x(t)$, and a vector of parameters **k** associated to the reactions and given by the dynamical rules of the system. The common case of continuous dynamics is the mass-action kinetic law [13], where the state vector $x(t) \in \mathbb{R}_{\geq 0}^m$ represents the concentration of species in the reaction network at time t, and the process vector function **v** is defined by:

$$v_i(t, k) = k_i \prod_{j=1}^{m} x_j^{a_{ij}}$$

for $i = 1, \ldots, n$, and $k = (k_1, ..., k_n)$ is a strictly positive vector whose coordinates are called reaction rate constants.

For (discrete or continuous) probabilistic dynamics, the process vector **v** represents the probability of occurrence of the reactions in the network.

A reaction network together with the discrete/continuous and deterministic/probabilistic kinetic law is called a *reaction system*. A reaction system is the most refined description of a reaction network because it describes how the local dynamics evolve.

3. Connecting the Description Levels: Chemical Organization Theory

The relational, stoichiometric and kinetic levels of representation present three increasingly precise ways of representing a reaction network and its processes. However, the gain in precision is compensated with an increase in the computational resources required to identify the properties of the reaction network. In Table 1, we summarize some important structural features and the computational resources required for identifying such structures at each level of representation.

Table 1. Table of scalability of properties depending on the level of representation. Each property is either not computable, or a level or scalability is associated. A property is more scalable if it can be computed for larger networks. Hence, Full, Moderate, and Hard scalability represent three levels of increasingly more complex computation, respectively.

Property-Type/Level	Relational	Stoichiometric	Kinetic
Topological Structure	Full	Full	Full
Phase Space Analysis	Uncomputable	Moderate	Hard
Time Evolution	Uncomputable	Moderate	Hard

Namely, the relational description is capable of identifying connectivity-related properties in the network by means of simple set-like operations, but is unable to describe properties of quantitative nature. The stoichiometric description allows for describing properties of quantitative nature by means of matrix algebra operations (which are computationally tractable for moderately large networks), but it is unable to describe the influence of kinetic parameters and the precise time-evolution of a reaction network. The kinetic description is able to fully represent the influence of parameters and the time-evolution of a reaction network, but such description requires solving a highly coupled and nonlinear system of (either discrete, stochastic, or differential) equations. These equations do not have an analytic solution in most cases. Hence, the exploration of the dynamics of a reaction network requires numerical solutions that become intractable for large reaction networks.

Since a full-featured understanding of the dynamics of a reaction network is intractable at the kinetic level, COT proposes that certain structural properties at the kinetic level can be traced at the relational or stoichiometric levels, and thus at a computationally affordable cost. Such structural traces are, technically speaking, necessary conditions for the desired properties of the reaction system. In particular, COT focuses on the connection between structural properties at the relational and stoichiometric levels with the long-term behavior of the reaction system.

To this end, COT introduces the crucial notion of organization:

Definition 7. *X is a (semi-)organization if and only if X is closed and (semi-)self-maintaining.*

An organization satisfies simultaneously the relational-level property of closure and the stoichiometric-level property of self-maintaining. By combining these two requirements, COT identifies the structural footprint of stable dynamics. Namely, a closed set of species entails a subnetwork whose processes do not produce new species, and within these closed dynamics, there are processes that allow self-maintenance of the quantity of species in the system. Therefore, as long as self-maintaining processes occur, the subnetwork $(X, \mathcal{R}_X)$ will be preserved in time.

In order to connect the notion of organization with the dynamical level, we introduce the following notions.

Definition 8. *Let $\mathcal{P}(\mathcal{M})$ be the power set of $\mathcal{M}$ and*

$$\phi(t) : \mathbb{R}_{\geq 0}^m \rightarrow \mathcal{P}(\mathcal{M}), \; \mathbf{x}(t) \mapsto \phi(\mathbf{x}(t)) \equiv \{s_i \in \mathcal{M} : x_i(t) > 0\}. \tag{5}$$

*For a state $\mathbf{x}(t) \in \mathbb{R}_{\geq 0}^m$, the set $\phi(\mathbf{x}(t))$ is the **abstraction** of $\mathbf{x}(t)$. For a given set of species $X \subseteq \mathcal{M}$, a state $\mathbf{x}(t) \in \mathbb{R}_{\geq 0}^m$ is an **instance** of X if and only if its abstraction equals X.*

The notions of abstraction and instance connect the representations of the reaction network with the reaction system, and organizations represent the abstractions of all the possible stable instances:

Theorem 1. *If $\mathbf{x}$ is a fixed-point of the ODE (4), i.e., $\mathbf{Sv}(\mathbf{x}, \mathbf{k}) = 0$, then the abstraction $\phi(\mathbf{x})$ is an organization [33].*

Fixed points entail the simplest dynamically stable instances of a reaction system, and are crucial for determining most important features of the dynamics of a system [38]. Thus, Theorem 1 provides a link between the long-term behavior of a reaction system and its underlying reaction network. In simple words, it proves that a necessary condition to be a fixed point at the kinetic level is to be an organization at the stoichiometric level (and thus a a semi-organization at the relational level). Moreover, in [39], Theorem 1 is extended to other stable asymptotic behaviors such as periodic orbits and limit cycles. In addition to these results, necessary conditions for the existence of adequate flux vectors are explored in [40], and algorithmic studies concerning the computation of the organizations of a reaction network are presented in [41–43].

4. Discussion: Reaction Networks and the Modeling of Systems

We now discuss some general aspects about using reaction networks as a language for the modeling of systems.

4.1. Reaction Networks as Universes and Organizations as Systems

In the reaction network formalism, we define species and reactions to specify how the entities we consider interact and transform. Therefore, it is important to stress that we do not start from the idea of a pre-existing system to be modeled. On the contrary, we start from a set of relevant entities, which can be of any nature (physical, cognitive, economic, etc.), and we determine a set of rules that specify how combinations of these entities transform into new combinations. Since, for a system, it is generally assumed that it is, to some extent, stable in time so it can be observed, and it holds certain properties that entail its qualitative identity, we propose that the more adequate notion of system is a subnetwork $(X, \mathcal{R}_X)$ such that X is an organization. Therefore, systems are conceived as self-maintaining entities that emerge from the universe of interactions [29,44]. This explains the notion of qualitative identity of a system from a dynamic perspective. Namely, a system is continuously changing its inner components and sub-processes, but the qualitative identity and unity are secured as long as the reaction network is structurally closed and its inner processes are self-maintaining. Hence, the reaction network $(\mathcal{M}, \mathcal{R})$ plays the role of universe of existence and interaction, and the organizations play the role of potentially observable systems in this universe.

The latter view allows for a recursive representation of the systems in a universe [45]. In case the whole reaction universe is an organization, we conceive it as the largest possible system, if not, several largest systems might exist. By looking inside the largest systems, we find smaller systems that are contained in the largest systems, and continue recursively until we arrive to the smallest organizations, which play the role of minimal observable entities.

Since we are free to chose our basic entities and processes, a fundamental step when modeling with reaction networks is to choose a *basic representational ontology* which includes the fundamental entities in the transformational universe, and then define the transformation rules among these basic entities. Next, we can extend the representation of such universe by incorporating either new entities and new reactions, or by replacing existing entities and reactions by a deeper or fine-grained representation of the replaced entities.

4.2. Inner and External Contexts

In the reaction network formalism, we identify two fundamentally different notions of context. The first is the epistemic (or external) context that corresponds to the choice of the subnetwork to be considered. The external context specifies what entities and interactions we consider. In this approach, we start from our universe $(\mathcal{M}, \mathcal{R})$ and analyze specific external contexts represented by subnetworks $(X, \mathcal{R}_X)$. The choice of an external context constrains the entities to be found and the transformations allowed to occur.

The second context operates within the external context. This behavioral (or inner) context specifies what processes **v** are allowed to occur and how they occur. The inner context **v** determines whether a subnetwork is self-maintaining, and thus an organization. If the inner context forbids the occurrence of self-maintaining processes, we have that $(X, \mathcal{R}_X)$ will not be stable in time, and thus not observable.

The previous observation implies an interesting dichotomy between structure (external context) and dynamics (inner context) in the study of systems. The importance of this dichotomy has not been widely acknowledged by the biochemistry-related community working with reaction networks models. It occurs for two main reasons. First, in most biochemical cases, the reaction network is meant to model a predefined system, thus the notion of external context is of virtual nature only. Secondly, biochemical processes are governed by deterministic physical laws. Hence, the inner context of the reaction network is fixed by deterministic principles of the biochemical domain [13].

However, by considering reaction networks as a language for modeling systems, species might not be biochemical or even physical entities. Hence, processes might not necessarily obey any determined set of rules, and thus external and inner contexts might become relevant and influence each other.

4.3. The Emergence of Systems and Meta-Systems

In the reaction network approach, the emergence of a stable system is a natural consequence of the dynamics. In fact, Theorem 1 states that stable dynamical regimes correspond to organizations in the reaction network. This is equivalent to say that the systems we observe are observable because they are stable enough in time to be observed, and that this stability is the consequence of a dynamical process.

Therefore, systems are stable structures of processes emerging from the transformational interactions, i.e., dynamics, occurring in the universe.

In COT, a reaction network is represented as a hierarchy of subnetworks. At the lowest level, we have the empty reaction network (0 species), and each subsequent level has subnetworks having more species. At the highest level, we find the set $\mathcal{M}$. This representation is known as the Hasse diagram (see Figure 1).

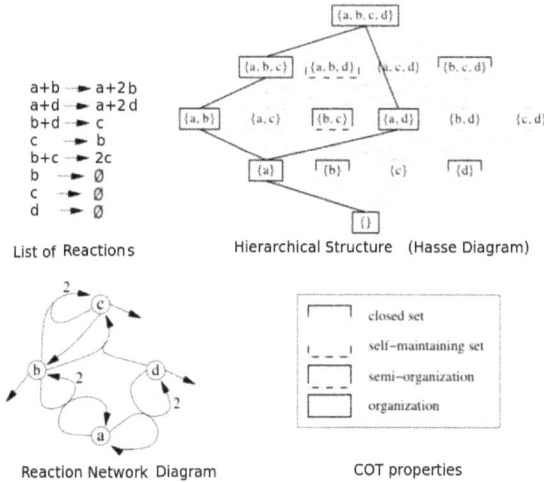

Figure 1. Example of hierarchy of reaction networks and their properties in COT. Inspired from [14].

This setting is convenient for explaining not only the emergence of systems from the interaction of species, but also the emergence of meta-systems from the interaction of systems. Since we conceive organizations as systems, we note that when two or more organizations interact, a new reaction network is formed. Since the new reaction network can have reactions that are not in any of the former organizations, the union of two (or more) organizations may or may not be an organization. In the example in Figure 1, we see that both $\{a, b\}$ and $\{b, d\}$ are organizations, but when they combine to become $\{a, b, d\}$, the new reaction $b + d \rightarrow c$ becomes active. Therefore, $\{a, b, d\}$ is not closed and thus is not an organization, but its closure $\{a, b, c, d\}$ is closed and self-maintaining, and thus an organization. This simple example illustrates that organizations form a hierarchy, and that this hierarchy can be used to explain how meta-systems can non-trivially emerge by the interaction of systems.

4.4. The Lack of Identity Problem and the Membrane Solution

One important drawback of the reaction network formalism is that the species do not have individual identity. This means that given a certain species type s, we have that all the species of this type are equivalent in the reaction network. This has proven to be problematic for modeling systems where species represent entities having mechanisms of memory or recognition [32]. Moreover, there is a way to construct virtual cells of interaction in an extended reaction network formalism known as membrane or P-systems [46]. In the P-systems framework, we allow different reaction networks to exist, and each of these reaction networks is enclosed in a membrane and thus is allowed to evolve separately from the other reaction networks. However, each reaction network is also allowed to exchange species with other reaction networks by means of a common space. Therefore, indirect communication between reaction networks is possible.

The use of membranes brings a new modeling dimension because we can attribute agency properties to reaction networks, and since membranes can be recursively defined, we can assign recursive layers of individual identity within an agent. Moreover, by properly labeling species according to the compartment they belong to, we can model interactions where mechanisms of recognition and even memory operate. Remarkably, it has been shown that when, for even the simplest cases where two reaction networks interact, it is possible that both networks co-stabilize in an organization, even though the two reaction networks are not stable on their own [40].

4.5. Resilience and Other Modern Systemic Notions

The reaction network approach provides a suitable landscape of concepts to formalize some modern systemic notions. As an example, we will elaborate on the notion of resilience. Resilience has been defined as *the ability of a system to cope with change* (There is a large number of definitions for the concept of resilience [47]. We use this definition due to its simplicity and generality.). By 'cope', authors generally mean 'to be able to maintain its qualitative identity', and 'change' means 'a perturbation.' However, the notion of qualitative identity, as well as the notion of perturbation, are generally applied in a non-formal manner. This leads to multiple interpretations of the notion of resilience. In our approach, the notions behind resilience can be properly defined. In fact, the qualitative identity of a system in this setting corresponds to an organization together with a self-maintaining process occurring in it, and the perturbation corresponds to three different types of change. The first is a change of state, and this means increasing or decreasing the values of the coordinates of the vector **x**. The second is a change of the inner context, and this means changing the set of possible processes **v** that can occur in the system. The third is a change of the external context, and this means adding or eliminating species and/or reactions in the system.

Since the notion of identity and change are properly defined, resilience can be formally studied using reaction networks. In a similar way, we propose that other notions introduced in the system theory literature such as robustness, adaptivity, etc. can be formalized using structural properties of a reaction network. We will not elaborate on the details of these notions here, but refer to [36] for a mathematical framework to formalize such notions.

5. Examples

We now overview three cases of non-biochemical systems that have been modeled using reaction networks.

5.1. Social System: Political Structure

The use of reaction networks as a language for modeling political systems was inspired by Luhmann's approach to sociology [48]. Luhmann introduced the notion of communication as the basis of societies' structuring and ordering [49]. The concept of communication is defined as the flow produced by the exchange of *social-symbols*. These symbols belong to different social structures. For example, for simple economical, legal and political structures, the communication flow is done through *money, justice* and *power*, respectively. In a general case, all of these structures overlap, and, hence, communications in one system may affect the others. Therefore, a social system emerges from these structures of communication.

In [31], Dittrich and Winter developed a reaction network that represents a *toy-model* of the political system based on Luhmann's concept of communication. They define 13 *communication species*, e.g., *social movement demands* '*SBFor*' (acronym from the German: Soziale Bewegung Forderung), *social movement members* '*SBMit*' (Soziale Bewegung Mitglieder), *potential collective binding decisions* '*KVEPot*' (Potenzielle Kollectiv Verbindliche Entscheidungen), etc., and a set of 20 reactions to model the interactions among these communications. For example, the reaction '*SBMit* + *KVEPot* → *SBFor*' models that *Demands from social movements can be stimulated by potential collectivly binding decisions*. The species '*SBMit*' corresponds to a social movement demand decision (e.g., *do not increase the tax*), that may be expressed as a protest, or by other actions, '*KVEPot*' might correspond to a potential law such as *increase the tax*, and '*SBFor*' corresponds to the communications that the social movement members discuss or spread (for example within social networks). For simplicity, we will not present the full set of species and reactions, but provide a diagram of the topology of the network and the hierarchy of organizations in Figure 2.

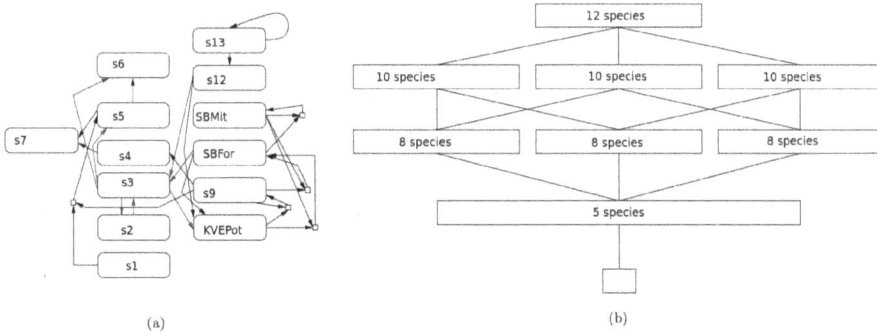

Figure 2. (**a**) Topology of the reaction network model of the political system. Labelled boxes represent species and arrows represent reactions. (**b**) The hierarchy of organizations.

In this network, the simplest organization represents a political system where there is formal political power of high and low levels, public opinion, thematic conflicts, and public force. The dynamics of self-maintaining networks is given by how public opinion influences thematic conflicts and how they get resolved by the political power. It resembles a monarchy-like system. More complex organizations in the hierarchy involve either social movements and social demands, political parties and their programs, or collectively binding decisions and their implementation. Further levels combine these structures in different ways, and the highest level at the hierarchy is the conjunction of all these cases together.

5.2. Decision System: Evolutionary Game Theory

In a game theoretical setting, an agent interacts with another agent by deciding a certain action on the basis of a set of possible actions and the payoffs of these actions. The payoffs depend on the decision of both agents, but no agent knows the decision of the other agent. For example, if we consider cooperative C and defecting D decisions, we have that a cooperative interaction requires two cooperative agents generating payoff for the cooperative payoff specified by P_C by the reaction $C + C \rightarrow C + C + 2P_C$, while the interaction of a cooperative decision with a defecting decision, $C + D \rightarrow C + D + P_D$, generates payoff for the defecting decision P_D only.

In the evolutionary game theory setting, agents are allowed to interact several times. They can eventually recognize and remember other agents as well as their past actions. From here, each agent develops a 'strategy' that sets how agents interact with each other depending on past interactions.

Since the interactions in evolutionary game theory are between agents, agent-based modeling is the dominant paradigm to represent these systems [50]. Unfortunately, it is very difficult to develop analytic results for agent-based models, and performing simulations to explore the parameter space is computationally very expensive when several strategies are in play. As an alternative, in [32], a reaction network model was developed to represent the evolutionary game theoretical setting of the prisoner's dilemma, and, in particular, the evolution of cooperation problem [51]. Species play the role of decisions and payoffs, and a reaction network is built from the payoff matrix of decisions (see Figure 3).

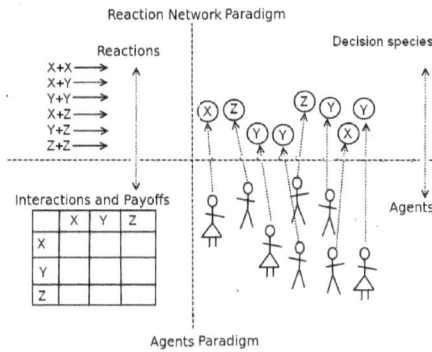

Figure 3. Paradigm change from agent-based (**bottom**) to reaction network (**top**) modeling. The interaction among agents corresponds to a vessel of decision species interacting, and the payoff matrix corresponds to a set of reactions which consumes a pair of decisions to produce the payoff of each decision and two new decisions determined by the strategies.

The reaction network model is able to fully reproduce the results obtained using agents for the evolutionary prisoner's dilemma. Interestingly, a formula that explains when cooperation is evolutionarily stable is obtained [32]. This formula is equivalent to the famous result obtained by Nowak in [52].

5.3. Ecological Systems

Ecological interactions among biological species can be modeled using reaction networks. As an example, Table 2 provides a simple model of a list of ecological interactions using reaction networks.

Table 2. Reactions associated to most common ecological interactions.

Reaction	Ecological Interaction
$prey + predator \rightarrow 2predator$	Depredation
$host + hosted \rightarrow 2hosted$	Parasitism
$host + hosted \rightarrow host + 2hosted$	Comensalism
$host + hosted \rightarrow host$	Amensalism
$Coop_1 + Coop_2 \rightarrow 2Coop_1 + 2Coop_2$	Mutualism

Since, in this setting, interactions are many-to-many directed relations, reaction networks allow a more complex representation of ecological interactions than traditional network models. Moreover, a more detailed account of the ecological concepts can be developed in certain ecosystems. For example, consider the mutualistic interaction between mychorrizae and plants [53]. Namely, plants y feed from mycelium x_r to grow roots y_r (r_1 in Equation (6)), mycorrhizae x feeds from the roots y_r to produce mycelium (r_2 in Equation (6)), and contributes to the production of mycelium x_r (r_3 in Equation (6)), which, in turn, increments the absorption capacities of plants y (r_4 in Equation (6)). Therefore, the following set of reactions:

$$r_1 = y + x_r \rightarrow y + y_r \text{ (Plant grow roots)},$$
$$r_2 = x + y_r \rightarrow x + x_r \text{ (Roots foster the growth of mychorrizea)},$$
$$r_3 = x \rightarrow x + x_r \text{ (Mychorrizea produces mycelium)},$$
$$r_4 = y + x_r \rightarrow 2y \text{ (Mycelium foster the growth of plants)},$$

(6)

provides a more complex model of a mutualistic relation than the one shown in Figure 2.

Therefore, reaction networks can be used to model the mechanisms of ecological interactions. Remarkably, COT provides a suitable conceptual landscape to formalize ecological notions. For example, organizations can be understood as sustainable ecosystems, and invasion of a particular ecological species x in an eco-system E can be modeled by adding species and reactions that represent the interaction mechanisms of the species x with the ecological species in the ecosystem E [30].

6. Conclusions

In this paper, we have proposed the language of reaction networks, and particularly its COT implementation, as a representational framework for systemic modeling. In particular, we focused on crucial notions in a system's theory such as the notion of the system itself, context, meta-system, resilience, etc., and presented a reaction network model for a political, agent-decision, and ecological systems.

Remarkably, this framework does not require the choice of a particular point of view, or field of knowledge, that serves as a reference for representing a system. On the contrary, it is required to identify a set of entities, that can be, in principle, of any nature (physical, biological, social, economical, etc.), and define a set of interactions of transformational nature (reactions) among them. The set of interactions is understood as a universe of basic processes, i.e., a reaction network. From here, a system corresponds to a sub-network such that its structural properties ensure its qualitative identity (closed) and observability (self-maintaining), i.e., organizations in the COT sense. Technically speaking, organizations characterize the global invariants of the local dynamics, and can be computed at a computationally tractable cost.

Since reactions are allowed to combine entities of different nature (see, for example, the interactions combining payoffs and decisions in Section 5.2), this approach is a priori interdisciplinary.

In order to advance the modeling of systems using COT, we envisage various challenges. First, this perspective requires proving its usefulness beyond the toy-model studies. We believe that a model of an ecological universe of interactions, extending the toy-model presented in Section 5.3, is a good option. Second, COT is still at an early stage of mathematical development. Several advances can be made applying lattice theoretical notions [54] to the hierarchy of organizations [14], and more profound studies in the topology of reaction networks could provide a more rich structure that the one presented in this paper. Such a richer structure could be used not only to better understand how systems emerge and combine, but also for improving algorithms regarding the computation of organizations [41]. In this vein, we present some advances in this issue [36].

Last but not least, we believe that it is fundamental to develop a semi-formal methodology that ensures that interdisciplinary reaction networks can be designed by combining the expertise from different fields. Indeed, previous reaction networks models of non-biochemical systems required an extensive dialogue among the disciplines involved in the problem in question (e.g., sociologist, economist, and biologist). The aim of such dialogue was to find an ontology (set of species) and the interaction mechanisms (sets of reactions) that combine the different perspectives into play. This exhaustive dialogue has been carried out due to the motivation of the members involved in the respective studies. However, a methodology to advance on such a dialogue could foster the application of this formalism at a wider scale.

We suggest two ways in which this potentially divergent process can be improved. The first is to provide semantic tools to construct the ontology given by the set of species. Researchers can be aided by current powerful language taxonomies and semantic tools that can help them to identify misunderstandings and ambiguous meanings. The second is the possibility to visualize the output of the COT analysis of a reaction network (e.g., see Figure 1b). In this way, researchers can modify in real-time the structure of the reaction network and observe how the structural properties of the reaction network depend on the local interactions.

The ultimate goal of using reaction networks as a language for modeling systems is to bring systemic thinking closer to the real world. We aim at not only scholars, but also decision-makers and the general public contributing with their knowledge and expertise, so multiple perspectives can be integrated in a single framework that brings forward a broader understanding of the emergent consequences of our local actions.

Acknowledgments: This work has been supported by the FONDECYT Grant 1150229 and the FONDECYT Postdoctoral Scholarship 3170122.

Author Contributions: All authors contributed equally to this work.

Conflicts of Interest: The authors declare no conflict of interest.

References

1. Bogdanov, A. *Essays in Tektology: The General Science of Organization*; Intersystems Publications: Seaside, CA, USA, 1984.
2. Von Bertalanffy, L. *General Systems Theory*; George Braziller: New York, NY, USA, 1968; Volume 41973, p. 40.
3. Vemuri, V. *Modeling of Complex Systems: An Introduction*; Academic Press: Cambridge, MA, USA, 2014.
4. Ashby, W.R. *An Introduction to Cybernetics*; Chapman & Hall: London, UK, 1956.
5. Laszlo, E.; Clark, J.W. *Introduction to Systems Philosophy*; Gordon and Breach: New York, NY, USA, 1972.
6. Goertzel, B. *Chaotic Logic: Language, Thought, and Reality from the Perspective of Complex Systems Science*; Springer Science & Business Media: Berlin, Germany, 2013; Volume 9.
7. Dittrich, P.; Ziegler, J.; Banzhaf, W. Artificial chemistries—A review. *Artif. Life* **2001**, *7*, 225–275.
8. Kitano, H. Systems biology: A brief overview. *Science* **2002**, *295*, 1662–1664.
9. Steuer, R.; Junker, B.H. Computational models of metabolism: Stability and regulation in metabolic networks. *Adv. Chem. Phys.* **2009**, *142*, 105–251.
10. Albrecht, M.; Kerren, A.; Klein, K.; Kohlbacher, O.; Mutzel, P.; Paul, W.; Schreiber, F.; Wybrow, M. *On Open Problems in Biological Network Visualization*; International Symposium on Graph Drawing; Springer: Berlin/Heidelberg, Germany, 2009; pp. 256–267.
11. Lee, D.Y.; Yun, H.; Park, S.; Lee, S.Y. MetaFluxNet: The management of metabolic reaction information and quantitative metabolic flux analysis. *Bioinformatics* **2003**, *19*, 2144–2146.
12. Schuster, S.; Fell, D.A.; Dandekar, T. A general definition of metabolic pathways useful for systematic organization and analysis of complex metabolic networks. *Nat. Biotechnol.* **2000**, *18*, 326–332.
13. Feinberg, M.; Horn, F.J. Dynamics of open chemical systems and the algebraic structure of the underlying reaction network. *Chem. Eng. Sci.* **1974**, *29*, 775–787.
14. Dittrich, P.; Di Fenizio, P.S. Chemical organisation theory. *Bull. Math. Biol.* **2007**, *69*, 1199–1231.
15. Klamt, S.; Gagneur, J.; von Kamp, A. Algorithmic approaches for computing elementary modes in large biochemical reaction networks. *IEE Proc. Syst. Biol.* **2005**, *152*, 249–255.
16. Matsumaru, N.; Centler, F.; di Fenizio, P.S.; Dittrich, P. Chemical organization theory applied to virus dynamics (Theorie chemischer organisationen angewendet auf infektionsmodelle). *IT Inf. Technol.* **2006**, *48*, 154–160.
17. Kreyssig, P.; Escuela, G.; Reynaert, B.; Veloz, T.; Ibrahim, B.; Dittrich, P. Cycles and the qualitative evolution of chemical systems. *PLoS ONE* **2012**, *7*, e45772.
18. Centler, F.; Dittrich, P. Chemical organizations in atmospheric photochemistries—A new method to analyze chemical reaction networks. *Planet. Space Sci.* **2007**, *55*, 413–428.
19. Matsumaru, N.; Centler, F.; di Fenizio, P.S.; Dittrich, P. Chemical organization theory as a theoretical base for chemical computing. In *Proceedings of the 2005 Workshop on Unconventional Computing: From Cellular Automata to Wetware*; Luniver Press: Beckington, UK, 2005; pp. 75–88.
20. Kaleta, C.; Richter, S.; Dittrich, P. Using chemical organization theory for model checking. *Bioinformatics* **2009**, *25*, 1915–1922.
21. Karp, R.M.; Miller, R.E. Parallel program schemata. *J. Comput. Syst. Sci.* **1969**, *3*, 147–195.
22. Murata, T. Petri nets: Properties, analysis and applications. *IEEE Proc.* **1989**, *77*, 541–580.
23. Crespi-Reghizzi, S.; Mandrioli, D. Commutative grammars. *Calcolo* **1976**, *13*, 173–189.
24. Heiner, M.; Koch, I.; Voss, K. Analysis and simulation of steady states in metabolic pathways with Petri nets. In Proceedings of the Workshop and Tutorial on Practical Use of Coloured Petri Nets and the CPN Tools (CPN'01), Aarhus, Denmark, 29–31 August 2001; pp. 15–34.

25. Veloz González, T.I. A Computational Study of Algebraic Chemistry. Master Thesis, Universidad de Chile, Santiago, Chile, 2010.
26. Heiner, M.; Gilbert, D.; Donaldson, R. Petri nets for systems and synthetic biology. In *International School on Formal Methods for the Design of Computer, Communication and Software Systems*; Springer: Berlin/Heidelberg, Germany, 2008; pp. 215–264.
27. Pascual, M.; Dunne, J.A. *Ecological Networks: Linking Structure to Dynamics in Food Webs*; Oxford University Press: Oxford, UK, 2006.
28. Borgatti, S.P.; Mehra, A.; Brass, D.J.; Labianca, G. Network analysis in the social sciences. *Science* **2009**, *323*, 892–895.
29. Razeto-Barry, P. Autopoiesis 40 years later. A review and a reformulation. *Orig. Life Evol. Biosph.* **2012**, *42*, 543–567.
30. Veloz, T. Teoría de organizaciones químicas: Un lenguaje formal para la autopoiesis y el medio ambiente. In *Autopoiesis: Un Concepto Vivo*; Universitas Nueva Civilización: Santiago, Chile 2014. (In Chile)
31. Dittrich, P.; Winter, L. Chemical organizations in a toy model of the political system. *Adv. Complex Syst.* **2008**, *11*, 609–627.
32. Veloz, T.; Razeto-Barry, P.; Dittrich, P.; Fajardo, A. Reaction networks and evolutionary game theory. *J. Math. Biol.* **2014**, *68*, 181–206.
33. Dittrich, P.; Winter, L. Reaction networks as a formal mechanism to explain social phenomena. In Proceedings of the 4th Iternational Workshop on Agent-Based Approaches in Economics and Social Complex Systems (AESCS 2005), Tokyo, Japan, 9–13 July 2005; pp. 9–13.
34. Feinberg, M. *Lectures on Chemical Reaction Networks*; Notes of lectures given at the Mathematics Research Center, University of Wisconsin: Madison, WI, USA, 1979.
35. Kreyssig, P.; Wozar, C.; Peter, S.; Veloz, T.; Ibrahim, B.; Dittrich, P. Effects of small particle numbers on long-term behaviour in discrete biochemical systems. *Bioinformatics* **2014**, *30*, i475–i481.
36. Veloz, T.; Razeto-Barry, P. Studying Structural Changes of a System using Reaction Networks. *Systems*, submitted for publication, 2016.
37. Horn, F.; Jackson, R. General mass action kinetics. *Arch. Ration. Mech. Anal.* **1972**, *47*, 81–116.
38. Strogatz, S.H. *Nonlinear Dynamics and Chaos: With Applications to Physics, Biology, Chemistry, and Engineering*; Westview Press: Boulder, CO, USA, 2014.
39. Peter, S.; Dittrich, P. On the relation between organizations and limit sets in chemical reaction systems. *Adv. Complex Syst.* **2011**, *14*, 77–96.
40. Peter, S.; Veloz, T.; Dittrich, P. Feasibility of Organizations—A Refinement of Chemical Organization Theory with Application to P Systems. In Proceedings of the 11th International Conference on Membrane Computing (CMC11), Jena, Germany, 24–27 August 2010; p. 369.
41. Centler, F.; Kaleta, C.; di Fenizio, P.S.; Dittrich, P. Computing chemical organizations in biological networks. *Bioinformatics* **2008**, *24*, 1611–1618.
42. Centler, F.; Kaleta, C.; di Fenizio, P.S.; Dittrich, P. A parallel algorithm to compute chemical organizations in biological networks. *Bioinformatics* **2010**, *26*, 1788–1789.
43. Veloz, T.; Reynaert, B.; Rojas, D.; Dittrich, P. A decomposition theorem in chemical organizations. In Proceedings of the European Conference in Artificial Life, Paris, France, 8–12 August 2011.
44. Maturana, H.; Varela, F. *Autopoiesis and Cognition: The Realization of the Living*; Springer: Dordrecht, The Netherlands, 1985.
45. Mesarovic, M.D.; Macko, D.; Takahara, Y. *Theory of Hierarchical, Multilevel, Systems*; Elsevier: Amsterdam, The Netherlands, 2000; Volume 68.
46. Păun, G. Computing with membranes. *J. Comput. Syst. Sci.* **2000**, *61*, 108–143.
47. Folke, C. Resilience: The emergence of a perspective for social–ecological systems analyses. *Glob. Environ. Chang.* **2006**, *16*, 253–267.
48. Luhmann, N. *The Differentiation of Society*; Columbia University Press: New York, NY, USA, 1983.
49. Razeto-Barry, P.; Cienfuegos, J. La paradoja de la probabilidad de lo improbable y el pensamiento evolutivo de Niklas Luhmann. *Convergencia* **2011**, *18*, 13–38.
50. Axelrod, R. *The Complexity of Cooperation*; Princeton University Press: Princeton, NJ, USA, 1997.
51. Axelrod, R.M. *The Evolution of Cooperation*; Basic Books: New York, NY, USA, 2006.
52. Nowak, M.A. Five rules for the evolution of cooperation. *Science* **2006**, *314*, 1560–1563.

53. Harley, J.L. *The Biology of Mycorrhiza*; Leonard Hill (Books) Ltd.: London, UK, 1959.
54. Birkhoff, G.; Birkhoff, G.; Birkhoff, G.; Mathématicien, E.U.; Birkhoff, G. *Lattice Theory*; American Mathematical Society: New York, NY, USA, 1948; Volume 25.

systems

MDPI

Article

Emergence at the Fundamental Systems Level: Existence Conditions for Iterative Specifications

Bernard P. Zeigler [1,*] and Alexandre Muzy [2]

[1]   Co-Director of the Arizona Center for Integrative Modeling and Simulation (ACIMS), University of Arizona and Chief Scientist, RTSync Corp. 12500 Park Potomac Ave. #905-S, Potomac, MD 20854, USA
[2]   CNRS, I3S, Université Côte d'Azur, 06900 Sophia Antipolis, France; alexandre.muzy@cnrs.fr
*   Correspondence: zeigler@rtsync.com

Academic Editors: Gianfranco Minati, Eliano Pessa and Ignazio Licata
Received: 17 August 2016; Accepted: 25 October 2016; Published: 9 November 2016

Abstract: Conditions under which compositions of component systems form a well-defined system-of-systems are here formulated at a fundamental level. Statement of what defines a well-defined composition and sufficient conditions guaranteeing such a result offers insight into exemplars that can be found in special cases such as differential equation and discrete event systems. For any given global state of a composition, two requirements can be stated informally as: (1) the system can leave this state, i.e., there is at least one trajectory defined that starts from the state; and (2) the trajectory evolves over time without getting stuck at a point in time. Considered for every global state, these conditions determine whether the resultant is a well-defined system and, if so, whether it is non-deterministic or deterministic. We formulate these questions within the framework of iterative specifications for mathematical system models that are shown to be behaviorally equivalent to the Discrete Event System Specification (DEVS) formalism. This formalization supports definitions and proofs of the afore-mentioned conditions. Implications are drawn at the fundamental level of existence where the emergence of a system from an assemblage of components can be characterized. We focus on systems with feedback coupling where existence and uniqueness of solutions is problematic.

Keywords: emergence; uniqueness; existence of solutions; input/output system; system specifications; Discrete Event System Specification

1. Introduction

Emergence has been characterized as taking place in strong and week forms. Mittal [1] pointed out that strong emergent behavior results in generation of new knowledge about the system representing previously unperceived complex interactions. This can occur in the form of one or more of new abstraction levels and linguistic descriptions, new hierarchical structures and couplings, new component behaviors, and new feedback loops. Once understood and curated, the behavior returns to the weak form, as it is no longer intriguing, and then can begin to be treated in regularized fashion. Emergent behavior is likely an inherent feature of any complex system model because abstracting a continuous real-world system (e.g., any complex natural system) to a constructed system-model must leave gaps of representation that may diverge in unanticipated directions. In [2] philosophically, following Ashby [3] and Foo and Zeigler [4], the perceived global behavior (holism) of a model might be characterized as: Components (reductionism) + interactions (computation) + higher-order effects where the latter can be considered as the source of emergent behaviors [5,6].

The Discrete Event Systems Specification (DEVS) formalism has been advocated as an advantageous vehicle for researching such structure–behavior relationships because it provides the components and couplings for models of complex systems and supports dynamic structure for genuine adaption and evolution. Furthermore, DEVS enables fundamental emergence modeling because it operationalizes the closure-under-coupling conditions that form the basis of well-defined resultants of system composition especially where feedback coupling prevails [7,8].

In this paper, we formulate conditions under which a composition of component systems form a well-defined system-of-systems at a fundamental level. Formal statement of what defines a well-defined composition and sufficient conditions guaranteeing such a result offer insight into exemplars that can be found in special cases such as differential equation and discrete event systems. Informally stated, we show that for any given global state of a composition, two requirements can be stated as: (1) the system can leave this state, i.e., there is at least one trajectory defined that starts from the state; and (2) the trajectory evolves over time without getting stuck at a point in time. Considered for every global state, these conditions determine whether the resultant is a well-defined system and if so, whether it is non-deterministic or deterministic. We formulate these questions within the framework of iterative specifications for mathematical system models that are shown to be behaviorally equivalent to the Discrete Event System Specification (DEVS) formalism. This formalization supports definitions and proofs of the afore-mentioned conditions and allows us to exhibit examples and counter-examples of condition satisfaction. Drawing on Turing machine halting decidability, we investigate the probability of legitimacy for randomly constructed DEVS models. We close with implications for further research on the emergence of new classes of well-defined systems.

Let us start with a well-known concept, the Turing Machine (TM) (cf. [9]). Usually, it is presented in a holistic, unitary manner but as shown in Figure 1b, we can decompose it into two stand-alone independent systems: the TM Control ($S1$) and the Tape System, ($S2$). Foo and Zeigler [4] argued that the re-composition of the two parts was an easily understood example of emergence wherein each standalone system has very limited power but their composition has universal computation capabilities. Examining this in more depth, the Tape system shown in Figure 1a is the dumber of the two, serving a memory with a slave mentality, it gets a symbol (sym) and a move (mv) instruction as input, writes the symbol to the tape square under the head, moves the head according to the instruction, and outputs the symbol found at the new head location. The power of the Tape system derives from its physicality—its ability to store and retrieve a potentially infinite amount of data—but this can only be exploited by a device that can properly interface with it. The TM Control by contrast has only finite memory but its capacity to make decisions (i.e., use its transition table to jump to a new state and produce state-dependent output) makes it the smarter executive. The composition of the two exhibits "weak emergence" in that the resultant system behavior is of a higher order of complexity than those of the components (logically undecidable versus finitely decidable), the behavior that results can be shown explicitly to be a direct consequence of the component behaviors and their essential feedback coupling—cross-connecting their outputs to inputs as shown by the arrows of Figure 1b. We are going to use this example to discuss the general issues in dealing with such compositions.

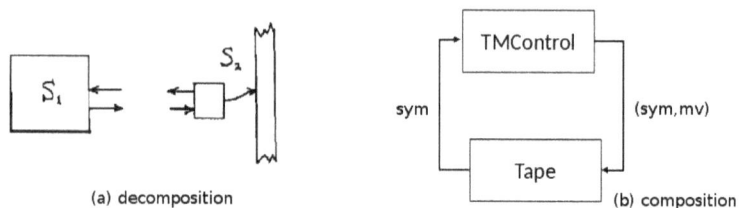

Figure 1. Turing Machine decomposition (**a**) and modular composition (**b**).

The interdependent form of the interaction between the TM control and the tape system illustrates a pattern found in numerous information technology and process control systems and recognized early on by simulation language developers [10] (though not necessarily in the modular form described here). In this interaction, each component system alternates between two phases, active and passive. When one system is active the other is passive—only one can be active at any time. The active system does two actions: (1) it sends an input to the passive system that activates it (puts it into the active phase); and (2) it transits to the passive phase to await subsequent re-activation. For example, in the re-composed Turing Machine, the TM control starts a cycle of interaction by sending a symbol and move instruction to the tape system then waiting passively for a new scanned symbol to arrive. The tape system waits passively for the sym, mv pair. When it arrives, it executes the instruction and sends the symbol now under the head to the waiting control.

Such active–passive compositions provide a class of systems from which we can draw intuition and examples for generalizations about system emergence at the fundamental level. We will employ the modeling and simulation framework based on system theory formulated in [10] especially focusing on its concepts of iterative specification and the Discrete Event Systems Specification (DEVS) formalism. Special cases of memory-less systems and the pattern of active–passive compositions are discussed to exemplify the conditions resulting ill-definition, deterministic, and non-deterministic as well as probabilistic systems. We provide sufficient conditions, meaningful especially for feedback coupled assemblages, under which iterative system specifications can be composed to create a well-defined resultant and that moreover can be simulated in the DEVS formalism.

However, to address more fundamental issues, we need to start with a more primitive and perhaps more intuitive notion of a system. As in Figure 2, consider a concept of system with states, transitions, and times associated with transitions. For example, there are transitions from state $S1$ to state $S3$ and from $S3$ to $S4$ which each takes 1 time unit and there is a cycle of transitions involving $S4, \ldots, S7$ each of which take zero time. There is a self-transition involving $S2$ which consumes an infinite amount of time (signifying that it is passive, remaining in that state forever.) This is distinguished from the absence of any transitions out of $S8$. A state trajectory is a sequence of states following along existing transitions, e.g., $S1, S3, S4$ is such a trajectory.

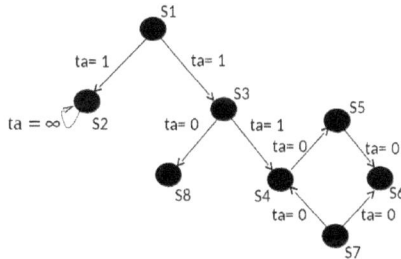

Figure 2. System with tuned transitions.

This example gives us a quick understanding of the conditions for system existence at the fundamental level.

We say that the system is:

- *not defined* at $S8$ because there is no trajectory emerging from it;
- *non-deterministic* at $S1$ because there are two distinct outbound transitions defined for it; and
- *deterministic* at $S2$ and $S4$ because there is only one outbound transition for each.

We say that the system is *well-defined* if it is defined at all its states. These conditions are relative to static properties, i.e., they relate to states not how the states follow one another over time. In contrast, state trajectories relate to dynamic and temporal properties. When moving along a trajectory, we keep adding the time advances to get the total traversal time, e.g., the time taken to go from $S2$ to $S4$ is 2. Here, a trajectory is said to be *progressive* in time if time always advances as we extend the trajectory. For example, the cycle of states $S4 \ldots S7$ is not progressive because as we keep adding the time advances the sum never increases. Conceptually, let us start a clock at 0 and, starting from a given state, we let the system evolve following existing transitions and advancing the clock according to the time advances on the transitions. If we then ask what the state of the system will be at some time later, we will always be able to answer if the system is well-defined and progressive. A well-defined system that is not progressive signifies that the system gets stuck in time and after some time, it becomes impossible to ask what the state of the system is in after that time. Zeno's paradox offers a well-known metaphor where the time advances diminish so that time accumulates to a point rather than continue to progress and offers an example showing that the pathology does not necessarily involve a finite cycle. Our concept of progressiveness generalizes the concept of legitimacy for DEVS [10] and deals with the "zenoness" property which has been much studied in the literature [11]. We return to it in more detail later.

Thus, we have laid the conceptual groundwork in which a system has to be well-defined (static condition) and progressive (temporal dynamic condition) if it is to have achieved independent existence when emerging from a composition of components.

2. Formal Background

Hereafter are presented basic and coupled specifications.

2.1. Basic Discrete Event System Specification

As presented in [12], a *Discrete Event System Specification (DEVS)* is a structure

$$M = < X, S, Y, \delta_{int}, \delta_{ext}, \lambda, ta >$$

where X is the *set of input values*, S is a *set of states*, Y is the *set of output values*, $\delta int : S \rightarrow S$ is the *internal transition function*, $\delta_{ext} : Q \times X \rightarrow S$ is the *external transition function*, where $Q = \{(s, e) | s \in S, 0 \le e \le ta(s)\}$ is the *total state* where e is the *time elapsed* since last transition, $\lambda : S \rightarrow Y$ is the *output function*, $ta : S \rightarrow \mathbb{R}_\infty^+$ is the *time advance function*.

In the Turing machine example of Figure 1, the tape system and control engine are each atomic DEVS. For the tape system, a state is a triple $(tape, pos, mv)$ where the tape is an infinite sequence of zeros and ones (symbols), pos represents the position of the head, and mv is a specification for moving left or right. An external transition accepts a symbol, move pair, writes the symbol in the square of the current head position and stores the move for the subsequent internal transition that executed the specified move. For the control engine, a state is a pair (st, sym) where st is a control state and sym is a stored symbol. An external transition stores the received symbol for subsequent use. An internal transition applies the TM transition table to the (st, sym) pair and transitions to the specified control state.

The core of this concept simplifies when the system is input-free (or autonomous, i.e., not responding to inputs) and outputs are not considered. Then, we have a transition system

$$M = < S, \delta, ta >$$

where $\delta \subseteq S \times S$ and $ta : S \times S \rightarrow \mathbb{R}_\infty^+$.

Here, we are allowing the transition system to be non-deterministic so that rather than δ and ta being functions they are presented as relations.

For example, in Figure 2, we have $S = \{S1, S2, S3, S4, S5, S6, S7\}$, $\delta = \{(S1, S3), (S3, S4), ...\}$, and $ta(S1, S3) = 1$, $ta(S3, S4) = 1, ...$

In this formal version of the informal statements above, the transition system, M, is

- *not defined* at S8 because there is no transition pair with S8 as the left member in δ;
- *non-deterministic* at S1 because it is a left member of two transition pairs $(S1, S2)$ and $(S1, S3)$; and
- *deterministic* at S2 and S4 because there is only one transition pair involving each one as a left member.

This gives concrete and formal expression to the earlier definition of these concepts.

2.2. Coupled DEVS Models

DEVS models can be coupled to form coupled models that themselves are DEVS models manifesting closure under coupling. To keep the presentation as straightforward as possible in this paper we will limit the discussion to coupling of two components without external inputs or outputs. This is illustrated in Figure 1b where the coupling recipe maps the output of the control engine (*symbol, move pair*) to the input of the tape system and likewise, the output of the tape system (*symbol*) to the input of the control engine. This mapping is assumed to take zero time (any delay in a real manifestation can be modeled by inserting delay components). To sketch the way the resultant of coupling is computed and show that it is expressed as a DEVS, let $M = < X, S, Y, \delta_{int}, \delta_{ext}, \lambda, ta >$ and $M' = < X', S', Y', \delta'_{int}, \delta'_{ext}, \lambda', ta' >$.

Then, the resultant is a DEVS:

$$M_{coup} = < Q_{coup}, \delta_{coup}, ta_{coup} >$$

where $Q_{coup} = Q \times Q'$, $ta_{coup} : Q_{coup} \to R_\infty^+$, $\delta_{coup} : Q_{coup} \to Q_{coup}$ and $ta((s,e),(s',e')) = \min\{ta(s) - e, ta'(s') - e'\}$.

To explain, the next event will occur according the time advance and will be driven by the imminent component (whose next event time is the minimum of the two—neglecting ties for simplicity here). For illustration, let t^* be the time advance and let the second component be imminent. This component generates its output and sends it to the first via the coupling which reacts to it using its external transition function establishing its next state $\delta_{ext}(s, t^*, \lambda'(s'))$; at the same time, the imminent component applies its internal transition function to establish its next state $\delta'_{int}(s')$. Thus, we have $\delta_{coup}((s,e),(s',e')) = (\delta_{ext}(s, t^*, \lambda'(s')), \delta'_{int}(s'))$.

Having briefly reviewed the basics of DEVS, we are ready to consider the general case of iterative specification of which DEVS is an example.

3. Iterative System Specifications

We briefly review an approach to Iterative Specification of Systems that was introduced in [12] and provide more detail in Appendix B. I/O systems describe system behavior with a global transition function that determines the final state given the initial state and the applied input segment. Since input segments are left segmentable and closed under composition, we are able to generate the state and output values along the entire input interval. However, such an approach is not very practical. What we need is a way to generate state and output trajectories in an iterative way going from one state along the trajectory to the next.

Iterative specification of systems is a general scheme for defining systems by iterative applications of *generator segments*. These are elementary segments from which all input segments of a system can be generated. Having such a concept we define a *generator state transition function* and iteratively apply generator segments to the generator state transition function. The results produced by the generator state transitions constitute the state trajectory for the input segment resulting from the composition of the generator segments. The general scheme of iterative specification forms a basis for more specialized types of specifications of systems. System specification formalisms are special forms of iterative specifications with their special type of generator segments and generator state transition functions.

Consider (Z, T) the set of all segments $\{\omega :< 0, t_1 > \to Z \mid t_1 \in T\}$. Here, the notation, $\omega :< 0, t_1 > \to Z\}$ means that ω is a mapping from an interval of time base, T to a set of values Z. For a subset Γ of (Z, T), the *concatenation closure* of Γ is denoted Γ^+. Example generators are bounded continuous segments, and constant segments of variable length generating bounded piecewise continuous segments piecewise constant segments, respectively. Unfortunately, if Γ generates Ω, we *cannot* expect each $\omega \in \Omega$ to have a unique decomposition by Γ. A single representative, or *canonical* decomposition can be computed using a *Maximal Length Segmentation* (MLS). First we find ω_1, the longest generator in Γ that is also a left segment of ω. This process is repeated with what remains of ω after ω_1 is removed, generating ω_2, and so on. If the process stops after n repetitions, then $\omega = \omega_1 \omega_2 \ldots \omega_n$. We say that Γ is an *admissible* set of generators for Ω if Γ generates Ω and for each $\omega \in \Omega$, a unique MLS decomposition of ω by Γ exists.

The following is the basis for further analysis in this paper:

Theorem 1. *Sufficient Conditions for Admissibility. If Γ satisfies the following conditions, it admissibly generates Γ^+:*

1. *Existence of longest initial segments:* $\omega \in \Gamma^+ \Rightarrow max\{t \mid \omega_{t>} \in \Gamma\}$ *exists*
2. *Closure under right segmentation:* $\omega \in \Gamma \Rightarrow \omega_{<\tau} \in \Gamma$ *for all $\tau \in dom(\omega)$*

An *iterative specification* of a system is a structure

$$G = < T, X, \Omega_G, Y, Q, \delta_G, \lambda >$$

where $T, X, Y,$ and Q have the same interpretation as for I/O systems, Ω_G is an admissible set of input segment generators, $\delta_G : Q \times \Omega_G \to Q$ is the single segment state transition function, and $\lambda : Q \times X \to Y$ is the output function.

An iterative specification $G = < T, X, \Omega_G, Y, Q, \delta_G, \lambda >$ specifies a time invariant system, $S_G = < T, X, \Omega_G^+, Y, Q, \delta_G^+, \lambda >$. The system is well-defined if δ_G^+, the extension of δ_G, has the composition property, i.e., $\delta_G^+(q, \omega_1 \bullet \omega_2) = \delta_G^+(\delta_G^+(q, \omega_1), \omega_2)$, for all $\omega_1, \omega_2 \in \Omega_G^+$.

For more details, see Appendix B.

3.1. DEVS Simulation of Iterative Specification

As reviewed in Appendix B, the notion of iterative specification was introduced to characterize diverse classes of systems such as differential equation systems and discrete time systems. With the motivation of including discrete event systems under the same umbrella as more familiar systems, DEVS was defined using iterative specification. Here we show that the converse is also true, namely, that DEVS can directly represent iterative specifications. Given an iterative specification $G = < T, X, \Omega_G, Y, Q, \delta_G, \lambda >$, we construct a DEVS model $M = < X, S, Y, \delta_{int}, \delta_{ext}, \lambda, ta >$ that can simulate it in a step-by-step manner moving from one input segment to the next. The basic idea is that we build the construction around an encoding of the input segments of G into the event segments of M as illustrated in Figure 3. This is based on the MLS which gives a unique decomposition of input segments to G into generator subsegments. The encoding maps the generator subsegments. ω_i in the order they occur into corresponding events which contain all the information of the segment itself. To do the simulation, the model M stores the state of G. It also uses its external transition function to store its version of G's input generator subsegment when it receives it. M then simulates G by using its internal transition function to apply its version of G's transition function to update its state maintaining correspondence with the state of G.

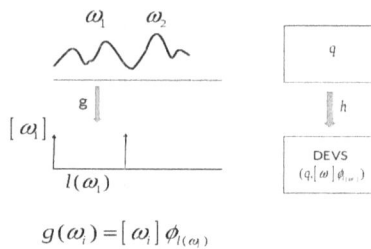

$$g(\omega_i) = [\omega_i] \, \phi_{l(\omega_i)}$$

Figure 3. DEVS Simulation of an Iterative Specification.

Details of the construction are provided in Appendix C. We note that the key to the proof is that iterative specifications can be expressed within the explicit event-like constraints of the DEVS formalism, itself defined through an iterative specification. Thus DEVS can be viewed as the computational basis for system classes that can be specified in iterative form satisfying all the requirements for admissibility.

3.2. Coupled Iterative Specification

Although closure of coupling holds for DEVS, the generalization to iterative specification does not immediately follow. The problem is illustrated in Figure 4 where two iterative system specifications, ISP1 and ISP2, are cross-coupled such as exemplified by the Turing machine example in Figure 1. The cross-coupling introduces two constraints shown by the equalities in the figure, namely, the input of ISP2, $\omega 2$ must equal the output of ISP1, $\rho 1$, and the input of ISP1, $\omega 1$ must equal the output of ISP2, $\rho 2$. Here we are referring to input and output trajectories over time as suggested graphically in Figure 4. Since each system imposes its own constraints on its input/output relation, the conjunction of the four constraints (two coupling-imposed, two system-imposed) may have zero, one, or multiple solutions.

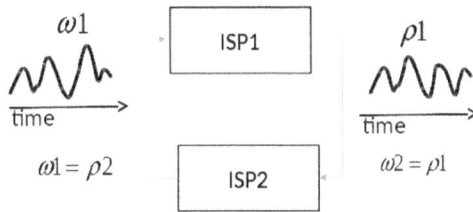

Figure 4. Input and output trajectories of coupled systems.

Definition 1. *A Coupled Iterative Specification is a network specification of components and coupling where the components are iterative specifications at the I/O System level.*

As indicated, in the sequel, we deal with the simplified case of two coupled components. However, the results can be readily generalized with use of more complex notation. We begin with a definition of the relation of input and output segments that a system imposes on its interface. We need this definition to describe the interaction of systems brought on through the coupling of outputs to inputs.

Definition 2. *The I/O Relation of an iterative specification is inherited from the I/O Relation of the system that it specifies. Likewise, the set of I/O Functions of an iterative specification is inherited from the system it specifies. Formally, let $S_G = < T, X, \Omega_G^+, Y, Q, \delta_G^+, \lambda >$ be the system specified by $G = < T, X, \Omega_G, Y, Q, \delta_G, \lambda >$. Then the I/O Functions associated with G is $\beta : Q \times \Omega_G^+ \to \Omega_G^+$ where for $q \in Q, \omega \in \Omega_G^+$, $\beta(q, \omega) = \lambda(\delta_G^+(q, \omega))$.*

Let $\beta 1$ and $\beta 2$ represent the I/O functions of the iterative specifications, ISP1 and ISP2, respectively. Applying the coupling constraints expressed in the equalities above, we make the definition:

Definition 3. *A pair of output trajectories $(\rho 1, \rho 2)$ is a consistent output trajectory for the state pair $(q1, q2)$ if $\rho 1 = \beta 1(q1, \rho 2)$ and $\rho 2 = \beta 2(q2, \rho 1)$.*

Definition 4. *A Coupled Iterative Specification has unique solutions if there is a function, $F: Q1 \times Q2 \to \Omega$; with $F(q1, q2) = (\rho 1, \rho 2)$, where there is exactly one consistent pair $(\rho 1, \rho 2)$ with infinite domain for every initial state $(q1, q2)$. Infinite domain is needed for convenience in applying segmentation.*

Definition 5. *A Coupled Iterative Specification is admissible if it has unique solutions.*

The following theorems are proved in Appendix D.

Theorem 2. *An admissible Coupled Iterative Specification specifies a* well-defined *Iterative Specification at the I/O System level.*

Theorem 3. *The set of Iterative Specifications is closed under admissible coupling.*

Theorem 4. *DEVS coupled model can component-wise simulate a coupled Iterative Specification.*

Proof. The coupled model has components, which are DEVS representations of the individual Iterative Specifications (according to Theorem D in Appendix D) and also a coordinator as shown in Figure 5.

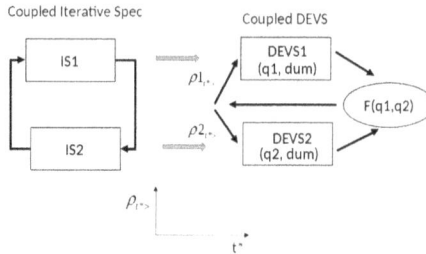

Figure 5. DEVS Component-wise simulation of Coupled Iterative Specifications.

The coordinator receives the current states of the components and applies the *F* function to compute unique consistent output segments. After segmentation using the MLS as in the proof of Theorem D in Appendix D, it packages each as a single event in a DEVS segment as shown. Each DEVS component computes the state of its Iterative Specification at the end of the segment as in Theorem D. Then it sends this state to the coordinator and the cycle repeats. This completes an informal version of the proof which would formally proceed by induction. □

The solution function *F* represents an idealization of the fixed point solutions required for Differential Equation System Specification (DESS) and the local solution approaches of Discrete Time System Specification (DTSS) and Quantized DEVS [12]. Generalized Discrete Event System Specification (GDEVS) [13] polynomial representation of trajectories is the closest realization but the approach opens the door to realization by other trajectory prediction methods.

3.3. Special Case: Memoryless Systems

Consider the case where each component's output does not depend on its state but only on its input.

Let gr represent the ground state in which the device is always found (all states are represented by this state since output does not depend on state.) In the following $R1$ and $R2$ are the I/O relations of systems 1 and 2, respectively. In this case, they take special forms:

$$\rho 1 = \beta 1(gr, \omega 1) \Leftrightarrow (\omega 1, \rho 1) \in R1$$
$$\rho 2 = \beta 2(gr, \omega 2) \Leftrightarrow (\omega 2, \rho 2) \in R2$$
$$\omega 2 = \rho 1$$
$$\omega 1 = \rho 2$$
$$\rho 2 = \beta 2(q2, \omega 2) \Leftrightarrow (\rho 1, \omega 1) \in R2 \Leftrightarrow (\omega 1, \rho 1) \in R2^{-1}$$

i.e., $(\rho 2, \rho 1)$ is consistent $\Leftrightarrow (\rho 2, \rho 1) \in R1 \cap R2^{-1}$

Let *f* and *g* be defined in the following way:

$$f(\rho2) = \beta1(gr, \rho2)$$
$$g(\rho1) = \beta2(gr, \rho1)$$

So for any $\rho1$, $\rho1 = f(\rho2) = f(g(\rho1))$ and $g = f^{-1}$ (considered as a relation).
Finally, $F(q1, q2) = (\rho1, f^{-1}(\rho1))$ has:

- no solutions if f^{-1} does not exist, yielding no resultant;
- a unique solution for every input if f^{-1} exists, yielding a deterministic resultant; and
- multiple solutions for a given segment ρ if $f^{-1}(\rho)$ is multivalued, yielding a non-deterministic resultant.

For examples, consider an *adder* that always adds 1 to its input, i.e.,

$$\beta(gr, \rho) = f(\rho) = \rho + 1$$
$$i.e., \forall t \in T, \beta(gr, \rho)(t) = \rho(t) + 1$$

Cross-coupling a pair of adders does not yield a well-defined resultant because $f^{-1}(\rho) = \rho - 1 \neq f$. However, coupling an adder to a *subtracter*, its inverse, $\beta(gr, \rho) = \rho - 1$, yields a well-defined deterministic system.

For other examples, consider combinatorial elements whose output is a logic function of the input and consider a pair of gates of the same type connected in a feedback loop:

- A NOT gate has an inverse $0 \rightarrow 1, 1 \rightarrow 0$, so the composition has two solutions one for each of two complementary assignments, i.e., $F(gr, gr) = \{(1,0), (0,1)\}$ yielding a non-deterministic system.
- An AND gate with one of its input held to 0 always maps the other input into 0 so has a solution only for inputs of 0 to each component, i.e., $F(gr, gr) = (0,0)$, yielding a deterministic system.
- An AND gate with a stuck-at input of 1 is the identity mapping and has multiple solutions i.e., $F(gr, gr) = \{(0,0), (1,1)\}$, yielding a non-deterministic system.

3.4. Active-Passive Systems

We now show that active–passive systems as described earlier offer a class of systems for which the iterative specifications of components satisfy the admissibility conditions specified in Theorem 4. As in Figure 6a, consider a pair of cross-coupled systems each having input generators that represent null and non-null segments.

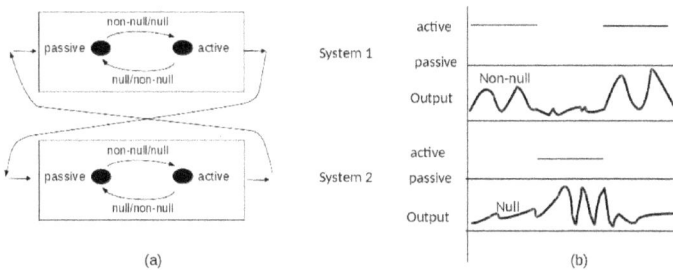

(a)   (b)

Figure 6. Active–Passive Example of Admissible Progressive Coupled Iterative Specifications.

A null generator represents the output of a passive system whereas a non-null generator represents the output of an active system. For example in the TM case, a non-null generator is a segment that represents transmission of a symbol by the tape unit and of a symbol, move pair for the TM control. As in Figure 6b let *S1* and *S2* start as active and passive, respectively. Then, they output non-null and null generators, respectively. Since the null generator has infinite extent, the end-time of the non-null generator determines the time of next event, t^* (as defined earlier) and we apply the transition functions in Figure 6a to find that at t^* the systems have reversed phases with *S1* in passive and *S2* in active.

Definition 6. *Define a triple* $< Q, \delta_G, \lambda >$ *by* $Q = \{active, passive\} \times \{active, passive\}$ *with* $F(active, passive) = (non\text{-}null, null)$ $\delta_G(active, passive) = (\delta(active, null_{t^*>}) = \delta(passive, non\text{-}null_{t^*>})) = (passive, active)$ *with* $t^*(active, passive) =$ *endtime of the non-null generator segment* $\lambda(active, passive) = (non\text{-}null, null)$

Let the non-null and null generators stand for sets of concrete segments that are equivalent with respect to state transitions and outputs. Then, such a scheme can define a deterministic or non-deterministic resultant depending on the number of consistent pairs of *(non-null, null)* generator segments are possible from the *(active, passive)* state. Furthermore, mutatis-mutandis for the state *(passive, active)*.

4. Temporal Progress: Legitimacy, Zenoness

Although we have shown how a DEVS coupled model can simulate a coupling of iterative specifications, it does not guarantee that such a model is legitimate [12]. Indeed, backing up, legitimacy of the simulating DEVS is reflective of the temporal progress character of the original coupling of specifications. We see that the time advance has a value provided that the function F is defined for current state. However, as in the case of zenoness, a sequence of such values may accumulate at a finite point thus not allowing the system to progress beyond this point. Consequently, we need to extend the requirements for coupled iterative specifications to include temporal progress in order to fully characterize the resultant composition of systems. Thus, we can extend the definitions:

Definition 7. *An Iterative Specification is* progressive *if every sequence of time advances diverges in sum.*

Definition 8. *A Coupled Iterative Specification is* progressively admissible *if its components are each progressive and the resultant Iterative Specification at the I/O System level is progressive.*

Definition 9. *A progressive Iterative Specification can be simulated by a* legitimate *DEVS.*

Definition 10. *A progressively admissible Coupled Iterative Specification can be component-wise simulated by a legitimate DEVS.*

Zeigler et al. [12] showed that the question of whether a DEVS is legitimate is undecidable. The approach was to express a Turing Machine as a DEVS and showing that solving the halting problem is equivalent to establishing legitimacy for this DEVS. We can transfer this approach to the formulation of the TM in Figure 1 by setting the time advances of both components to zero except for a halt state which is a passive state (and has an infinite time advance). Then a TM DEVS is legitimate just in case the TM it implements ever halts. Since there is no algorithm to decide whether an arbitrary TM will halt or not the same is true for legitimacy. Since DEVS is an iterative specification, the problem of determining whether an Iterative Specification is progressive is also undecidable. Likewise, the question of whether a coupled iterative specification is progressive is undecidable.

Fundamental Systems Existence: Probabilistic Characterization of Halting

The existence of an iteratively specified system requires temporal progress and there is no algorithm to guarantee in a finite time that such progress is true. The underlying point is that, given an arbitrary TM, we have to simulate it step-by-step to determine if it will halt. Importantly, if a TM has not halted after some time, however long, this gives us no further information on its potential halting in the future. However, while this is true for individual TMs, the situation may be different for the statistics of subclasses of TMs. In other words, for a given class of TMs, the probability of an instance halting may be found to increase or decrease for the future given that it has not yet halted.

We performed an empirical study of randomly generated TMs described in Appendix E. The results seem to show that TMs break into two classes—those that halt within a small number of steps (e.g., 16) and those that do not halt. Indeed, the data suggest that approximately one-third of the sampled TMs halt, while two-thirds do not. For those that halt, the probability of halting at the first step is greatest and then decreases exponentially and so is well-described by a geometric distribution with probability of success approximately $1/6$. In other words the probability of halting is $1/6$ at each step given the TM has not halted earlier for the set that will eventually halt.

Summarizing, the existence of an iteratively specified system is not algorithmically decidable but there may be useful probabilistic formulations that can be applied to sub-classes of such specifications. For example, it seems to be that with high probability the halting of a two-symbol, three-state TM can be decided within 16 simulation steps. Therefore, it might be tractable for an assemblage of components to find a way to solve its composition problems in a finite time for certain classes of components couplings.

5. Discussion: Future Research

The discussion presented here was limited to compositions having two components with cross-coupled connections. However, this scope was sufficient to expose the essential effect of feedback in creating the need for consistent assignments of input/output pairs and the additional—distinct— requirement for time progression. Further development can extend the scope to arbitrary compositions generalizing the statement of consistent assignments to finite and perhaps, infinite sets of components.

Further, when extending to arbitrary composition sizes, the analysis revealing fundamental conditions for the emergent existence of systems from component systems suggests new classes of systems that can be defined using iterative specifications. For example, we might consider the active–passive systems in a more general setting. First allowing any finite number of them in a coupling can satisfy the requirement for at most one active component at any time by restricting the coupling to a single influencer for each component. In fixed coupling exhibiting feedback, this amounts to a cyclic formation in which activity will travel around the cycle from one component to the next. Another common pattern is a single component (root) influencing a set of components (branches) each of which influences only the root. Several possibilities for interaction of the root with the branches can result in admissible compositions. For example, the root might select a single branch to activate to satisfy the single active component requirements. Alternatively, it might activate all branches and count the number of branches as they successively go passive, becoming active when all the branches have become passive. These alternative configuration require knowledge of the protocol underlying the interaction but this only means that proofs of emergent well-defined resultants will be conditioned to narrower sub-classes of systems rather than to broad classes such as all differential equation systems.

After much progress in understanding how dynamic structure can be managed in DEVS-based simulations [14–16], a framework that incorporates several insights has recently been developed [17]. This framework, formulated in the DEVS formalism, can be extended to iterative specifications and investigated for conditions of existence as done here. The extension will probably increase the complexity required to state and establish the conditions but would, if successful, bring us closer to understanding the emergence of systems from assemblages of components in the real world.

Appendix A. Turing Machine as a DEVS

Appendix A.1. Tape System

A state is a triple $(tape, pos, mv)$ where $tape : \mathbb{I} \rightarrow \{0,1\}$, $pos \in \mathbb{I}$, $mv \in \{-1,1\}$ and $\mathbb{I}$ is the integers; in other words, the tape is an infinite sequence of bits, where *pos* represents the position of the head, and *mv* is a specification for moving left or right. An internal transition moves the head as specified; an external transition accepts a symbol, move pair, stores the symbol in the current position of the head and stores the move for subsequent execution. The slot for storing the move can also be null, which indicates that a move has taken place.

Each function can be described as follows:

$$\delta_{int}(tape, pos, mv) = (tape, move(pos, mv), null)$$
$$\delta_{ext}((tape, pos, null), e, (sym, mv)) = (store(tape, pos, sym), pos, mv)$$
$$ta(tape, pos, mv) = 1$$
$$ta(tape, pos, null) = \infty$$
$$\lambda(tape, pos, mv) = getSymbol(tape, pos)$$

where

$$move(pos.mv) = pos + mv$$
$$store(tape, pos, sym) = tape' \text{ where } tape'(pos) = sym, \ tape'(i) = tape(i)$$
$$getSymbol(tape, pos) = tape(pos)$$

Appendix A.2. TM Control

A state is a pair (st, sym) where st is a control state and sym is a stored symbol. An internal transition applies the TM transition table to the (st, sym) pair and transitions to the specified control state. An external transition stores the received symbol for subsequent use.

Each function can be described as follows:

$$\delta_{int}(st, sym) = (TMState(st, sym), null)$$
$$\delta_{ext}((st, null), e, sym) = (st, sym)$$
$$ta(st, sym) = 1$$
$$ta(st, null) = \infty$$
$$\lambda(st, sym) = TMOutput(st, sym)$$

where

$$TMState(st, sym) = st'$$
$$TMOutput(st, sym) = sym'$$

Appendix B. Iterative Specification of Systems

The following is extracted from [12] Chapter 5.

Appendix B.1. Generator Segments

Consider (Z,T) the set of all segments $\{\omega :< 0, t_1 > \rightarrow Z \mid t_1 \in T\}$, which is a semigroup under concatenation. For a subset Γ of (Z,T), we designate by Γ^+ the *concatenation closure* of Γ (also called the *semigroup generated* by Γ). Example generators are bounded continuous segments, and constant segments of variable length generating bounded piecewise continuous segments piecewise constant segments, respectively. Unfortunately, if Γ generates Ω (a subset of segments), we *cannot* expect each $\omega \in \Omega$ to have a unique decomposition by Γ; that is, there may be distinct decompositions $\omega_1, \omega_2, \ldots, \omega_n$ and $\omega'_1, \omega'_2, \ldots, \omega'_n$ such that $\omega 1 \bullet \omega 2 \bullet \cdots \bullet \omega_n = \omega$ and $\omega 1' \bullet \omega'_2 \bullet \cdots \bullet \omega'_n = \omega$.

A single representative, or *canonical* decomposition can be computed using a *maximal length segmentation*. First, we find ω_1, the longest generator in Γ that is also a left segment of ω. This process is repeated with what remains of ω after ω_1 is removed, generating ω_2, and so on. If the process stops after n repetitions, then $\omega = \omega_1 \omega_2 \ldots \omega_n$.

It is not necessarily the case that MLS decompositions exist (i.e., that the just-mentioned processes will stop after a finite number of repetitions). Thus, we are interested in checkable conditions on a set of generators that will guarantee that each segment generated has an MLS decomposition. Fortunately, a segment can have at most one MLS decomposition. We say that Γ is an *admissible* set of generators for Ω if Γ generates Ω and for each $\omega \in \Omega$, a unique MLS decomposition of ω by Γ exists. (We also say that Γ admissibly generates Ω.) The following is proved in [12]:

Theorem B1. *Sufficient Conditions for Admissibility. If Γ satisfies the following conditions, it admissibly generates Γ^+:*

1. *Existence of longest initial segments:* $\omega \in \Gamma^+ \Rightarrow max\{t \mid \omega_{t>} \in \Gamma\}$ *exists*
2. *Closure under right segmentation:* $\omega \in \Gamma \Rightarrow \omega_{<t} \in \Gamma$ *for all* $\tau \in dom(\omega)$

Appendix B.2. Generator State Transition Systems

Having established a terminating process for obtaining MLS decompositions, we wish to use these decompositions to help us generate a transition function given only that function's action on the generators. In other words, let Ω_G be an admissible generating set for Ω and suppose we have defined a function $\delta_G : Q \times \Omega_G \to Q$, which we call a *single segment transition function*.

Let $\omega_1, \omega_2, \ldots, \omega_n$ be the MLS decomposition of ω. Having δ defined for each segment ω_i, we wish to piece together these parts to obtain compound transition associated with ω itself.

An *iterative specification* of a system is a structure

$$G = < T, X, \Omega_G, Y, Q, \delta, \lambda >,$$

where T, X, Y, and Q have the same interpretation as for I/O systems; Ω_G is the set of input segment generators; $\delta_G : Q \times \Omega_G \to Q$ is the single segment state transition function; $\lambda : Q \times X \to Y$ is the output function, with the restriction that $\Omega_G \subseteq (X, T)$; and, most important, Ω_G is an admissible set of generators and $\delta_G^+ : Q \times \Omega_G^+ \to Q$ has the composition property.

An iterative specification $G = < T, X, \Omega_G, Y, Q, \delta_G, \lambda >$ specifies a time invariant system, $S_G = < T, X, \Omega_G^+, Y, Q, \delta_G^+, \lambda >$ where δ_G^+ is the extension of δ_G which is well-defined according to the following:

Theorem B2. *Sufficient Conditions for Iterative Specification. Let* $G = < T, X, \Omega_G, Y, Q, \delta_G, \lambda >$ *be a structure as just defined. Then, if the following conditions hold, G is an iterative specification and* $S_G = < T, X, \Omega_G^+, Y, Q, \delta_G^+, \lambda >$ *is a system.*

1. *Existence of longest prefix segments:* $\omega \in \Omega_G^+ \Rightarrow max\{t \mid omega_{t>} \in \Omega_G\}$ *exists*
2. *Closure under right segmentation:* $\omega \in \Omega_G \Rightarrow \omega_{<t} \in \Omega_G$ *for* $t \in dom(\omega)$
3. *Closure under left segmentation:* $\omega \in \Omega_G \Rightarrow \omega_{t>} \in \Omega_G$ *for* $t \in dom(\omega)$
4. *Consistency of composition:* $\delta_G^+(q, \omega_1 \bullet \omega_2 \bullet \cdots \bullet \omega_n) = \delta_G(\delta_G(\ldots \delta_G(\delta_G(q, \omega_1), \omega_2), \ldots), \omega_n)$

Appendix C. DEVS Atomic Model Simulation of an Iterative Specification

Theorem C1. *A DEVS atomic model can simulate an Iterative Specification.*

Proof. As shown in Figure 3 and in Figure 5, given an iterative specification $G = < T, X, \Omega_G, Y, Q, \delta_G, \lambda >$, we construct a DEVS model $M = < X, S, Y, \delta_{int}, \delta_{ext}, \lambda, ta >$, such that

$$g : \Omega_G \to \Omega_{DEVS} \text{ using mls}$$
$$g(\omega) = [\omega]_{l(\omega)>} = [\omega] \phi_{l(\omega)>}$$
$$h : Q \to Q \times \Omega_G$$
$$h(q) = (q, dummy)$$

According to the definition of the simulating DEVS:

$$S = Q \times \Omega_G$$
$$\delta_{int}(q, \omega) = (\delta_G(q, \omega), dummy)$$
$$ta(q, \omega) = l(\omega)$$
$$\delta_{ext}((q, \omega), e, \omega') = (\delta_G(q, \omega_{e>}), \omega')$$
$$\lambda(q, dummy) = \lambda_G(q)$$

In the Iterative Specification:

$$\delta_G^+(q, \omega\omega') = \delta_G(\delta_G(q, \omega), \omega')$$

In the DEVS, let δ_{DEVS} be the transition function of the system specified by the DEVS. We want to show that $h(\delta_G^+(q, \omega\omega')) = \delta_{DEVS}(h(q), g(\omega\omega'))$, i.e.,

$$\delta_{DEVS}(h(q), g(\omega\omega'))$$
$$= \delta_{DEVS}((q, dummy), g(\omega)g(\omega'))$$
$$= \delta_{DEVS}((q, dummy), [\omega]_{l(\omega)>}, [\omega']_{l(\omega')>})$$
$$= \delta_{DEVS}(\delta_{ext}((q, dummy), [\omega]), \phi_{l(\omega)>}), \omega'_{l(\omega')>})$$
$$= \delta_{DEVS}((q, [\omega]), \phi_{l(\omega)>}), \omega'_{l(\omega')>})$$
$$= \delta_{DEVS}(\delta_{int}(q, [\omega]), \phi_{l(\omega)>}), \omega'_{l(\omega')>})$$
$$= \delta_{DEVS}((\delta_G(q, \omega), dummy), \omega'_{l(\omega')>})$$
$$= (\delta_G(\delta_G(q, \omega), \omega'), dummy)$$

Thus,

$$h(\delta_G^+(q, \omega\omega'))$$
$$= h(\delta_G(\delta_G(q, \omega), \omega'))$$
$$= (\delta_G(\delta_G(q, \omega), \omega'), dummy)$$
$$= \delta_{DEVS}(h(q), g(\omega\omega'))$$

□

Appendix D. Coupled Iterative Specification at the I/O System level

Theorem D1. *An admissible Coupled Iterative Specification specifies a well-defined Iterative Specification at the I/O System level.*

Proof. Given two iterative specifications, G_i and their state sets, Q_i $i = 1, 2$, let $Q = Q_1 \times Q_2$, eventually the state set of the iterative specification to be constructed. For any pair $(q1,q2)$ in Q, let $(\rho1, \rho2)$ be a consistent output trajectory for $(q1, q2)$, i.e., $\rho1 = \beta(q1, \rho2)$ and $\rho2 = \beta(q2, \rho1)$ and $F(q1, q2) = (\rho1, \rho2)$.

Define an autonomous Iterative Specification $< Q, \delta_G, \lambda >$ by

$$\delta_G(q1, q2) = (\delta(q1, \rho2_{t^*(q1,q2)>}), \delta(q2, \rho1_{t^*(q1,q2)>}))$$

where $t^*(q1, q2) = \min\{t_1^*, t_2^*\}$ and t_1^*, t_2^* are the times of the MLS for $\rho1$ and $\rho2$.

In other words, since each of the component iterative specifications have maximum length segmentations we take the time of next update of the constructed specification to be determined by earliest of the times of these segmentations for the consistent pair of output trajectories. This allows us to define a step-wise transition for the constructed transition function. Closure under composition for this transition function can be established using induction on the number of generators in the segments under consideration. Similarly, the output function is defined by: $\lambda(q1, q2) = (\lambda(q1), \lambda(q2))$. □

Theorem D2. *The set of Iterative Specifications is closed under admissible coupling.*

Proof. This theorem is a corollary of the previous theorem. □

Appendix E. A Statistical Experiment Sampling from 2-Symbol, 3-State TMs

We performed a statistical experiment sampling from two-symbol, three-state TMs. Each sampled TM fills in the xs in Table E1 with values from the sets shown, where one of the rows is chosen at random and its next state change to the halt state (e.g., the first row can be $(A, -1, 0)$ meaning jump to state A, move back, and print 0). The number of TMs in this class is $(2 \times 2 \times 3)^{2 \times 3} = 12^6$. We generated 1000 samples randomly and simulated them until they halted or for a maximum of 1000 time steps.

Table E1. Template for two-symbol, three-state TMs.

State	Symbol	Next State {A, B, C}	Move {1, −1}	Print Symbol {0, 1}
A	0	x	x	x
A	1	x	x	x
B	0	x	x	x
B	1	x	x	x
C	0	x	x	x
C	1	x	x	x

The results are shown in the first two columns of Table E2. They seem to show that TMs break into two classes—those that halt within a small number of steps (e.g., 16) and those that do not halt. Indeed, the data in the table suggest that approximately one-third of the sampled TMs halt, while two-thirds do not. For those that halt, the probability of halting at the first step is greatest and then decreases exponentially (cf. Figure E1). The third row shows the probabilities computed from a geometric success model using probability of success as 0.161. The agreement of the observed and predicted distributions are in close agreement. This suggests that for the set that will eventually halt, the probability of a TM halting at each step is approximately 1/6, given that it has not halted earlier.

Table E2. Empirical Frequency Distribution for halting step vs. Geometric Model.

Halts at N	Frequency	Geometric Model
1	0.161	0.161
2	0.1	0.09
3	0.042	0.039
4	0.023	0.022
5	0.016	0.015
6	0.005	0.00488
7	0.004	0.00390
11	0.001	0.000990
16	0.001	0.000985
10000	0.647	

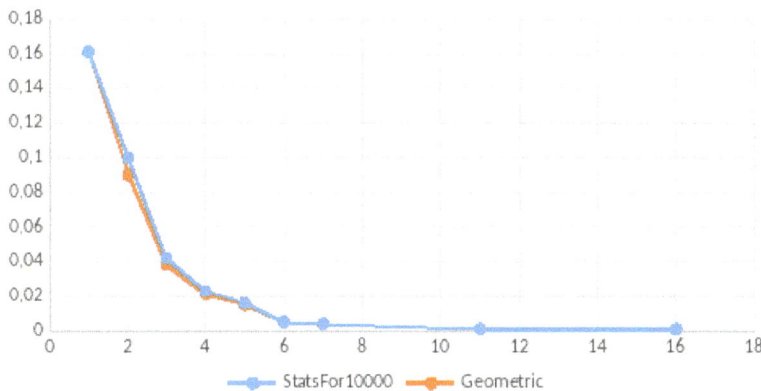

Figure E1. Empirical frequency vs. geometric model.

Summarizing, the existence of an iteratively specified system is not algorithmically decidable but there may be useful probabilistic formulations that can be applied to sub-classes of such specifications. For example, it seems to be that with high probability the halting of a two-symbol, three-state TM can be decided within 16 simulation steps. Therefore, it might be tractable for an assemblage of components to find a way to solve its composition problems in a finite time for certain classes of components couplings.

References

1. Mittal, S. Emergence in stigmergic and complex adaptive systems: A formal discrete event systems perspective. *Cogn. Syst. Res.* **2013**, *21*, 22–39.
2. Mittal, S.; Rainey, L. Harnessing emergence: The control and design of emergent behavior in system of systems engineering. In Proceedings of the Conference on Summer Computer Simulation, SummerSim'15, Chicago, IL, USA, 26–29 July 2015; pp. 1–10.
3. Ashby, W. *An Introduction to Cybernetics*; University Paperbacks: Methuen, MA, USA; London, UK, 1964.
4. Foo, N.Y.; Zeigler, B.P. Emergence and computation. *Int. J. Gen. Syst.* **1985**, *10*, 163–168.
5. Kubik, A. Toward a formalization of emergence. *Artif. Life* **2003**, *9*, 41–65.
6. Szabo, C.; Teo, Y.M. Formalization of weakemergence in multiagent systems. *ACM Trans. Model. Comput. Simul.* **2015**, *26*, 6:1–6:25.
7. Ören, T.I.; Zeigler, B.P. System Theoretic Foundations of Modeling and Simulation: A Historic Perspective and the Legacy of A. Wayne Wymore. *Simul. Trans. Soc. Model. Simul.* **2012**, *88*, 1033–1046.
8. Zeigler, B.; Muzy, A. Some Modeling & Simulation Perspectives on Emergence in System-of-Systems. In Proceedings of the SpringSim2016, Virginia Beach, VA, USA, 23–26 April 2016; pp. 11:1–11:4.
9. Wikipedia. Available online: https://en.wikipedia.org/wiki/Turing_machine (accessed on 2 November 2016).
10. Uhrmacher, A.M. Dynamic structures in modeling and simulation: A reactive approach. *ACM Trans. Model. Comput. Simul.* **2001**, *11*, 206–232.
11. Nutaro, J. *Building Software for Simulation: Theory and Algorithms with Applications in C++*; Wiley: Hoboken, NY, USA, 2011.
12. Zeigler, B.; Praehofer, H.; Kim, T. *Theory of Modeling and Simulation: Integrating Discrete Event and Continuous Complex Dynamic Systems*; Academic Press: New York, NY, USA, 2000.
13. Giambiasi, N.; Escude, B.; Ghosh, S. GDEVS: A generalized discrete event specification for accurate modeling of dynamic systems. In Proceedings of the Autonomous Decentralized Systems, Dallas, TX, USA, 26–28 March 2001; pp. 464–469.
14. Barros, F.J. Modeling and Simulation of Dynamic Structure Heterogeneous Flow Systems. *SIMULATION Trans. Soc. Model. Simul. Int.* **2002**, *78*, 18–27.
15. Barros, F.J. Dynamic Structure Multi-Paradigm Modeling and Simulation. *ACM Trans. Model. Comput. Simul.* **2003**, *13*, 259–275.
16. Steiniger, A.; Uhrmacher, A.M. Intensional couplings in variable-structure models: An exploration based on multilevel-DEVS. *ACM Trans. Model. Comput. Simul.* **2016**, *26*, 9:1–9:27.
17. Muzy, A.; Zeigler, B.P. Specification of dynamic structure discrete event systems using single point encapsulated control functions. *Int. J. Model. Simul. Sci. Comput.* **2014**, *5*, 1450012.

systems

MDPI

Article

Temporal Modeling of Neural Net Input/Output Behaviors: The Case of XOR

Bernard P. Zeigler [1,*] and Alexandre Muzy [2]

[1] Co-Director of the Arizona Center for Integrative Modeling and Simulation (ACIMS),
     University of Arizona and Chief Scientist, RTSync Corp., 12500 Park Potomac Ave. #905-S,
     Potomac, MD 20854, USA
[2] CNRS, I3S, Université Côte d'Azur, 06900 Sophia Antipolis, France; alexandre.muzy@cnrs.fr
[*] Correspondence: zeigler@rtsync.com

Academic Editors: Gianfranco Minati, Eliano Pessa and Ignazio Licata
Received: 17 November 2016; Accepted: 18 January 2017; Published: 25 January 2017

Abstract: In the context of the modeling and simulation of neural nets, we formulate definitions for the behavioral realization of memoryless functions. The definitions of realization are substantively different for deterministic and stochastic systems constructed of neuron-inspired components. In contrast to earlier generations of neural net models, third generation spiking neural nets exhibit important temporal and dynamic properties, and random neural nets provide alternative probabilistic approaches. Our definitions of realization are based on the Discrete Event System Specification (DEVS) formalism that fundamentally include temporal and probabilistic characteristics of neuron system inputs, state, and outputs. The realizations that we construct—in particular for the *Exclusive Or* (XOR) logic gate—provide insight into the temporal and probabilistic characteristics that real neural systems might display. Our results provide a solid system-theoretical foundation and simulation modeling framework for the high-performance computational support of such applications.

Keywords: neural nets; spiking neurons; xor problem; input/output system; system specifications; Discrete Event System Specification

---

1. Introduction

Bridging the gap between neural circuits and overall behavior is facilitated by an intermediate level of neural computations that occur in individual and populations of neurons [1]. The computations performed by Artificial Neural Nets (ANN) can be viewed as a very special, but currently popular, instantiation of such a concept [2]. However, such models map vectors to vectors without considering the immediate history of recent inputs nor the time base on which such inputs occur in real counterparts [3–5]. In reality, however, time matters because the interplay of the nervous system and the environment occurs via time-varying signals. Recently, third-generation neural nets which feature temporal behavior, including processing of individual spikes, have gained recognition [4].

Computing the XOR function has received special attention as a simple example of resisting implementation by the simplest ANNs with direct input to output mappings [6], and requiring ANNs having a hidden mediating layer [7,8]. From a systems perspective, the XOR function—and indeed all functions computed by ANNs—are memoryless functions not requiring states for their definition [2,9,10]. It is known that Spiking Neural Nets (SNN)—which employ spiking neurons as computational units—account for the precise firing times of neurons for information coding, and are computationally more powerful than earlier neural networks [11,12]. Discrete Event System Specification (DEVS) models have been developed for formal representations of spiking neurons in end-to-end nervous system architectures from a simulation perspective [10,13,14]. Therefore, it is of interest to examine the properties of DEVS realizations that employ dynamic features that are distinctive to SNNs, in contrast to their static neuronal counterparts.

Although typically considered as deterministic systems, Gelenbe introduced a stochastic model of ANN that provided a markedly different implementation [15]. With the advent of increasingly complex simulations of brain systems [13] the time is ripe for reconsideration of the forms of behavior displayed by neural nets. In this paper, we employ systems theory and a modeling and simulation framework [16] to provide some formal definitions of neural input/output (I/O) realizations and how they are applied in deterministic and probabilistic systems. We formulate definitions for the behavioral realization of memoryless functions with particular reference to the XOR logic gate. The definitions of realization are substantively different for deterministic and stochastic systems constructed from neuron-inspired components. In contrast to ANNs that can compute functions such as XOR, our definitions of realizations fundamentally include temporal and probabilistic characteristics of their inputs, state, and outputs. The realizations of the XOR function that we describe provide insight into the temporal and probabilistic characteristics that real neural systems might display.

In the following sections, we review system specifications and concepts for their input/output (I/O) behaviors that allow us to provide definitions for systems implementation of memoryless functions. This allows us to consider the temporal characteristics of neural nets in relation to the functions they implement. In particular, we formulate a deterministic DEVS version of the neural net model defined by Gelenbe [15], and show how this model implements the XOR function. In this context, we discuss timing considerations related to the arrival of pulses, coincidence of pulses, end-to-end time of computation, and time before new inputs can be submitted. We close this section by showing how these concepts apply directly to characterize the I/O behaviors of Spiking Neural Networks (SNN). We then derive a Markov Continuous Time model [17] from the deterministic version, and point out the distinct characteristics of the probabilistic system implementation of XOR. We conclude with implications about the characteristics of real-brain computational behaviors suggested by contrasting the ANN perspective and the systems-based formulation developed here. We note that Gelenbe and colleagues have generated a huge amount of literature on the extensions and applications of random neural networks. As just described, the focus of this paper is not on DEVS modeling of such networks in general. However, some aspects related to I/O behavior will be discussed in the conclusions as potential for future research.

2. System Specification and I/O Behaviors

Inputs/outputs and their logical/temporal relationships represent the I/O behavior of a system. A major subject of systems theory deals with a hierarchy of system specifications [16] which defines levels at which a system may be known or specified. Among the most relevant is the Level 2 specification (i.e., the I/O Function level specification), which specifies the collection of input/output pairs constituting the allowed behavior, partitioned according to the initial state the system is in when the input is applied. We review the concepts of input/output behavior and their relation to the internal system specification in greater depth.

For a more in-depth consideration of input/output behavior, we start with the top of Figure 1, which illustrates an input/output (I/O) segment pair. The input segment represents messages with content x and y arriving at times t_1 and t_2, respectively. Similarly, the output segment represents messages with contents z and z', at times t_3 and t_4, respectively.

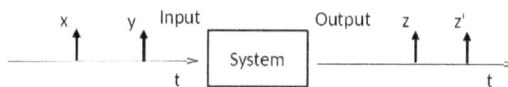

Figure 1. Representing an input/output pair.

To illustrate the specification of behavior at the I/O level, we consider a simple system—an adder—all it does is add values received on its input ports and transmit their sum as output. However simple this basic *adding* operation is, there are still many possibilities to consider to characterize its I/O behavior, such as which input values (arriving at different times) are paired to produce an output value, and the order in which the inputs must arrive to be placed in such a pairing. Figure 2 portrays two possibilities, each described as a DEVS model at the I/O system level of the specification hierarchy. In Figure 2a, after the first inputs of contents x and y have arrived, their values are saved, and subsequent inputs refresh these saved values. The output message of content z is generated after the arrival of an input, and its value is the sum of the saved values. In Figure 2b, starting from the initial state, both contents of messages must arrive before an output is generated (from their most recent values), and the system is reset to its initial state after the output is generated. This example shows that even for a simple function, such as adding two values, there can be considerable complexity involved in the specification of behavior when the temporal pattern of the messages bearing such values is considered. Two implications are immediate. One is that there may be considerable incompleteness and/or ambiguity in a semi-formal specification where explicit temporal considerations are often not made. The second implication follows from the first: an approach is desirable to represent the effects of timing in as unambiguous a manner as possible.

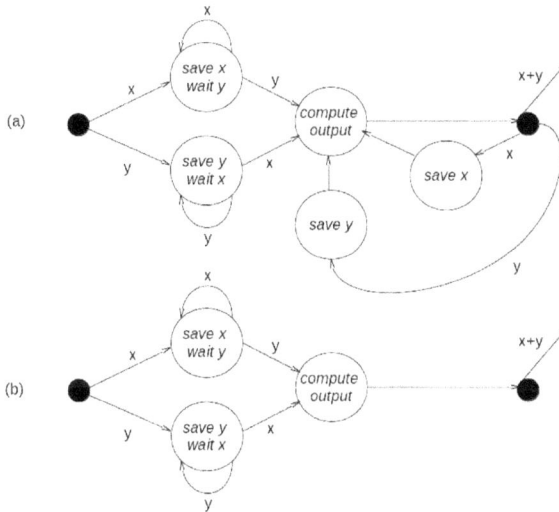

Figure 2. Variants of behavior and corresponding input/output (I/O) pairs, with (**a**) saving input values when they arrived; or (**b**) resetting to initial state once the output is computed. White circles indicate states, black circles initial states, and arrows transitions.

3. Systems Implementation of a Memoryless Function

Let $f : X \rightarrow Y$ be a memoryless function; i.e., it has no time or state dependence [16]. Still, as we have just seen, a system that implements this function may have dynamics and state dependence. Thus, the relationship between a memoryless function and a system that somehow displays that behavior needs to be clearly defined. From the perspective of the hierarchy of systems specifications [16], the relationship involves (1) mapping the input/output behavior of the system to the definition of the function; and (2) working at the state transition level correctly. Additional system specification levels may be brought to bear as needed. Recognizing that the basic relationship is that of simulation between two systems [16], we will keep the discussion quite restricted to limit the complexities.

The first thing we need to do is represent the injection of inputs to the function by events arriving to the system. Let us say that the order of the arguments does not count. This is the case for the *XOR function*. Therefore, we will consider segments of zero, one, or two pulses as input segments, and expect segments of zero or one pulses as outputs. In other words, we are using a very simple decoding of an event segment into the number of events in its time interval. While simplistic, this concept still allows arbitrary event times for the arguments, and therefore consideration of important timing issues. Such issues concern spacing between arguments and time for a computation to be completed. Figure 3 sketches this approach and corresponding *deterministic system for f* with two input ports *P1* and *P2* receiving contents *P* and an output port *P3* sending a content *P*.

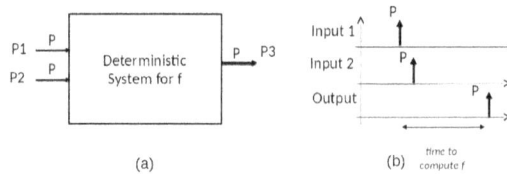

Figure 3. Deterministic system realization of memoryless function: (**a**) Input/Output Black Box; (**b**) Input and Output Trajectories.

Appendix A gives the formal structure of a DEVS basic model, and Appendix B gives our working definition of the simulation relation to be used in the sequel. Having a somewhat formal definition of what it means for a discrete event model to display a behavior equivalent to computing a memoryless function, we turn toward discussing DEVS models that can exhibit such behaviors for the XOR function.

4. DEVS Deterministic Representation of Gelenbe Neuron

Figure 4a shows a DEVS model that captures the spirit of the Gelenbe stochastic neuron (as shown in [15]) in deterministic form. We first introduce the deterministic model to prepare the ground for discussion of the stochastic neuron in Section 7. Positive pulse arrivals increment the state up to the maximum, while negative pulses decrement the state, stopping at zero. Non-zero states down-transition in a time *tfire*, a parameter. The DEVS model is given as:

$$DEVS = (X, Y, S, \delta_{ext}, \delta_{int}, \lambda, ta)$$

where,

$X = \{P^+, P^-\}$ is the *set of positive and negative input pulses*,
$Y = \{P\}$ is the *set of plain pulse outputs*,
$S = \{0, 1, 2\}$ is the *set of non-negative integer states*,
$\delta_{ext}(s, e, P^+) = s + 1$ is the *external transition increasing the state by 1 when receiving a positive pulse*,
$\delta_{ext}(s, e, P^+, P^+) = s + 2$ is the *external transition increasing the state by 2 when simultaneously receiving two positive pulses*,
$\delta_{ext}(s, e, P^-) = max(s - 1, 0)$ is the *external transition decreasing the state by 1 (except at zero) when receiving a negative pulse*,
$\delta_{int}(s > 0) = max(s - 1, 0)$ is the *non-zero states internal transition function decreasing the state by one (except at zero)*,
$\lambda(s > 0) = P$ is the *non-zero states output a pulse*,
$\lambda(s) = \phi$ is the *output sending non-event for states below threshold*,
$ta(s) = tfire$ is the *time advance, tfire, for states above 0*, and
$ta(0) = +\infty$ is the *infinity time advance for zero passive state*.

See Appendix A for definitions symbols.

Figure 4b shows an input/state/output trajectory in which two successive positive pulses cause successive increases in the state to 2, which transitions to 1 after *tfire*, and outputs a pulse. Note that the second positive pulse arrives before the elapsed time has reached *tfire*, and increases the state. This effectively cancels and reschedules the internal transition back to 0. Figure 4c shows the case where the second pulse comes after firing has happened. Thus, here we have an explicit example of the temporal effects discussed above. Two pulses arriving close enough to each other (within *tfire*) will effectively be considered as coincident. In contrast, if the second pulse arrives too late (outside the *tfire* window), it will *not* be considered as coincident, but will establish its own firing window.

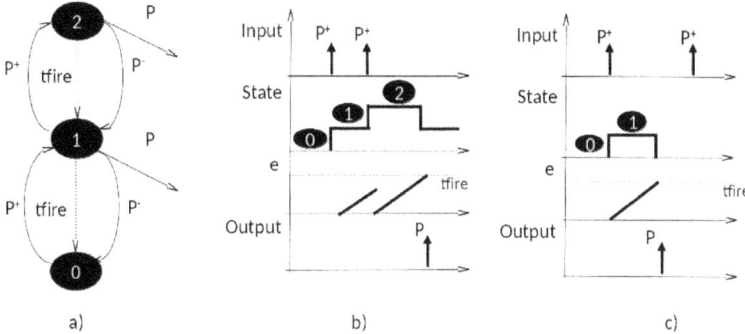

Figure 4. Two-state deterministic Discrete Event System Specification (DEVS) model of Gelenbe neuron, with (**a**) DEVS state graph; (**b**) Closely Spaced Inputs; and (**c**) Widely Spaced Inputs. The *time elapsed since the last transition* is indicated as $e \in \mathbb{R}_0^{+,\infty}$.

To implement the two logic functions *Or* and *And*, we introduce a second parameter into the model—the *threshold*. Now, states greater or equal to the threshold will transition to zero state in a time *tfire* and output a pulse. The threshold is set to 1 for the *Or*, and to 2 for the *And* function. Thus, any pulse arriving alone is enough to output a pulse for *Or*, while 2 pulses must arrive to enable a pulse for the *And*. However, there is an issue with the time advance needed for state 1 in the *And* case (due to an arrival of a first positive pulse). If this time advance is 0, then there is no time for a second pulse to arrive after a first. If it is *infinity*, then the model waits forever for a second pulse to arrive. We introduce a third parameter, *tdecay*, to establish a finite non-zero window after receiving the first pulse for a second one to arrive and be counted as coincident with the first. The revised DEVS model is:

$$DEVS = (X, Y, S, \delta_{ext}, \delta_{int}, \lambda, ta)$$

where,

$X = \{P^+, P^-\}$ is the *set of positive and negative input pulses,*
$Y = \{P\}$ is the *set of plain pulse outputs,*
$S = \{0, 1, 2\}$ is the *set of non-negative integer states,*
$\delta_{ext}(s, e, P^+) = s + 1$ is the *external transition increasing the state by 1 when receiving a positive pulse,*
$\delta_{ext}(s, e, P^+, P^+) = s + 2$ is the *external transition increasing the state by 2 when simultaneously receiving two positive pulses,*
$\delta_{ext}(s, e, P^-) = floor(s - 1, 0)$ is the *external transition decreasing the state by 1 (except at zero) when receiving a negative pulse,*
$\delta_{int}(s > 0) = floor(s - 1, 0)$ is the *non-zero states internal transition function decreasing the state by one (except at zero),*
$\lambda(s \geq Thresh) = P$ is the *output sending a pulse for states above or equal threshold,*
$\lambda(s) = \phi$ is the *output sending non-event for states below threshold,*

$ta(s \geq Thresh) = tfire$ is the *time advance, tfire, for states above or equal threshold,* and
$ta(s < Thresh) = tdecay$ is the *time advance, tfire, for states below threshold.*

5. Realization of the XOR Function

We can use the *And* and *Or* models as components in a coupled model, as shown in Figure 5a to implement the XOR function. However as we will see in a moment, we need the response of the *And* to be slower than that of the *Or* to enable the correct response to a pair of pulses. So, we let *tfireOr* and *tfireAnd* be the time advances of the *Or* and *And* response in the above threshold states. As in Figure 5b,c, pulses arriving at the input ports *P1* and *P2* are mapped in positive pulses by the external coupling that sends them as inputs to both components. When a single pulse arrives within the *tdecay* window, only the *Or* responds and outputs a pulse. When a pair of pulses arrive within *tdecay* window, the *And* detects them and produces a pulse after *tfireAnd*. The internal coupling from And to Or maps this pulse into a double negative pulse at the input of the *Or*. Meanwhile, the *Or* is holding in State 2 from the pair of positive pulses it has received from the input. So long as the *tfireOr* is greater than *tfireAnd*, the double negative pulse will arrive quickly enough to the *Or* model to reduce its state to zero, thereby suppressing its response. In this way, XOR behavior is correctly realized.

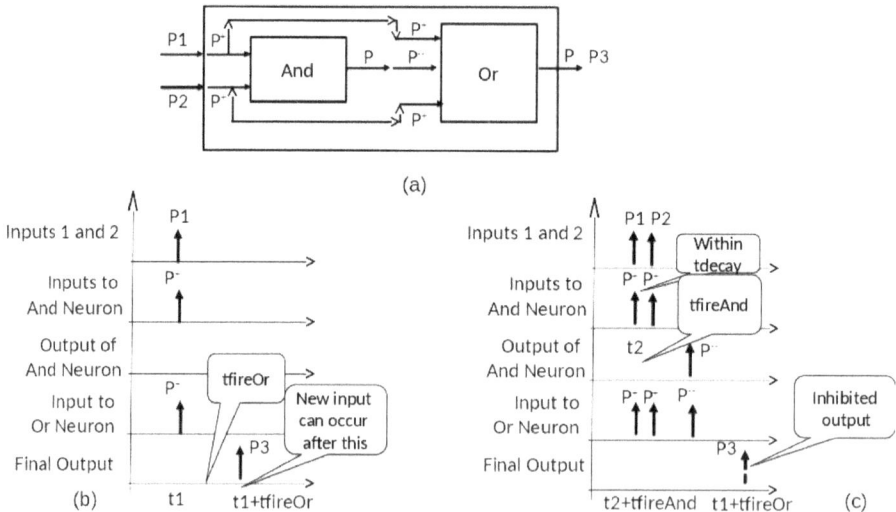

Figure 5. Coupled Model for XOR Implementation, with (**a**) XOR Network Description; (**b**) Single Input Pulse; and (**c**) Double Input Pulse. Note that t1 in (**c**) is the same time in (**b**), representing the time an inhibited pulse would have arrived.

Assertion: The coupled model of Figure 5 with *tfireAnd*<*tfireOr*<*tdecay* realizes the XOR function in the following sense:

1. When there are no input pulses, there are no output pulses,
2. When a single input pulse arrives and is not followed within *tfireAnd* by a second pulse, then an output pulse is produced after *tfireOr* of the input pulse arrival time.
3. When the pair of input pulses arrive within *tfireAnd* of each other, then no output pulse is produced.

Thus, the *computation time* is *tfireOr*, since that is the longest time after the arrival of the input arguments (first pulse or second pulse in the pair of pulses case) that we have to wait to see if there is an output pulse. Another metric could also be considered which starts the clock when the first argument arrives, rather than when all arguments arrive.

On the other hand, the *time for the system to return to its initial state*—and we can send in new arguments for computation—may be longer than the computation time. Indeed, the *Or* component returns to the zero state after outputting a pulse at *tfireOr* in both single and double pulse input cases. However, in the first case, the *And* component—having been put into a non-zero state—only relaxes back to zero after *tdecay*. Since *tdecay* is greater than *tfireOr*, the initial state return time is *tdecay*.

6. Characterization of SNN I/O Behaviors and Computations

Reference [5] provides a comprehensive review of SNNs, concluding that they have significant potential for solving complicated time-dependent pattern recognition problems because of their inclusion of temporal and dynamic behavior. Maass [11,18,19] and Schmitt [12] characterized third generation SNNs which employ spiking neurons as computational units, accounting for the precise firing times of neurons for information coding, and showed that such networks are computationally more powerful than earlier neural networks. Among other results, they showed that a spiking neuron cannot be simulated by a Boolean network (in particular, a disjunctive composition of conjunctive components with fixed degree). Furthermore, SNNs have the ability to approximate any continuous function [4]. As far as realization of SNN's in DEVS, the reader may refer to reference [20] for a generic model of a discrete event neuron in an end-to-end nervous system architecture, and to [21] for a complete formal representation of Maass' SNN from a DEVS simulation perspective. Thus, while there is no question concerning the general ability of SNNs to compute the XOR function, it is of interest to examine the properties of a particular realization—especially one that employs dynamic features that are distinctive to SNNs vice their static neuronal counterparts. Here we draw on the approach of Booij [22], who exhibited an architecture for SNN computation of XOR directly, as opposed to one that relies on the training of a generic net. Like Booij, we change the input and output argument coding to restrict inputs and outputs to particular locations on the timeline. Employing earlier convention, Booij requires inputs to occur at fixed positions, such as either at 0 or 6, and outputs to occur at 10 or 16. Such tight specifications enable a device to be designed that employs synaptic delays and weights to be manipulated to rise above or stay below a threshold, as required. However, the result is highly sensitive to noise, in that any slight change in input position can upset the delicate balancing of delay and weight effects. In contrast, we employ a coding that enables the inputs to have much greater freedom of location while fundamentally employing synaptic delays (although we reduce the essential computation to a static Boolean computation).

As before, we consider the XOR function,

$$f : X \rightarrow X$$

where $X = \{0, 1, 2\}$, $f(x) = x + 1 (mod 2)$.

However, we slightly distinguish the decoding of domain and range. Let $g1 : DEVS(\rho) \rightarrow \{0, 1, 2\}$ specify the decoding of segments to domain of f. Let $g2 : DEVS(\rho) \rightarrow \{0, 1, 2\}$ specify decoding of segments to range of f. With $\beta_q : DEVS(\rho) \rightarrow DEVS(\rho)$, the I/O function of state q, we require, $\forall \omega \in \Omega_x, g2(\beta_q(\omega)) = f(g1(\omega))$. In this example, for $L > 0$, we define $g1(\omega) = $ *number of pulses in* ω *that arrive earlier than L*, and $g2(\omega) = $ *number of pulses in* ω *that arrive earlier than 2L*; i.e., we require f (*number of pulses in* ω *that arrive earlier than L*) = *number of pulses in* $\beta_q(\omega)$ *that arrive earlier than 2L*.

The basic SNN component is shown in Figure 6a, which has two delay elements feeding an OR gate with weights shown. The delay element is behaviorally equivalent to a synaptic delay in an SNN, and is described as DEVS:

$$Delay(d) = < \{p\}, \{passive, active\}, \delta_{ext}, \delta_{int}, ta, \lambda >$$

where $\delta_{ext}(passive, p, e) = active$, $\delta_{int}(active) = passive$, $ta(passive) = +\infty$, $ta(active) = d$, $\lambda(active) = p$.

The device rests passively until becoming active when receiving a pulse. It remains active for a time d (the delay parameter), and then outputs a pulse and reverts to passive. The top delay element in Figure 6a has delay, L, and is activated by a bias pulse at time 0 to start the computation. If an input pulse arrives any time before time L, it inhibits the output of the OR; otherwise, a pulse is emitted at time L. Thus, the net of Figure 6a can be called an L-arrival detector, since it detects whether a pulse arrives before L and outputs its decision at L. Two such sub-nets are employed in Figure 6b to construct the XOR solution employing SNN equivalent components. After the initial bias, the incoming pulses, $P1$ and $P2$, each arrive early or late relative to L, as detected by the L-detectors. Moreover, any pulses output at the L-arrival detectors are synchronized so that they can be processed by a straightforward XOR gate of the kind constructed earlier (i.e., without concern for timing of arrival). We feed the result into an L-arrival detector in order to report the output back in the form of a pulse that will appear earlier than $2L$ if we start the output bias at L. Thus, the computation time for this implementation is $2L$, and as is its time to resubmission. Indeed, it can function like a computer logic circuit with clock cycle L. Note that unlike Booij's solution, the solution is not sensitive to exact placement of the pulses, and realistic delays in the gates can be accommodated by delaying the onset of the second bias and reducing the output L-detector's delay.

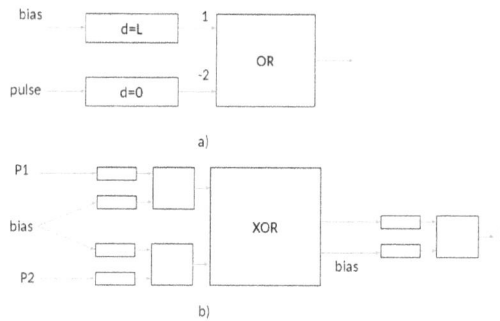

Figure 6. Implementation of the XOR using Spiking Neural Net (SNN) equivalent components described in DEVS: (**a**) L-Arrival Component; (**b**) XOR Coupled Model.

7. Probabilistic System Implementation of XOR

Gelenbe's implementation of the XOR [15] differs quite radically from the deterministic one just given. The concept of what it means for a probabilistic system to realize a memoryless function differs from that given above for a deterministic one.

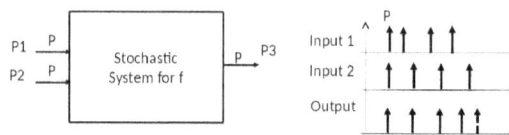

Figure 7. Stochastic system realization of a memoryless function.

As illustrated in Figure 7, each argument of the function is represented by an infinite stream of pulses. A stream is modeled as a *Poisson stochastic process* with a specified rate. An argument value of zero is represented by a null stream (i.e., a rate of zero). We will set the rate equal 1 for a stream representing an argument value of 1. The output of the function is represented similarly as a stream of pulses with a rate representing the value. However, rather than a point, we use an interval on the real line to represent the output value. In the XOR, Gelenbe's implementation uses an interval $[0, \alpha)$ to represent 0, with $[\alpha, 1]$ representing 1.

Furthermore, the approach to distinguishing the presence of a single input stream from a pair of such streams—the essence of the problem—is also radically different. The approach formulates the DEVS neuron of Figure 4 as a Continuous Time Markov model (CTM) [17] shown in Figure 8, and exploits its steady state properties in response to different levels of positive and negative input rates. In Figure 8, the CTM on the left has input ports P^+ and P^- and output port P. In non-zero states, it transitions to the next lower state with rate *FireRate*, which is set to the inverse of *tfire*, interpreted as the mean time advance for such transitions in Figure 4. The Markov Matrix model [17] on the right is obtained by replacing the P^+, P^- and P ports by rates *posInputRate* and *negInputRate*, resp. Further, the output port P is replaced by the *OutpuRate*, which is computed as the *FireRate* multiplied by the probability of firing (i.e., being in a non-zero state.)

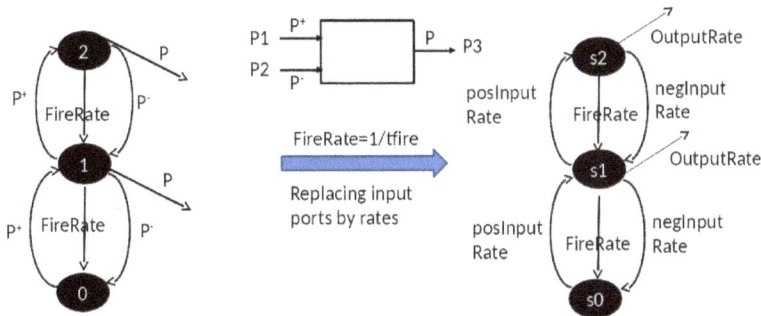

Figure 8. Mapping DEVS Neuron Continuous Time Markov (CTM) model to Markov Matrix model.

As in Figure 9, each input stream splits into two equal streams of positive and negative pulses by external coupling to two components, each of which is a copy of the CTM model of Figure 8. The difference between the components is that the first component receives only positive pulses, while the second component receives both positive and negative streams. Note that whenever two equal streams with the same polarity converge at a component, they effectively act as a single stream of twice the rate. However, when streams of opposite polarity converge at a component, the result is a little more complex, as we now show.

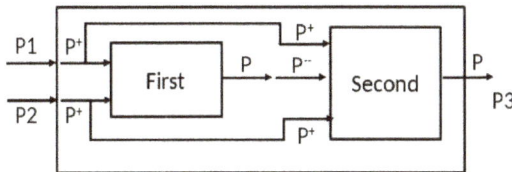

Figure 9. Stochastic coupled model implementation of XOR.

Now let us consider the two input argument cases. Case 1: one null stream, one non-null stream (representing arguments $(0, 1)$ or $(1, 0)$); Case 2: two non-null streams (representing $(1, 1)$). In this set-up, Appendix C describes how the first component *saturates* (fires at its maximum rate) when it receives the stream of positive pulses at either the basic or combined intensities. Therefore, it transmits a stream of positive pulses at the same rate in both Cases 1 and 2. Now, the second component differs from the first in that it receives the (constant) output of the first one. Therefore, it reacts differently in the two cases – its output rate is smaller when the negative pulse input rate is larger (i.e., it is inhibited in inverse relation to the strength of the negative stream). Thus, the output rate is lower in Case 2, when there are two input streams of pulses, than in Case 1, when only one is present. However, since the output rates are not exactly 0 and 1, there needs to be a dividing point (viz., α as above), to make the decision about which case holds. Appendix C shows how α can be chosen so that the output rate of the overall model is below α when two input streams are present, and above α when only one (or none) is present, as required to implement XOR.

8. Discussion

Discussing the proposition of deep neural nets (DNN) as the primary focus of artificial general intelligence, Smith asserts that largely as used, DNNs map vectors to vectors without considering the immediate history of recent inputs nor the time base on which such inputs occur in real counterparts [2]. Note that this is not to minimize the potentially increased ability of relatively simple ANN models to support efficient learning methods [5]. We do not address learnability in this paper. In reality however, time matters because the interplay of the nervous system and the environment occurs via time-varying signals. To be considered seriously as Artificial General Intelligence (AGI), a neural net application will have to work with time-varying inputs to produce time-varying outputs: the world exists in time, and the reaction of a system exhibiting AGI also has to include time [2,23]. Third-generation SNNs have been shown to employ temporal and dynamic properties in new forms of applications that point to such future AGI applications [4,5]. Our results provide a solid system-theoretical foundation and simulation modeling framework for high-performance computational support of such applications.

Although typically considered as deterministic systems, Gelenbe introduced a stochastic model of ANN that provided a markedly different implementation [15]. Based on his use of the XOR logic gate, we formulated definitions for the behavioral realization of memoryless functions, with particular reference to the XOR gate. The definitions of realization turned out to be substantively different for deterministic and stochastic systems constructed of neuron-inspired components. Our definitions of realizations fundamentally include temporal and probabilistic characteristics of their inputs, state, and outputs. Moreover, the realizations of the XOR function that we constructed provide insight into the temporal and probabilistic characteristics that real neural systems might display.

Considering the temporal characteristics of neural nets in relation to functions they implement, we formulated a deterministic DEVS version of Gelenbe's neural net model, and showed how this model implements the XOR function. Here, we considered timing related to the arrival of pulses, coincidence of pulses, end-to-end time of computation, and time before new inputs can be submitted. We went on to apply the same framework to the realization of memoryless functions by SNNs, illustrating how the formulation allowed for different input/output coding conventions that enabled the computation to exploit the synaptic delay features of SNNs. We then derived a Markov Continuous Time model [17] from the deterministic version, and pointed out the distinct characteristics of the probabilistic system implementation of XOR. We conclude with implications about the characteristics of real-brain computational behaviors suggested by contrasting the ANN perspective and systems-based formulation developed here.

System state and timing considerations we discussed include:

1. *Time dispersion of pulses*—the input arguments are encoded in pulses over a time base, where inter-arrival times make a difference in the output.
2. *Coincidence of pulses*—in particular, whether pulses represent arguments from the same submitted input or subsequent submission depends on their spacing in time.
3. *End-to-end computation time*—the total processing time in a multi-component concurrent system depends on relative phasing as well as component timings, and may be poorly estimated by summing up of individual execution cycles.
4. *Time for return to ground state*—the time that must elapse before a system that has performed a computation is ready to receive new inputs may be longer than its computation time, as it requires all components to return to their ground states.

Although present in third-generation models, system state and timing considerations are abstracted away by neural networks typified by DNNs that are idealizations of intelligent computation; consequently, they may miss the mark in two aspects:

1. As static recognizers of memoryless patterns, DNNs may become ultra-capable (analogous to *AlphaGo* progress [24]), but as representative of human cognition, they may vastly overemphasize that one dimension and correspondingly underestimate intelligent computational capabilities in humans and animals in other respects.
2. As models of real neural processing, DNNs do not operate within the system temporal framework discussed here, and therefore may prove impractical in real-time applications which impose time and energy consumption constraints such as those just discussed [25].

It is instructive to compare the computation-relevant characteristics of the deterministic and stochastic versions of the DEVS neuron models we discussed. The deterministic version delivers directly interpretable outputs within a specific processing time. The Gelenbe stochastic version formulates inputs and outputs as indefinitely extending streams modelled by Poisson processes. Practically speaking, obtaining results requires measurement over a sufficiently extended period to obtain statistical validity and/or to enable a Bayesian or Maximum Likelihood detector to make a confidence-dependent decision. On the other hand, a probabilistic version of the DEVS neuron can be formulated that retains the direct input/output encoding, but can also give probability estimates for erroneous output. Some of these models have been explored [9,21], while others explicitly connecting to leaky integrate-and-fire neurons are under active investigation [26]. Possible applications of DEVS modeling to the extensive literature on Gelenbe networks are considered in Appendix D. Along these lines, we note that both the deterministic and probabilistic implementations of XOR use the negative inputs in an essential (although different) manner to identify the $(1, 1)$ input argument and inhibit the output produced when it occurs (note that the use of negative synaptic weights is also essential in the SSN implementation, although in a somewhat different form). This suggests research to show that XOR cannot be computed without use of negative inputs, which would establish a theoretical reason for why inhibition is fundamentally needed for leaky integrate-and-fire neuron models—a reason that is distinct from the hidden layer requirement uncovered by Rumelhart [7].

Although not within the scope of this paper, the DEVS framework for I/O behavior realization would seem to be applicable to the issue of spike coding by neurons. A reviewer pointed to the recent work of Yoon [27], which provides new insights by considering neurons as analog-to-digital converters. Indeed, encoding continuous-time signals into spikes using a form of sigma-delta modulation would fit the DEVS framework which accommodates both continuous and discrete event segments. Future research could seek to characterize properties of I/O functions that map continuous segments to discrete event segments [16,20].

Author Contributions: B.P. Zeigler developed the different models in interaction with Alexandre Muzy who provided his expertise in mathematical system-theory and neural net modeling.

Conflicts of Interest: The authors declare no conflict of interest.

Appendix A. Discrete Event System Specification (DEVS) Basic Model

A *basic Discrete Event System Specification (DEVS)* is a mathematical structure

$$DEVS = (X, Y, S, \delta_{ext}, \delta_{int}, \lambda, ta),$$

where X is the *set of input events*, Y is the *set of output events*, S is the *set of partial states*, $\delta_{ext} : Q \times X \to S$ is the *external transition function* with $Q = \{(s,e) \mid s \in S, 0 \le e \le ta(s)\}$ the set of total states with e the *elapsed time* since the last transition, $\delta_{int} : S \to S$ is the *internal transition function*, $\lambda : S \to Y$ is the *output function*, and $ta : S \to \mathbb{R}_\infty^{0,+}$ is the *time advance function*.

Figure A1 depicts simple trajectories of a DEVS. The latter starts in initial state s_0 at time t_0, and schedules an internal event occurring after time advance $ta(s_0)$, where value y_0 is output, and state changes to $s_1 = \delta_{int}(s_0)$. At time t_2, an external event of value x_0 occurs, changing the state to $s_2 = \delta_{ext}(s_1, e_1, x_0)$ with e_1 the elapsed time since the last transition. Then, an internal event is scheduled at time advance $ta(s_1)$, and so on.

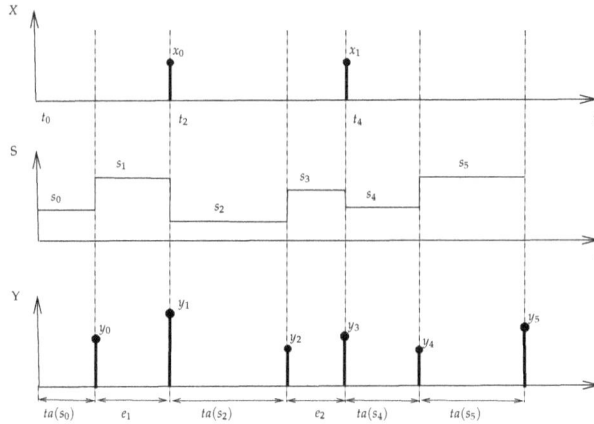

Figure A1. Simple DEVS trajectories.

Appendix B. Simulation Relation

Consider a function having the same domain and range,

$$f : X \rightarrow X$$

For example, an XOR function where $X = \{0, 1, 2\}$, $f(x) = x + 1 \pmod{2}$. Let,

$$g : \Omega_X \rightarrow X$$

That is, specify decoding of segments to domain and range of f. Let

$$\beta_q : \Omega_X \rightarrow \Omega_X$$

be the *I/O Function of state q* mapping input segments to output segments.

If $\beta_q(\omega) = \rho$, we require $g(\rho) = f(g(\omega))$; i.e., input segment ω mapped to output segment ρ when decoded is required to satisfy f. That is, $g(\beta_q(\omega)) = f(g(\omega))$. Applying the requirement to DEVS segments of pulses, let $g : DEVS(p) \rightarrow \{0, 1, 2\}$; i.e., $g(\omega)$ = *number of pulses in ω* requiring $\beta_q : DEVS(p) \rightarrow DEVS(p)$. That is, number of pulses in $\beta_q(\omega) = f(\text{number of pulses in } \omega)$.

Appendix C. Behavior of the Markov Model

We first reduce the infinite state Matrix model to a two-state version that is equivalent with respect to the output pulse rate in steady state. As in Figure C1, all the non-zero states are lumped into a single firing state, *sFire*, and we will interpret each of the probabilities in terms of the original rates as follows:

1. There is only one way to transition from *s0* to *sFire*, and that is by going from *s0* to *s1* in the original model, which happens with *posInputRate*. Therefore, P01 = *posInputRate*.
2. Similarly, there is only one way to transition from *sFire* to *s0*, and this happens with *negInputRate* + *FireRate*. Therefore, P10 = *negInputRate* + *FireRate*.

3. The probability of remaining in the *sFire*, $P11 = 1 - P10$ (these must sum to 1).
4. Similarly, $P00 = 1 - P01$.

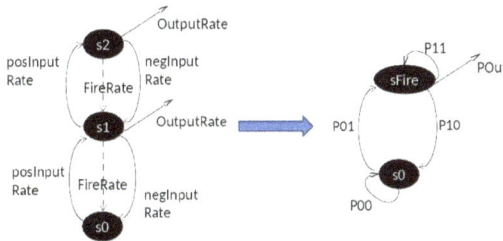

Figure C1. Reduction to two-state Markov Matrix model.

Now, in the reduced model, the steady state probabilities are easy to compute in terms of the transition probabilities. Indeed, the probability of being in the firing state, $P_{Fire} = \frac{P01}{P01 + P10} = 1$ for $P01 \approx 1$.

Additionally, the rate of producing output pulses

$$OutputRate = P_{Fire} \times FireRate = \frac{posInputRate}{posInputRate + negInputRate + FireRate} \times FireRate$$

The case of saturation occurs when the positive input rate is "very large" compared to the rates that lower the state, especially when the negative input rate is 0, so that $P_{Fire} = 1$ and $OutputRate = FireRate$.

Thus, the output of the first component saturates at the maximum, *FireRate*. This is input to the second component so that we have for it:

$$
\begin{aligned}
OutputRate &= P_{Fire} \times FireRate \\
&= \frac{FireRate}{FireRate + negInputRate + FireRate} \times FireRate \\
&= \frac{FireRate}{2 + \dfrac{negInputRate}{FireRate}}
\end{aligned}
$$

So, we see that the output rate is inversely related to the negative pulse input which, by design, the second component receives, but not the first.

Appendix D. Possible Applications of DEVS Modeling to Random Neural Networks

Random neural network (RNN)—a probabilistic model inspired by neuronal stochastic spiking behavior—have received much examination. Here we focus on two main extensions: synchronous interaction and spike classes. Gelenbe developed an extension of the RNN [28–33] to the case when synchronous interactions can occur, modeling synchronous firing by large ensembles of cells. Included are recurrent networks having both conventional excitatory–inhibitory interactions and synchronous interactions. Although modeling the ability to propagate information very quickly over relatively large distances in neuronal networks, the work focuses on developing a related learning algorithm. Synchronous interactions take the form of a joint excitation by a pair of cells on a third cell. One can assign $Q(i, j, m)$ as the probability that when cell i fires, then if cell j is excited, it will also fire immediately, with an excitatory spike being sent to cell m. This synchronous behavior can be extended to an arbitrary number of cells that can simultaneously fire. DEVS modeling includes zero time advance possibility to capture such behavior, as was illustrated in application to physical action-at-a-distance by Zeigler [14]. The standard RNN approach has been concerned with equilibrium analysis, and it may be interesting to see how the DEVS equivalent modeling can throw light on the plausibility of such zero time advances and any difference they would make in the temporal I/O behavior of interest to us here.

RNNs with multiple spike classes of signals were introduced to represent interconnected neurons which simultaneously process multiple streams of data, such as the color information of images or networks which simultaneously process streams of data from multiple sensors. One network was used to generate a synthetic texture that imitates the original image. To exchange spikes of different types, neurons have potentials that generate corresponding excitatory spikes in a manner similar to the single potential case. Inhibitory spikes are of only one type, and affect class potential in proportion to their levels. DEVS models can represent such neurons, but there seems to be no evidence for biological plausibility of such a structure. It would be interesting to see if the structure and behavior manifested by multi-class RNNs can be realized by groups of ordinary neurons in the roles of spike processing classes; e.g., interacting cell assemblies specifically tuned to red, green, and blue color wavelengths.

References

1. Carandini, M. From circuits to behavior: A bridge too far? *Nat. Neurosci.* **2012**, *15*, 507–509.
2. Smith, L.S. Deep neural networks: The only show in town? In Proceeedings of the Workshop on Can Deep Neural Networks (DNNs) Provide the Basis for Articial General Intelligence (AGI) at AGI 2016, New York, NY, USA, 16–19 July 2016.
3. Goertzel, B. Are There Deep Reasons Underlying the Pathologies of Today's Deep Learning Algorithms? In *Artificial General Intelligence*; Springer International Publishing: Cham, Switzerland, 2015; pp. 70–79.
4. Paugam-Moisy, H.; Bohte, S. Computing with Spiking Neuron Networks. In *Handbook of Natural Computing*; Kok, J., Heskes, T., Eds.; Springer: Berlin/Heidelberg, Germany, 2009.
5. Ghosh-Dastidar, S.; Hojjat, A. Spiking neural networks. *Int. J. Neural Syst.* **2009**, *19*, 295–308.
6. Minsky, M.; Papert, S. *Perceptrons*; MIT Press: Cambridge, MA, USA, 1969.
7. Rumelhart, D.E.; Hinton, G.E.; Williams, R.J. *Parallel Distributed Processing: Explorations in the Microstructure of Cognition. Vol. 1: Foundations*; Rumelhart, D.E., McClelland, J.L., Eds.; MIT Press: Cambridge, MA, USA, 1986; pp. 318–362.
8. Bland, R. *Learning XOR: Exploring the Space of a Classic Problem*; Computing Science Technical Report; Department of Computing Science and Mathematics, University of Stirling: Stirling, Scotland, June 1998.
9. Toma, S.; Capocchi, L.; Federici, D. A New DEVS-Based Generic Artificial Neural Network Modeling Approach. In Proceedings of the EMSS 2011, Rome, Italy, 12 September 2011.
10. Pessa, E. Neural Network Models: Usefulness and Limitations. In *Nature-Inspired Computing: Concepts, Methodologies, Tools, and Applications*; IGI Global: Hershey, PA, USA, 2017; pp. 368–395.
11. Maass, W. Lower bounds for the computational power of spiking neural networks. *Neural Comput.* **1996**, *8*, 1–40.
12. Schmitt, M. On computing Boolean functions by a spiking neuron. *Ann. Math. Artif. Intell.* **1998**, *24*, 181–191.
13. Brette, R.; Rudolph, M.; Carnevale, T.; Hines, M.; Beeman, D.; Bower, J.M.; Diesmann, M.; Morrison, A.; Goodman, P.H.; Harris, F.C., Jr.; et al. Simulation of networks of spiking neurons: A review of tools and strategies. *J. Comput. Neurosci.* **2007**, *23*, 349–398.
14. Zeigler, B.P. Cellular Space Models: New Formalism for Simulation and Science. In *The Philosophy of Logical Mechanism: Essays in Honor of Arthur W. Burks*; Salmon, M.H., Ed.; Springer: Dordrecht, The Netherlands, 1990; pp. 41–64.
15. Gelenbe, E. Random Neural Networks with Negative and Positive Signals and Product Form Solution. *Neural Comput.* **1989**, *1*, 502–510.
16. Zeigler, B.P.; Kim, T.G.; Praehofer, H. *Theory of Modeling and Simulation: Integrating Discrete Event and Continuous Complex Dynamic Systems*, 2nd ed.; Academic Press: Boston, MA, USA, 2000.
17. Zeigler, B.P.; Nutaro, J.; Seo, C. Combining DEVS and Model-Checking: Concepts and Tools for Integrating Simulation and Analysis. *Int. J. Process Model. Simul.* **2016**, in press.
18. Maass, W. Fast sigmoidal networks via spiking neurons. *Neural Comput.* **1997**, *9*, 279–304.

19. Maass, W. Networks of Spiking Neurons: The Third Generation of Neural Network Models. *Neural Netw.* **1996**, *10*, 1659–1671.

20. Zeigler, B.P. Discrete Event Abstraction: An Emerging Paradigm For Modeling Complex Adaptive Systems. In *Perspectives on Adaptation in Natural and Artificial Systems*; Booker, L., Forrest, S., Mitchell, M., Riolo, R., Eds.; Oxford University Press: New York, NY, USA, 2005; pp. 119–141.

21. Mayerhofer, R.; Affenzeller, M.; Fried A.; Praehofer, H. DEVS Simulation of Spiking Neural Networks. In Proceedings of the Euro-Pean Meeting on Cybernetics and Systems, Vienna, Austria, 30 March–1 April 2002.

22. Booij, O. Temporal Pattern Classification using Spiking Neural Networks. Master's Thesis, Universiteit van Amsterdam, Amsterdam, The Netherlands, August 2004.

23. Maass, W.; Natschlager, T.; Markram, H. Real-time computing without stable states: A new framework for neural computation based on perturbations. *Neural Comput.* **2002**, *14*, 2531–2560.

24. Koch, C. How the Computer Beat the Go Master, Scientific American. 2016. Available online: https://www.scientificamerican.com/article/how-the-computer-beat-the-go-master/ (accessed on 14 January 2017).

25. Hu, X.; Zeigler, B.P. Linking Information and Energy—Activity-based Energy-Aware Information Processing. *Simul. Trans. Soc. Model. Simul. Int.* **2013**, *89*, 435–450.

26. Muzy, A.; Zeigler, B.P.; Grammont, F. Iterative Specification of Input-Output Dynamic Systems and Implications for Spiky Neuronal Networks. *IEEE Syst. J.* **2016**. Available online: http://www.i3s.unice.fr/ muzy/Publications/neuron.pdf (accessed on 14 January 2017).

27. Yoon, Y.C. LIF and Simplified SRM Neurons Encode Signals Into Spikes via a Form of Asynchronous Pulse Sigma-Delta Modulation. *IEEE Trans. Neural Netw. Learn. Syst.* **2016**, *PP*, 1–14.

28. Gelenbe, E. G-networks: A unifying model for neural and queueing networks. *Ann. Oper. Res.* **1994**, *48*, 433–461.

29. Gelenbe, E.; Fourneau, J.M. Random Neural Networks with Multiple Classes of Signals. *Neural Comput.* **1999**, *11*, 953–963.

30. Gelenbe, E. The first decade of G-networks. *Eur. J. Oper. Res.* **2000**, *126*, 231–232.

31. Gelenbe, E. G-networks: Multiple classes of positive customers, signals, and product form results. In *IFIP International Symposium on Computer Performance Modeling, Measurement and Evaluation*; Springer: Berlin/Heidelberg, Germany, 2002.

32. Gelenbe, E.; Timotheou, S. Random Neural Networks with Synchronized Interactions. *Neural Comput.* **2008**, *20*, 2308–2324.

33. Gelenbe, E.; Timotheou, S. Synchronized Interactions in Spiked Neuronal Networks. *Comput. J.* **2008** , *51*, 723–730.